Study Guide and Solutions Manual for

FUNDAMENTALS OF ORGANIC CHEMISTRY

THIRD EDITION

Susan McMurry
Cornell University

Brooks/Cole Publishing Company
Pacific Grove, California

I(T)P™ The trademark ITP is used under license.

Brooks/Cole Publishing Company
A Division of Wadsworth, Inc.

© 1994 by Wadsworth, Inc., Belmont, California 94002. All rights reserved. No part of this book may be reproduced, stored in a retrieval system, or transcribed, in any form or by any means—electronic, mechanical, photocopying, recording, or otherwise—without the prior written permission of the publisher, Brooks/Cole Publishing Company, Pacific Grove, California 93950, a division of Wadsworth, Inc.

Printed in the United States of America

10 9 8 7 6 5 4

ISBN 0-534-21212-3

Sponsoring Editor: Audra C. Silverie
Editorial Assistant: Beth Wilbur
Production Coordinator: Dorothy Bell
Cover Design: Vernon T. Boes
Cover Photo: Shattil/Rozinski—Stock Imagery, Inc.
Printing and Binding: Malloy Lithographing, Inc.

Preface

If you're in a typical organic chemistry course, you go to lecture and take notes, read the text, work some problems, and take tests. For many of you, performing this ritual is all that's necessary to succeed in organic chemistry. Many others of you, however, follow all the correct steps but still feel bewildered by the course. In such a situation, supplementary material is often needed.

This book has been written with two functions in mind. First, it gives an overview of the course, both in chapters and in appendices. Second, it furnishes solutions to the problems presented in the text. The first of these functions might be described as "the big picture," and the second as "the details." Understanding both the big picture and the details is necessary if organic chemistry is to be more than the memorization of unrelated facts.

How to use this book

This *Study Guide and Solutions Manual* can't perform miracles if you don't read the textbook, *Fundamentals of Organic Chemistry*. In all cases, step one is to go to class, take notes, and read the text. At this point, the Study Guide should be helpful.

Study the Chapter Outline. The outline should help you to see how the topics in the chapter are related. Many people are able to learn the facts in each chapter, but are unable to recognize the principles underlying them. The outline will make clearer both the relationships between reactions and how these reactions are related to larger concepts.

Solve the problems in the text. Initially, don't use the *Solutions Manual* to help you with the problems. After completing the problems, consult the *Solutions Manual* to see if your answer is correct and if your method of solution is logical and systematic. If you're confused by a problem, carefully read the solution; then try to solve similar problems on your own.

Check the Study Guide at the end of each chapter in the *Solutions Manual*. All the skills you should have acquired after studying the chapter are listed here, along with the numbers of the problems that reinforce each skill. If a particular type of problem is difficult for you, work related problems until you feel confident.

Before a test or final exam, look at the appendices in the Study Guide and use them as a self-test to see if you know the relevant information. Many of these appendices summarize or tabulate information that has been presented over several chapters. Especially helpful before an exam are the following sections: Reagents Used in Organic Synthesis, Summary of Functional Group Preparations, and Summary of General Reaction Mechanisms. Other appendices present interesting chemical facts and tables.

For most people, understanding organic chemistry takes a long time — sometimes longer than the duration of an organic chemistry course. I hope that the combination of *Fundamentals of Organic Chemistry* plus this *Study Guide and Solutions Manual* makes the study of organic chemistry easier and more rewarding for you.

Acknowledgements: I would like to thank John McMurry for his advice and encouragement during this enjoyable project. I also thank David and Paul McMurry for their understanding and patience during the months I was busy with this book.

Contents

Solutions to Problems:

    Chapter 1   Structure and Bonding   1
    Chapter 2   The Nature of Organic Compounds: Alkanes   15
    Chapter 3   Alkenes: The Nature of Organic Reactions   36
    Chapter 4   Alkenes and Alkynes   50
    Chapter 5   Aromatic Compounds   68
    Chapter 6   Stereochemistry   88
    Chapter 7   Alkyl Halides   106
    Chapter 8   Alcohols, Ethers, and Phenols   123
    Chapter 9   Aldehydes and Ketones: Nucleophilic Addition Reactions   142
    Chapter 10   Carboxylic Acids and Derivatives   161
    Chapter 11   Carbonyl Alpha-Substitution and Condensation Reactions   184
    Chapter 12   Amines   202
    Chapter 13   Structure Determination   219
    Chapter 14   Biomolecules; Carbohydrates   235
    Chapter 15   Biomolecules: Amino Acids, Peptides, and Proteins   252
    Chapter 16   Biomolecules: Lipids and Nucleic Acids   268
    Chapter 17   The Organic Chemistry of Metabolic Pathways   283

Appendices:

    Summary of General Reaction Mechanisms   292
    Summary of Functional Group Preparations   296
    Reagents Used in Organic Synthesis   300
    Organic Name Reactions   305
    Glossary   308
    Symbols and Abbreviations   334
    Proton NMR Chemical Shifts   337
    Infrared Absorption Frequencies   338
    Nobel Prizes in Chemistry   341
    Top 40 Organic Chemicals   348

Chapter 1 – Structure and Bonding

1.1 The elements of the periodic table are organized into groups that are based on the number of outer-shell electrons each element has. For example, an element in group 1A has one outer-shell electron, and an element in group 5A has five outer-shell electrons. To find the number of outer-shell electrons for a given element, use the periodic table to find the element's group.

    a) Potassium is a member of group 1A and thus has one outer-shell electron.
    b) Calcium (group 2A) has two outer-shell electrons.
    c) Aluminum (group 3A) has three outer-shell electrons.

1.2   a) To find the ground-state electronic configuration of an element, first locate its atomic number. For boron, the atomic number is 5; boron thus has 5 protons and *5 electrons*. Next, assign the electrons to the proper energy levels (shown in Figure 1.3), starting with the lowest level:

    Boron   2p ↑ __ __
                  2s ↑↓
                  1s ↑↓

Remember that only two electrons can occupy the same orbital and that they must be of opposite spin.

A different way to represent the ground-state electron configuration is to simply write down the occupied orbitals and to indicate the number of electrons in each orbital. For example, the electron configuration of boron is $1s^2 2s^2 2p$.

Often, we are interested only in the electrons in the outer shell. We can then represent all filled levels by the symbol for the noble gas having the same levels filled. In the case of boron, the filled $1s$ energy level is represented by [He], and the *valence shell configuration* is symbolized by $[He]2s^2 2p$.

  b) Let's consider an element with many electrons. Phosphorus, with an atomic number of 15, has *15 electrons*. Assigning these to energy levels:

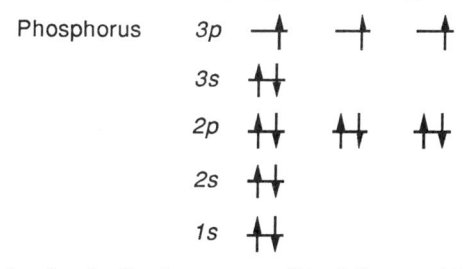

    Phosphorus   3p ↑ ↑ ↑
                         3s ↑↓
                         2p ↑↓ ↑↓ ↑↓
                         2s ↑↓
                         1s ↑↓

Notice that the $3p$ electrons are all in different orbitals. According to *Hund's rule*, we must place one electron into each orbital of the same energy until all orbitals are half-filled.

The more concise way to represent the ground-state electron configuration for phosphorus is: $1s^2 2s^2 2p^6 3s^2 3p^3$
Valence-shell electron configuration: $[Ne]3s^2 3p^3$

c) Oxygen (atomic number 8)

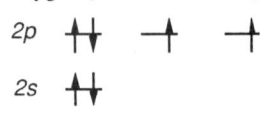

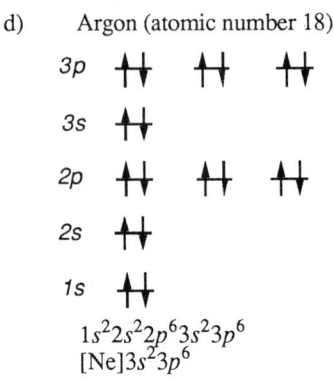

2p ↿⇂ ↿ ↿
2s ↿⇂
1s ↿⇂

$1s^2 2s^2 2p^4$
$[He]2s^2 2p^4$

d) Argon (atomic number 18)

3p ↿⇂ ↿⇂ ↿⇂
3s ↿⇂
2p ↿⇂ ↿⇂ ↿⇂
2s ↿⇂
1s ↿⇂

$1s^2 2s^2 2p^6 3s^2 3p^6$
$[Ne]3s^2 3p^6$

1.3

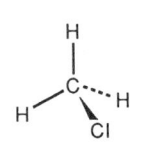

Chloromethane (structure: C bonded to three H and one Cl)

1.4 Use the periodic table to find the group to which an element belongs. For any element, the group number is the same as the number of valence shell electrons.

	Element	Group	Valence Shell Electrons
a)	Be	2A	2
b)	S	6A	6
c)	Br	7A	7

1.5 Elements on the left of the periodic table are electropositive; elements on the right of the periodic table are electronegative.

a) Oxygen is more electronegative than potassium.
b) Bromine is more electronegative than calcium.

1.6 a) Carbon (group 4A) has four electrons in its valence shell and forms four bonds to achieve the noble-gas configuration of neon. Hence, a likely formula is CCl_4.

	Element	Group	Likely Formula
b)	Al	3A	AlH_3
c)	C	4A	CH_2Cl_2
d)	Si	4A	SiF_4

Structure and Bonding   3

1.7 Writing Lewis structures of a molecule requires that you first determine the number of valence, or outer-shell, electrons for each atom in the molecule. For chloroform, we know that carbon has four valence electrons, hydrogen has one, and each chlorine has seven.

$\cdot \ddot{C} \cdot$ $4 \times 1 = 4$

$H \cdot$ $1 \times 1 = 1$

$:\ddot{Cl}\cdot$ $7 \times 3 = 21$

26 total valence electrons

Next, use two electrons for each single bond.

$$\begin{array}{c} H \\ Cl : \ddot{C} : Cl \\ \ddot{Cl} \end{array}$$

Finally, use the remaining electrons to achieve an noble-gas configuration for all atoms.

a) CHCl$_3$

$$\begin{array}{c} H \\ :\ddot{Cl}:\ddot{C}:\ddot{Cl}: \\ :\ddot{Cl}: \end{array} \qquad \begin{array}{c} H \\ | \\ Cl-C-Cl \\ | \\ Cl \end{array}$$

b) H$_2$S   Total valence electrons = 8

$$\begin{array}{c} H:\ddot{S}: \\ H \end{array} \qquad \begin{array}{c} H-S \\ | \\ H \end{array}$$

c) CH$_3$NH$_2$ Total valence electrons = 14

$$\begin{array}{c} H \ H \\ H:\ddot{C}:\ddot{N}:H \\ H \end{array} \qquad \begin{array}{c} H \ H \\ | \ \ | \\ H-C-N-H \\ | \\ H \end{array}$$

1.8 Bonds formed between an electropositive element and an electronegative element are ionic. Bonds formed between an element in the middle of the periodic table and another element are most often covalent, but exceptions can be found.

Ionic bonds: LiI, KBr, MgCl$_2$
Covalent bonds: CH$_4$, CH$_2$Cl$_2$, Cl$_2$

1.9

$$\begin{array}{c} H \ H \\ H:\ddot{C}:\ddot{C}:H \\ H \ H \end{array} \quad \text{Ethane} \quad \begin{array}{c} H \ H \\ | \ \ | \\ H-C-C-H \\ | \ \ | \\ H \ H \end{array}$$

Chapter 1

1.10

Cl—C(Cl)(Cl)—Cl   Tetrachloromethane

1.11 An electron in an sp^3 orbital is farther from the nucleus than an electron in a $1s$ orbital. Thus, a bond that uses an sp^3 orbital of carbon and a $1s$ orbital of hydrogen is longer than a bond that uses two $1s$ orbitals (H-H bond).

1.12

[Structure of propane with all carbons labeled sp^3]   Propane

All carbon atoms are tetrahedral, and all bond angles are approximately 109.5°.

1.13 The two carbons bond to each other by overlap of two sp^3 hybrid orbitals. Six sp^3 hybrid orbitals (three from each carbon) are left over, and they can bond with a maximum of six hydrogens. Thus, a formula such as C_2H_7 is not possible.

1.14

H:C:C::O: with H's   Acetaldehyde   H—C(H)(H)—C(H)=O

1.15

[Structure of propene showing C3 as sp^3, C2 and C1 as sp^2]   Propene

The C3-H bonds are sigma bonds formed by overlap of an sp^3 orbital of carbon 3 with an s orbital of hydrogen. Bond angles at C3 are approximately 109°.

The C2-H and C1-H bonds are sigma bonds formed by overlap of an sp^2 orbital of carbon with an s orbital of hydrogen.

The C2-C3 bond is a sigma bond formed by overlap of an sp^3 orbital of carbon 3 with an sp^2 orbital of carbon 2.

There are two C1-C2 bonds. One is a sigma bond formed by overlap of an sp^2 orbital of carbon 1 with an sp^2 orbital of carbon 2. The other is a pi bond formed by overlap of a $2p$ orbital of carbon 1 with a $2p$ orbital of carbon 2. All four atoms connected to the carbon-carbon double bond lie in the same plane, and all bond angles between these atoms are 120°.

Structure and Bonding 5

1.16

$$\text{H}-\underset{\underset{\text{H}}{|}}{\overset{\overset{\text{H}}{|}}{\underset{sp^2}{C}}}\overset{3}{\underset{sp^2}{=}}\underset{\underset{\text{H}}{|}}{\overset{\overset{\text{H}}{|}}{\underset{sp^2}{C}}}\overset{2}{\underset{sp^2}{-}}\underset{sp^2}{\overset{1}{C}}-\text{H}$$

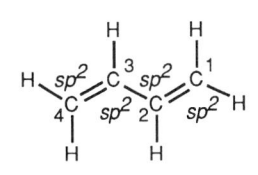

All atoms lie in the same plane, and all bond angles are approximately 120°

1.17

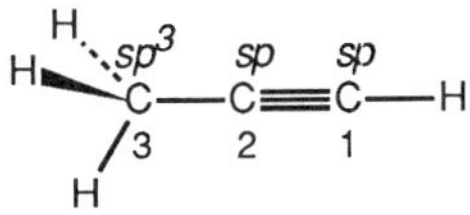

$$\underset{\underset{\text{H}}{|}}{\overset{\overset{\text{H}}{\underset{sp^3}{\diagdown}}}{\underset{3}{C}}}-\underset{2}{\overset{sp}{C}}\equiv\underset{1}{\overset{sp}{C}}-\text{H} \qquad \text{Propyne}$$

The C3-H bonds are sigma bonds formed by overlap of an sp^3 orbital of carbon 3 with an s orbital of hydrogen. Bond angles at C3 are approximately 109°.

The C1-H bond is a sigma bond formed by overlap of an sp orbital of carbon 1 with an s orbital of hydrogen.

The C2-C3 bond is a sigma bond formed by overlap of an sp orbital of carbon 2 with an sp^3 orbital of carbon 3.

There are three C1-C2 bonds. One is a sigma bond formed by overlap of an sp orbital of carbon 1 with an sp orbital of carbon 2. The other two bonds are pi bonds formed by overlap of two $2p$ orbitals of carbon 1 with two $2p$ orbitals of carbon 2.

The three carbon atoms of propyne lie on a straight line, with a bond angle of 180°.

1.18 Use Figure 1.17 to answer this problem. The larger the EN number, the more electronegative the element.

	More electronegative	Less electronegative	
a)	H (2.1)	Li	(1.0)
b)	Br (2.8)	Be	(1.6)
c)	Cl (3.0)	I	(2.5)

1.19 As in Problem 1.18, use Figure 1.17. Remember that the arrow points toward the more electronegative atom in the bond.

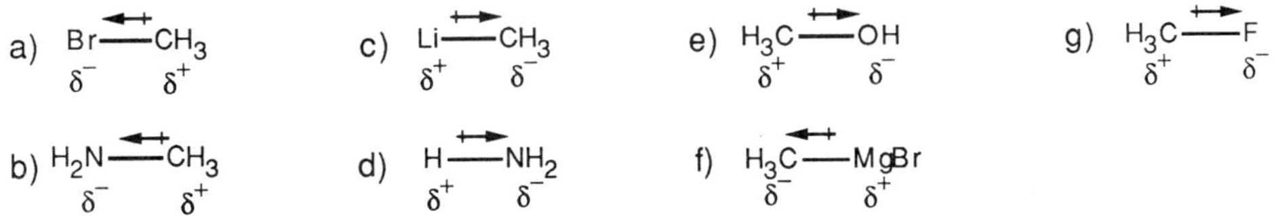

a) Br⟵CH₃
 δ⁻ δ⁺

b) H₂N⟵CH₃
 δ⁻ δ⁺

c) Li⟶CH₃
 δ⁺ δ⁻

d) H⟶NH₂
 δ⁺ δ⁻

e) H₃C⟶OH
 δ⁺ δ⁻

f) H₃C⟵MgBr
 δ⁻ δ⁺

g) H₃C⟶F
 δ⁺ δ⁻

6   Chapter 1

1.20   Use Figure 1.17 to locate the EN of each element. The larger the difference in EN, the more ionic the bond.

CCl$_4$	MgCl$_2$	TiCl$_3$	Cl$_2$O
Cl : EN = 3.0	Cl : EN = 3.0	Cl : EN = 3.0	Cl : EN = 3.0
C : EN = 2.5	Mg : EN = 1.2	Ti : EN = 1.5	O : EN = 3.5
Δ EN = 0.5	Δ EN = 1.8	Δ EN = 1.5	Δ EN = 0.5

Least ionic ⟶ Most ionic
CCl$_4$ and ClO$_2$, TiCl$_3$, MgCl$_2$

1.21   Since a lower pK_a value indicates a stronger acid, picric acid is stronger than formic acid.

1.22
Conjugate base		Acid	
stronger	$^-$NH$_2$	NH$_3$	weaker
weaker	$^-$OH	H$_2$O	stronger

The conjugate base of a strong acid is a weak base, and the conjugate base of a weak acid is a strong base. In line with this reasoning, water is a stronger acid than ammonia.

1.23

a)   H–CN + CH$_3$COO$^-$ Na$^+$ $\xrightarrow{?}$ Na$^+$ $^-$CN + CH$_3$COO–H
     pK_a = 9.2   pK_a = 4.7
     weaker acid   stronger acid

The lower the pK_a, the stronger the acid. Since CH$_3$COOH is the stronger acid and gives up a proton more readily than HCN, the reaction will not take place as written.

b)   CH$_3$CH$_2$O–H + Na$^+$ $^-$CN $\xrightarrow{?}$ CH$_3$CH$_2$O$^-$ Na$^+$ + H–CN
     pK_a = 16.0   pK_a = 9.2
     weaker acid   stronger acid

Using the same reasoning as in part (a), we can see that the above reaction will not take place.

1.24   A Lewis base has a non-bonding electron pair to share. A Lewis acid has a vacant orbital to accept an electron pair. Look for a lone electron pair when identifying a Lewis base.

Lewis acids: MgBr$_2$, B(CH$_3$)$_3$, $^+$CH$_3$

Lewis bases: CH$_3$CH$_2$ÖH, CH$_3$N̈HCH$_3$, CH$_3$P̈CH$_3$
                                                      |
                                                     CH$_3$

1.25 See Problem 1.1 if you need help.

Element	Group	Number of outer shell electrons
a) Oxygen	6A	6
b) Magnesium	2A	2
c) Fluorine	7A	7

1.26

Element	Atomic Number	Number of outer shell electrons
a) Li	3	$1s^2 2s$
b) Na	11	$1s^2 2s^2 2p^6 3s$
c) Al	13	$1s^2 2s^2 2p^6 3s^2 3p$
d) S	16	$1s^2 2s^2 2p^6 3s^2 3p^4$

1.27 a) $AlCl_3$ b) CF_2Cl_2 c) NI_3

1.28 Ionic bonds: BeF_2
Covalent bonds: SiH_4, CBr_4

1.29

a) H:C:::C:H

b) H:Al:H
       H

c)   H
   H:C:O:H
      H

d) H:C::C:Cl:
      H H

1.30

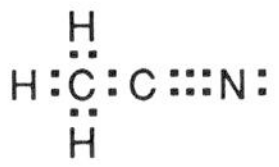

     H
H:C:C:::N: Acetonitrile
     H

Nitrogen has five electrons in its valence shell. Three are used in the carbon-nitrogen triple bond, and two are a nonbonding electron pair.

1.31

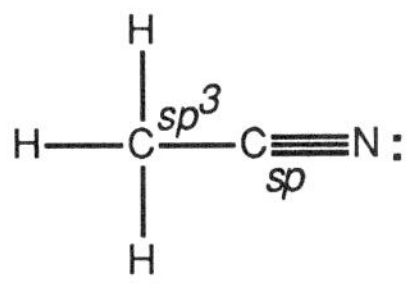

$$H-\underset{\underset{H}{|}}{\overset{\overset{H}{|}}{C}}{}^{sp^3}-\underset{sp}{C}\equiv N:\qquad \text{Acetonitrile}$$

1.32

(a) $CH_3-\ddot{O}-CH_3$

(b) $CH_3-\underset{H}{\overset{..}{N}}-H$

(c)   H
   :Cl-C-Cl:
        H

(d)   :O:
       ‖
   H_3C-C-CH_3

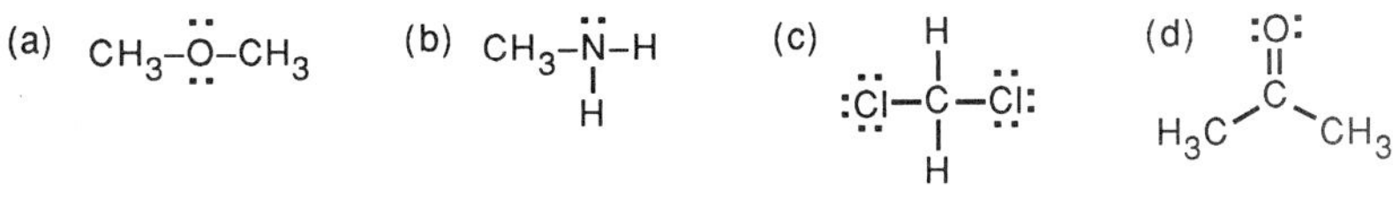

1.33

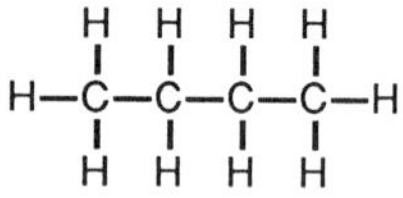

H-C(H)(H)-C(H)(H)-C(H)(H)-C(H)(H)-H   and   H-C(H)(H)-C(H)(H)-C(H)(H)-H with H-C(H)(H)-H branch on middle carbon

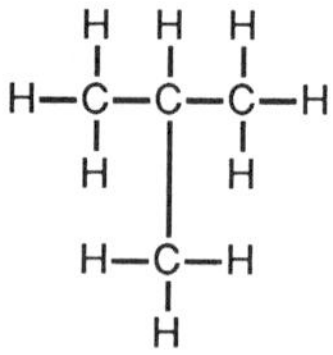

1.34

In order to work a problem of this sort, you must examine all possible structures that have the correct number of bonds. You must systematically consider all possible attachments, including those that have branches, rings and multiple bonds.

a) 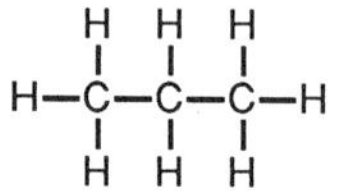 H-C(H)(H)-C(H)(H)-C(H)(H)-H

b) 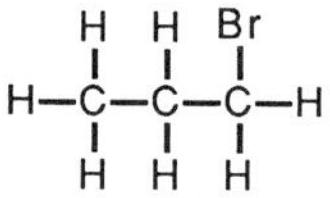 H-C(H)(H)-C(H)(H)-C(H)(Br)-H   and   H-C(H)(H)-C(H)(Br)-C(H)(H)-H

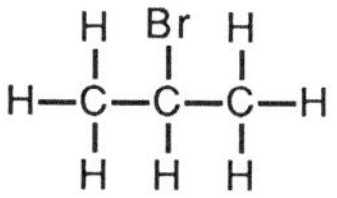

These are the only two possible structures with the formula C_3H_7Br.

c) 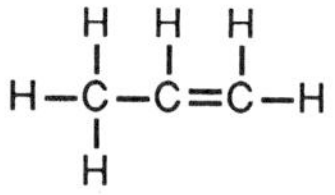 H-C(H)(H)-C(H)=C(H)-H   and   cyclopropane ring with H's

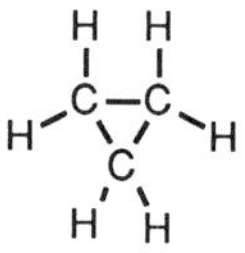

d) 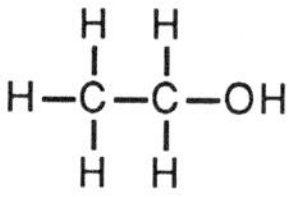 H-C(H)(H)-C(H)(H)-OH   and   H-C(H)(H)-O-C(H)(H)-H

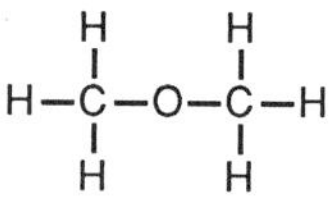

1.35

a) 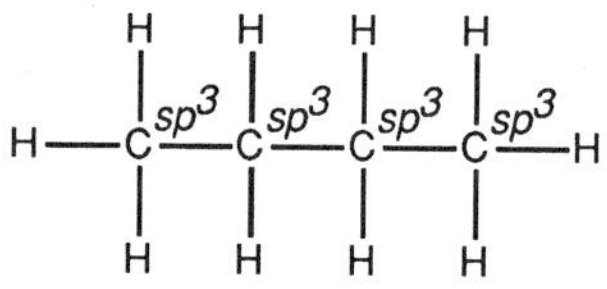 H-C(sp^3)(H)(H)-C(sp^3)(H)(H)-C(sp^3)(H)(H)-C(sp^3)(H)(H)-H

Butane

b) 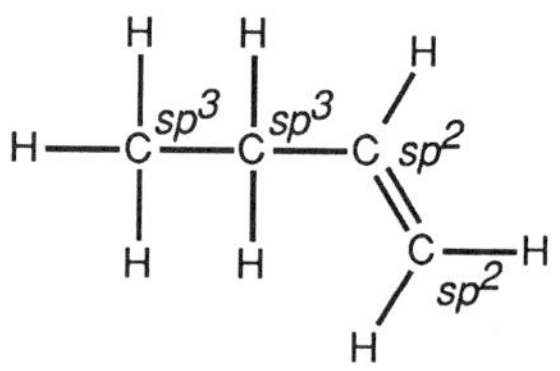 H-C(sp^3)(H)(H)-C(sp^3)(H)(H)-C(sp^2)(H)=C(sp^2)(H)(H)

1-Butene

c) 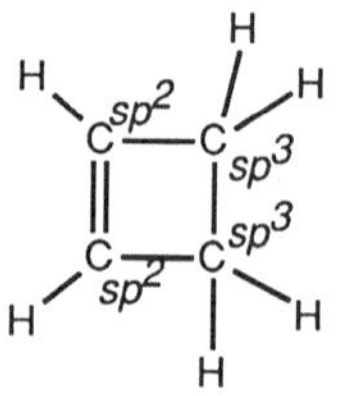 Cyclobutene with two sp^2 carbons (double bond) and two sp^3 carbons

Cyclobutene

d) 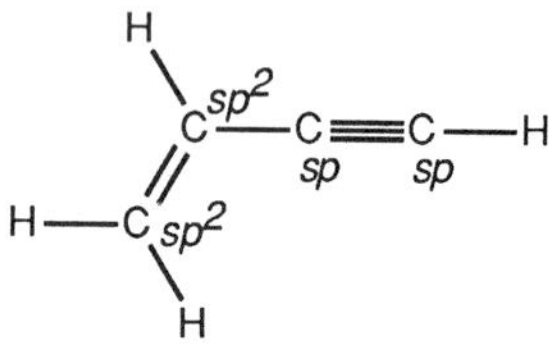 H-C(sp^2)(H)=C(sp^2)(H)-C(sp)≡C(sp)-H

1-Buten-3-yne

1.36

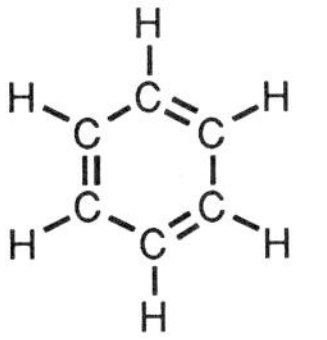

Benzene

All carbon atoms of benzene are sp^2 hybridized, and all bond angles of benzene are 120°. Benzene is a planar molecule.

1.37

a) 16 valence electrons

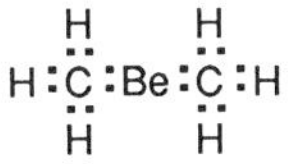

b) 26 valence electrons

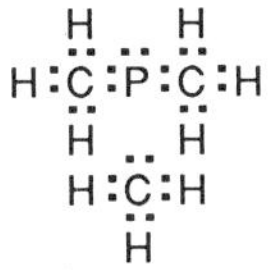

c) 32 valence electrons
(4 from Ti, 7 from each Cl)

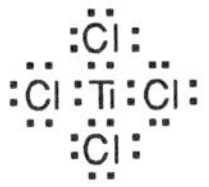

1.38-1.39

a) Br—Br

nonpolar

b)

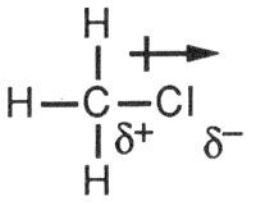

c)

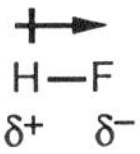

d)

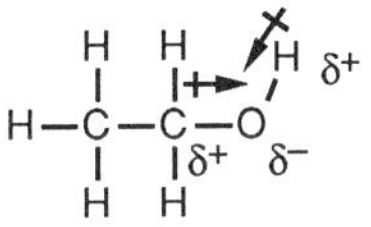

Molecules b-d are polar. Carbon-hydrogen bonds are only slightly polar.

1.40

a)

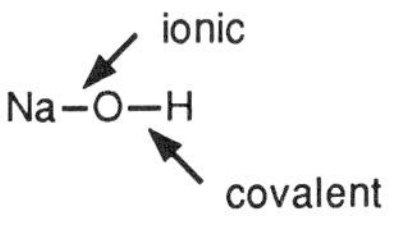

b)

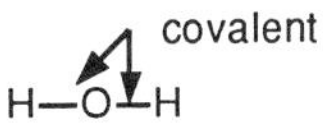

c)

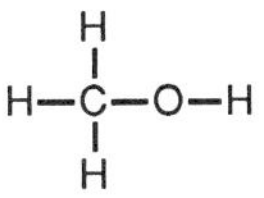

all covalent

d)

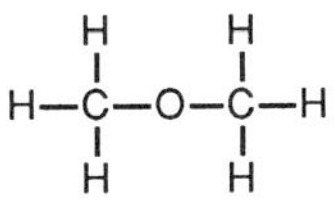

all covalent

e) F—F covalent

1.41

$Na^+ \; ^-O-\underset{H}{\overset{H}{\underset{|}{C}}}-H$ — ionic bond (indicated)

All other bonds are covalent.

1.42 The most electronegative element is underlined.

a) CH₂<u>F</u>Cl b) <u>F</u>CH₂CH₂CH₂Br c) H<u>O</u>CH₂CH₂NH₂ d) CH₃<u>O</u>CH₂Li

1.43-1.44

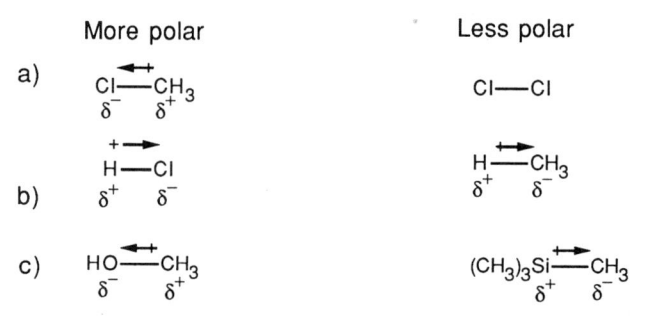

More polar

a) $\underset{\delta^-}{Cl}\text{—}\underset{\delta^+}{CH_3}$

b) $\underset{\delta^+}{H}\text{—}\underset{\delta^-}{Cl}$

c) $\underset{\delta^-}{HO}\text{—}\underset{\delta^+}{CH_3}$

Less polar

Cl—Cl

$\underset{\delta^+}{H}\text{—}\underset{\delta^-}{CH_3}$

$\underset{\delta^+}{(CH_3)_3Si}\text{—}\underset{\delta^-}{CH_3}$

1.45

a) CH₃S̈H

b) CH₃-N̈-CH₃
　　　　|
　　　　CH₃

c) CH₃CH₂B̈r:

d) CH₃C̈OH (with =O: double bond)

e) CH₃C(=O:)-C̈l:

1.46

Chloroform: Cl---C(H)(Cl)(Cl) with wedge/dash notation

Ethanol: H---C(CH₃)(H)(OH) with wedge/dash notation

1.47

$H_3C\overset{O}{\overset{\|}{C}}CH_3 + Na^{+\;-}:NH_2 \longrightarrow H_3C\overset{O}{\overset{\|}{C}}CH_2{:}^- \; Na^+ + NH_3$

pKₐ = 20　　　　　　　　　　　　　　　　　　　　　pKₐ = 36

stronger acid　　　　　　　　　　　　　　　　　　weaker acid

The above reaction will take place as written because acetone is a stronger acid than ammonia.

Structure and Bonding   11

1.48   Lewis acids: AlBr$_3$, HF
Lewis bases: CH$_3$CH$_2$NH$_2$, CH$_3$SCH$_3$

1.49   The reaction between methanol and bicarbonate does not take place in the indicated direction because methanol (pK_a = 15.5) is a weaker acid than bicarbonate (pK_a = 6.4).

1.50

a) CH$_3$OH + H$^+$ $\longrightarrow$ CH$_3$OH$_2^+$
   base   acid

b) CH$_3$OH + $^-$:NH$_2$ $\longrightarrow$ CH$_3$O:$^-$ + :NH$_3$
   acid   base

c) CH$_3$–C(=O:)–H + ZnCl$_2$ $\longrightarrow$ CH$_3$–C(=O$^+$–ZnCl$_2^-$)–H
   base   acid

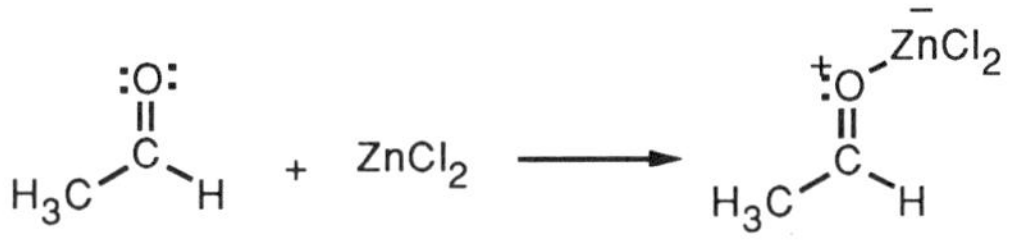

1.51

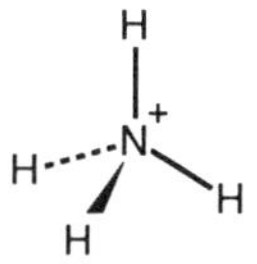

Ammonium ion (tetrahedral NH$_4^+$)

The nitrogen atom of the tetrahedral ammonium ion is sp^3 hybridized because, like the carbon atom of methane, nitrogen forms bonds to four different hydrogen atoms.

1.52   Ethane (tetrahedral carbons)

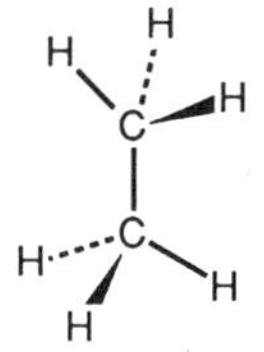

1.53

a) H–C(sp^3)H$_2$–C(sp^2)(=O)–OH

b) H–C(sp^2)H=C(sp^2)H–C(sp^2)(=O)–C(sp^3)H$_3$

c) H–C(sp^2)H=C(sp^2)H–C(sp)≡N

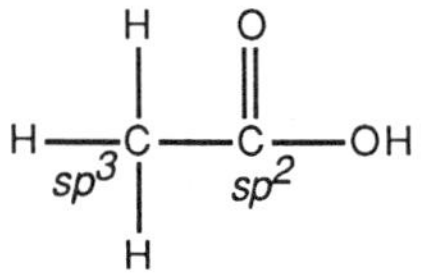

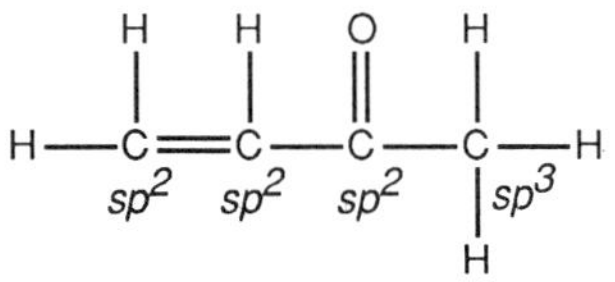

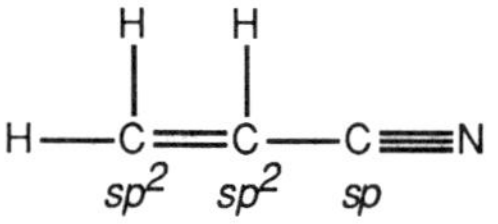

1.54 Carbon is most positive when it is bonded to the most electronegative atom.

Most negative carbon ⟶ Most positive carbon.

CH₃Li, CH₃-CH₃, CH₃-I, CH₃-NH₂, CH₃-OH, CH₃-F

1.55

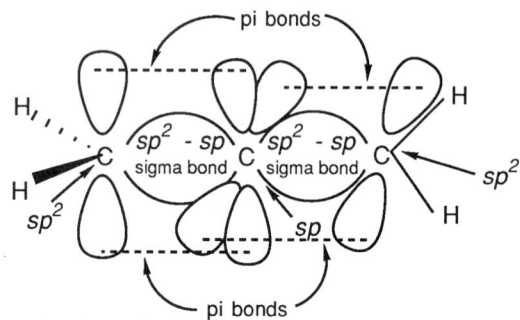

The central carbon of allene forms two sigma bonds and two pi bonds. The central carbon is *sp*-hybridized, and the two terminal carbons are *sp²*-hybridized. The carbon-carbon bond angle is 180°, indicating linear geometry for the carbons of allene.

1.56 The carbon atom of CO_2 is *sp*-hybridized. Allene and CO_2 are both linear molecules.

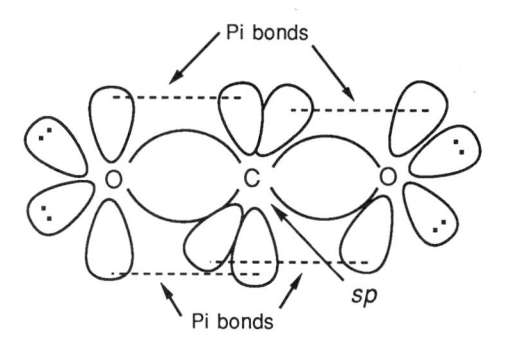

1.57 The carbon atom, which has three valence shell electrons, is *sp²* hybridized. A carbocation is planar and is *isoelectronic* with (has the same number of electrons as) a trivalent boron compound.

Chapter Outline

I. Atomic structure (Sections 1.1 - 1.2)
   A. Nucleus (Section 1.1)
      1. Protons
      2. Neutrons
   B. Electrons
      1. Shells
      2. Orbitals
   C. Electronic configuration of atoms (Section 1.2)
II. Chemical bonds (Sections 1.3 - 1.6)
   A. Development of chemical bonding theory (Section 1.3)
   B. Geometry of tetravalent carbon
   C. The ionic bond (Section 1.4)
      1. Electropositive elements
      2. Electronegative elements
   D. The covalent bond (Section 1.5)
      1. Lewis electron-dot structures
      2. Kekulé line-bond structures
   E. Formation of covalent bonds (Section 1.6)
      1. Bond strength and bond length
      2. Sigma (σ) bonds - head-on overlap of orbitals
III. Hybridization (Sections 1.7 - 1.11)
   A. Formation of sp^3 orbitals (Section 1.7)
      1. Excited-state configuration
      2. Single bonds and bond angles
      3. Structure of methane (Section 1.8)
      4. Structure of ethane (Section 1.9)
   B. Formation of sp^2 orbitals (Section 1.10)
      1. Double bonds
      2. Pi (π) - bonds - sideways overlap of bonds
      3. Structure of ethylene
   C. Formation of sp hybrids (Section 1.11)
      1. Triple bonds
      2. Structure of acetylene

14 Chapter 1

IV. Bond polarity and electronegativity (Section 1.12)
   A. Polar covalent bonds
   B. Electronegativity
   C. Inductive effects
V. Acids and Bases (Section 1.13)
   A. Bronsted-Lowry acids and bases
      1. Conjugate acids and conjugate bases
      2. Acidity constant
   B. Lewis acids and bases

Study Skills for Chapter 1

After studying this chapter, you should be able to:

1. Predict the number of valence-shell electrons and the electronegativity of an element (Problems 1.1, 1.4, 1.5, 1.18, 1.25, 1.42, 1.46)
2. Predict the ground-state electronic configuration of atoms (Problems 1.2, 1.26)
3. Draw simple organic compounds with the correct three-dimensional geometry (Problems 1.3, 1.10, 1.46, 1.52)
4. Write the likely formula of a simple compound (Problems 1.6, 1.27)
5. Draw Lewis electron-dot structures of simple compounds (Problems 1.7, 1.9, 1.14, 1.29, 1.30, 1.32, 1.37, 1.45, 1.56)
6. Draw line-bond structures of compounds (Problems 1.7, 1.9, 1.12, 1.14, 1.15, 1.16, 1.33, 1.34, 1.38)
7. Identify bonds as either ionic or covalent (Problems 1.8, 1.20, 1.28, 1.40, 1.41)
8. Predict and describe the hybridization of bonds in simple organic compounds (Problems 1.15, 1.16, 1.17, 1.31, 1.35, 1.36, 1.51, 1.53, 1.55, 1.56, 1.57)
9. Predict the direction of polarity of a bond (Problems 1.19, 1.39, 1.43, 1.44)
10. Predict the relative acidity of molecules (Problems 1.21, 1.22, 1.50)
11. Use pK_a values to predict the likelihood of a reaction taking place (Problems 1.23, 1.47 1.49)
12. Decide if a molecule is a Lewis acid or a Lewis base (Problem 1.24, 1.48)

Chapter 2 – The Nature of Organic Compounds: Alkanes

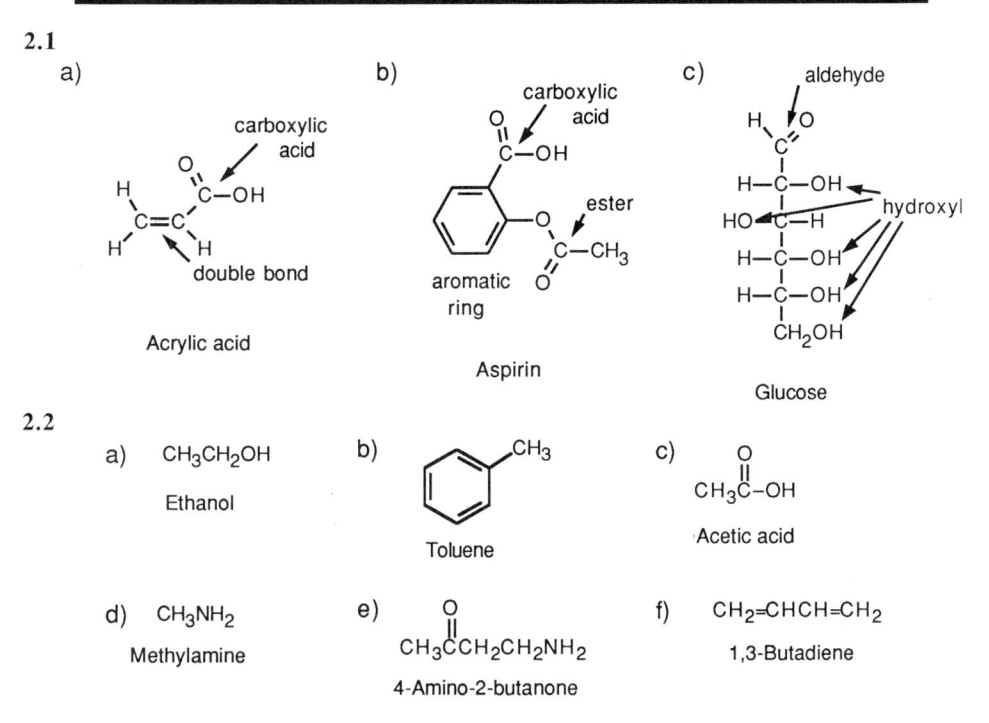

2.1

a) Acrylic acid: H₂C=CH–C(=O)–OH — labeled carboxylic acid, double bond

b) Aspirin — labeled carboxylic acid, aromatic ring, ester

c) Glucose:
```
      H   O
       \ //
        C
   H—C—OH
  HO—C—H
   H—C—OH
   H—C—OH
      CH₂OH
```
labeled aldehyde, hydroxyl

2.2

a) CH₃CH₂OH — Ethanol

b) Toluene (benzene ring with CH₃)

c) CH₃C(=O)–OH — Acetic acid

d) CH₃NH₂ — Methylamine

e) CH₃C(=O)CH₂CH₂NH₂ — 4-Amino-2-butanone

f) CH₂=CHCH=CH₂ — 1,3-Butadiene

Many other compounds containing these functional groups can be drawn.

2.3 We know that carbon forms four bonds and hydrogen forms one bond. Thus, if you draw all possible ways of connecting six carbons and add hydrogens so that all carbons have four bonds, you will arrive at the following formulas.

CH₃CH₂CH₂CH₂CH₂CH₃

CH₃CH₂CH₂CH(CH₃)CH₃

CH₃CH₂CH(CH₃)CH₂CH₃

CH₃CH₂C(CH₃)₂CH₃

CH₃CH(CH₃)CH(CH₃)CH₃

2.4 a) There are 18 isomers with the formula C₈H₁₈.

Octane:

CH₃CH₂CH₂CH₂CH₂CH₂CH₂CH₃

Chapter 2

Heptanes:

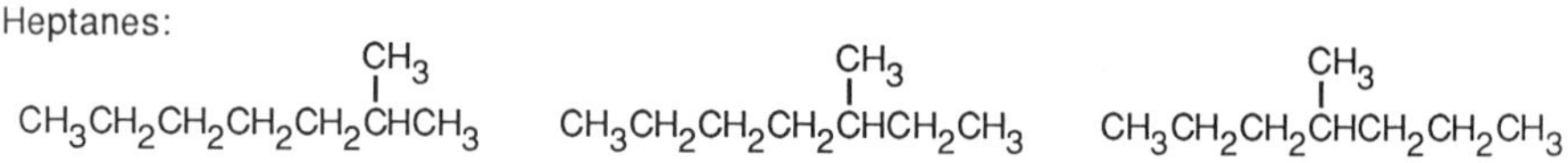

CH₃CH₂CH₂CH₂CH₂CH(CH₃)CH₃   CH₃CH₂CH₂CH₂CH(CH₃)CH₂CH₃   CH₃CH₂CH₂CH(CH₃)CH₂CH₂CH₃

Hexanes:

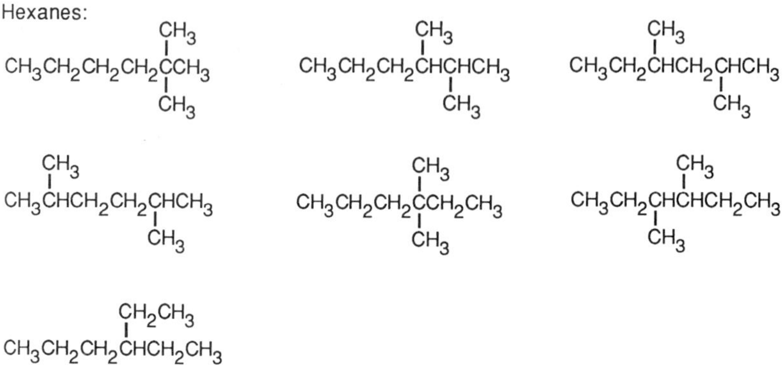

CH₃CH₂CH₂CH₂C(CH₃)₂CH₃   CH₃CH₂CH₂CH(CH₃)CH(CH₃)CH₃   CH₃CH₂CH(CH₃)CH₂CH(CH₃)CH₃

CH₃CH(CH₃)CH₂CH₂CH(CH₃)CH₃   CH₃CH₂CH₂C(CH₃)₂CH₂CH₃   CH₃CH₂CH(CH₃)CH(CH₃)CH₂CH₃

CH₃CH₂CH₂CH(CH₂CH₃)CH₂CH₃

Pentanes:

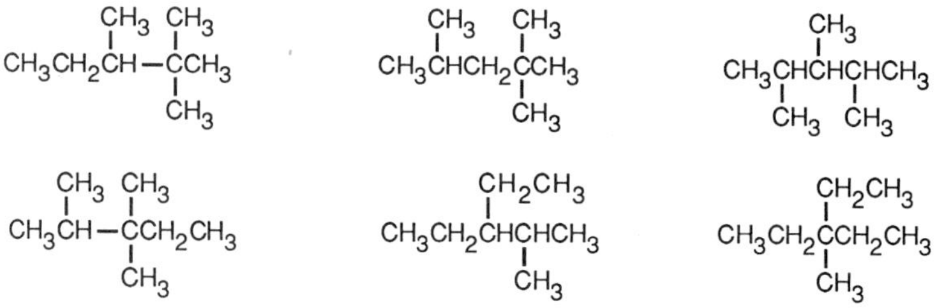

CH₃CH₂CH(CH₃)—C(CH₃)₂CH₃   CH₃CH(CH₃)CH₂C(CH₃)₂CH₃   CH₃CH(CH₃)CH(CH₃)CH(CH₃)CH₃

CH₃CH(CH₃)—C(CH₃)₂CH₂CH₃   CH₃CH₂CH(CH₂CH₃)CH(CH₃)CH₃   CH₃CH₂C(CH₂CH₃)(CH₃)CH₂CH₃

Butane:

CH₃C(CH₃)(CH₃)—C(CH₃)(CH₃)CH₃

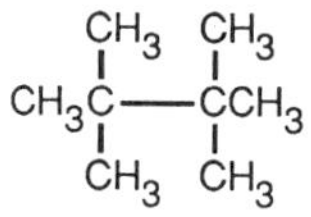

b) Many isomers of the formula C₄H₈O₂ containing different functional groups can be drawn. Here are three examples:

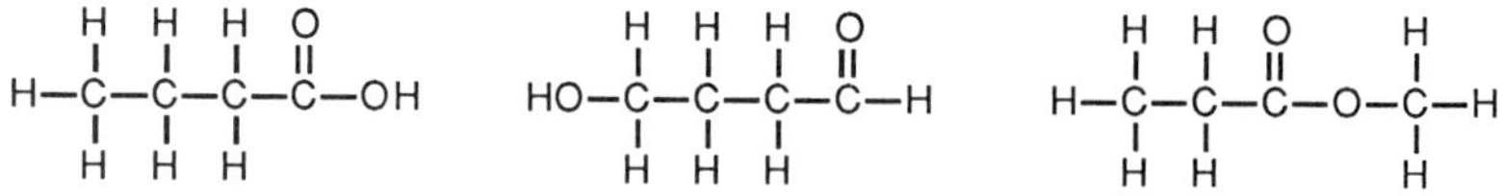

H−C(H)(H)−C(H)(H)−C(H)(H)−C(=O)−OH   HO−C(H)(H)−C(H)(H)−C(H)(H)−C(=O)−H   H−C(H)(H)−C(H)(H)−C(=O)−O−C(H)(H)−H

2.5

CH₃CH₂CH₂CH₂CH₂−⧯   CH₃CH₂CH₂CH(CH₃)−⧯   CH₃CH₂CH(CH₂CH₃)−⧯   CH₃CH₂CH(CH₃)CH₂−⧯

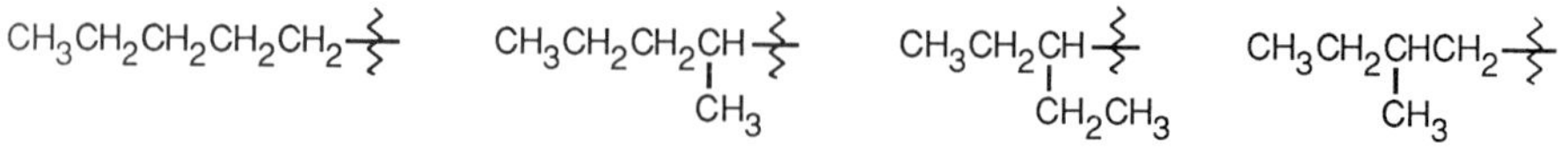

The Nature of Organic Compounds: Alkanes   17

$$CH_3CHCH_2CH_2\text{–}\xi \quad\quad CH_3CH_2\underset{\underset{CH_3}{|}}{\overset{\overset{CH_3}{|}}{C}}\text{–}\xi \quad\quad CH_3\underset{\underset{CH_3}{|}}{\overset{\overset{CH_3}{|}}{CH}}CH\text{–}\xi \quad\quad CH_3\underset{\underset{CH_3}{|}}{\overset{\overset{CH_3}{|}}{C}}CH_2\text{–}\xi$$
$$\underset{CH_3}{|}$$

2.6 Many other answers to these problems are acceptable.

a) $CH_3\underset{\underset{t\ CH_3}{|}}{\overset{\overset{CH_3\ \nearrow\ t}{|}}{CH}}CHCH_3$ b) $CH_3CH_2\boxed{\overset{CH_3}{\underset{|}{CH}}CH_3}$ c) $CH_3CH_2\underset{\underset{s\ CH_3}{\uparrow}}{\overset{\overset{q\ CH_3}{\nwarrow}}{C}}CH_3$

2.7

a) $\overset{p}{CH_3}$
$CH_3\underset{p\ \ t}{CH}\underset{s}{CH_2}\underset{s}{CH_2}\underset{p}{CH_3}$

b) $\overset{p\ \ t\ \ p}{CH_3CHCH_3}$
$CH_3CH_2\underset{p\ \ s\ \ t\ \ s\ \ p}{CH}CH_2CH_3$

c) $\overset{p}{CH_3} \quad \overset{p}{CH_3}$
$CH_3\underset{p\ \ t\ \ s}{CH}CH_2\text{—}\underset{\underset{p}{CH_3}}{\overset{\overset{}{|}}{C}}\text{—}\overset{p}{CH_3}$
$ |q\ p$

p = primary; s = secondary; t = tertiary; q = quaternary

2.8

a) $CH_3CH_2CH_2CH_2CH_3 \quad\quad CH_3CH_2\underset{|}{\overset{\overset{CH_3}{|}}{CH}}CH_3 \quad\quad CH_3\underset{\underset{CH_3}{|}}{\overset{\overset{CH_3}{|}}{C}}CH_3$

Pentane   2–Methylbutane   2,2–Dimethylpropane

b) $\boxed{\underset{65}{CH_3CH_2}\underset{4}{\overset{\overset{CH_3}{|}}{CH}}\underset{}{\overset{\overset{3}{}}{CH}}\text{–}CH_3}$
$\underset{21}{CH_2CH_3}$

The longest chain is a *hexane*.
The substituents are: 3–methyl, 4–methyl.
The IUPAC name is 3,4–dimethylhexane.

c) $CH_3\underset{|}{\overset{\overset{CH_3}{|}}{CH}}CH_2\underset{|}{\overset{\overset{CH_3}{|}}{CH}}CH_3 \quad\quad CH_3\underset{\underset{CH_3}{|}}{\overset{\overset{CH_3}{|}}{C}}\text{–}CH_2CH_2\underset{|}{\overset{\overset{CH_2CH_3}{|}}{CH}}CH_3$

2,4–Dimethylpentane   2,2,5–Trimethylheptane

2.9 a) First, draw the carbon structure of the parent hydrocarbon. Here, it is a nonane.

C–C–C–C–C–C–C–C–C

Then, add the substituents — methyl groups at C3 and C4.

$$\underset{987654321}{C\text{–}C\text{–}C\text{–}C\text{–}C\text{–}\underset{\underset{CH_3}{|}}{\overset{\overset{CH_3}{|}}{C}}\text{–}C\text{–}C\text{–}C}$$

18   Chapter 2

Finally, add hydrogens to complete the structure.

$$\underset{\underset{CH_3}{|}}{\overset{\overset{CH_3}{|}}{CH_3CH_2CH_2CH_2CH_2CHCHCH_2CH_3}}$$ 3,4-Dimethylnonane

b)   3-Ethyl-4,4-dimethylheptane $$\underset{\underset{CH_3}{|}\;\underset{CH_2CH_3}{|}}{\overset{\overset{CH_3}{|}}{CH_3CH_2CH_2C{-}CHCH_2CH_3}}$$

c)   2,2-Dimethyl-4-propyloctane $$\underset{\underset{CH_3CH_2CH_2}{|}\;\underset{CH_3}{|}}{\overset{\overset{CH_3}{|}}{CH_3CH_2CH_2CH_2CHCH_2CCH_3}}$$

d)   2,2,4-Trimethylpentane $$\underset{\underset{CH_3}{|}}{\overset{\overset{CH_3\;\;\;CH_3}{|\;\;\;\;\;|}}{CH_3CHCH_2CCH_3}}$$

2.10

Most stable conformation
(staggered)

Least stable conformation
(eclipsed)

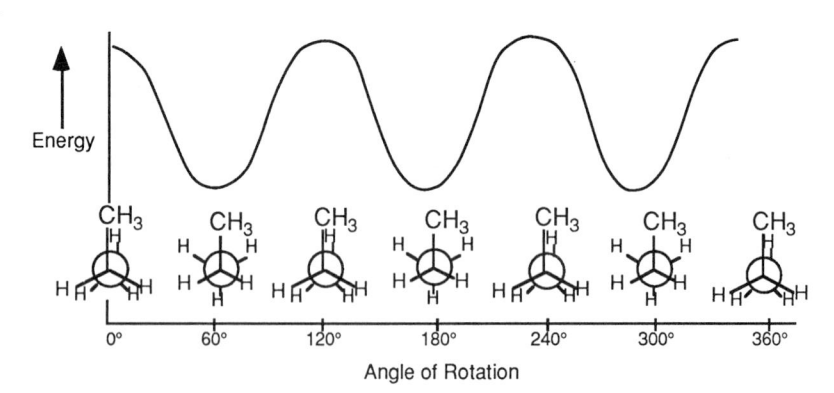

2.11

Energy

0°   60°   120°   180°   240°   300°   360°
Angle of Rotation

2.12

Staggered butane

Eclipsed butane

2.13
The first staggered conformation of butane (pictured above) is the most stable, because the large methyl groups are as far apart as possible.

2.14

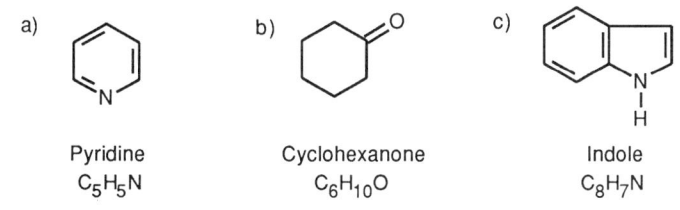

a) Pyridine C_5H_5N

b) Cyclohexanone $C_6H_{10}O$

c) Indole C_8H_7N

Although a particular shorthand structure has only one molecular formula, a particular molecular formula may represent many structures.

2.15
a) C_4H_8

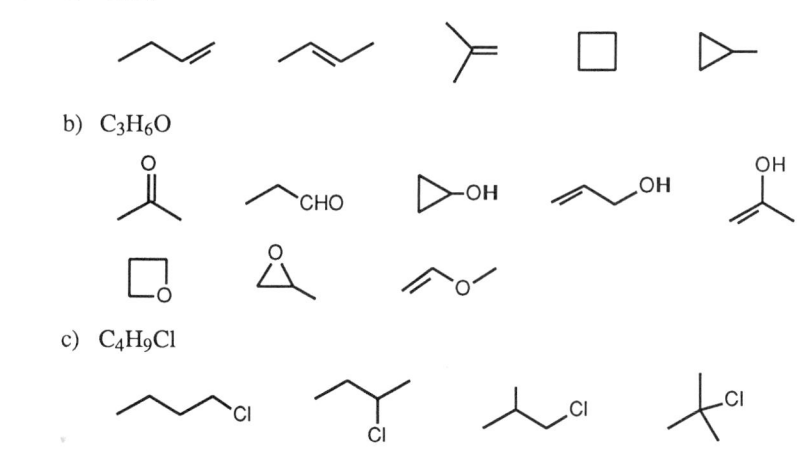

b) C_3H_6O

c) C_4H_9Cl

2.16

a) 1,4–Dimethylcyclohexane

b) 1–Ethyl–3–methylcyclopentane

c) Isopropylcyclobutane

2.17

a) 1-*tert*-Butyl-2-methylcyclopentane

b) 1,1–Dimethylcyclobutane

c) 1-Ethyl-4-isopropylcyclohexane

2.18

cis–1–Chloro–3–methylcyclopentane

2.19

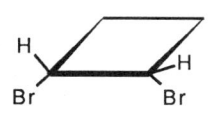

cis–1,2–Dibromocyclobutane

trans–1,2–Dibromocyclobutane

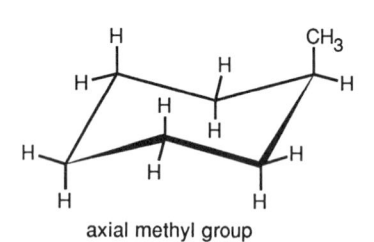

2.20

axial methyl group

equatorial methyl group

The Nature of Organic Compounds: Alkanes 21

2.21

The conformation having bromine in the equatorial position is more stable.

2.22 Make a model of *cis*–1,2–dichlorocyclohexane. Notice that all *cis* substituents are on the same side of the ring and that two adjacent *cis* substituents have an axial-equatorial relationship. Now perform a ring-flip on the cyclohexane.

After the ring-flip, the relationship of the two substituents is still axial-equatorial. No two adjacent *cis* substituents can be converted to being both axial or both equatorial without breaking bonds.

2.23 For a *trans*–1,2–disubstituted cyclohexane, two adjacent substituents must be either both axial or both equatorial.

A ring-flip converts two adjacent axial substituents into equatorial substituents, and *vice versa*. As in Problem 2.22, no two adjacent *trans* substituents can be converted to an axial-equatorial relationship without bond breaking.

22 Chapter 2

2.24

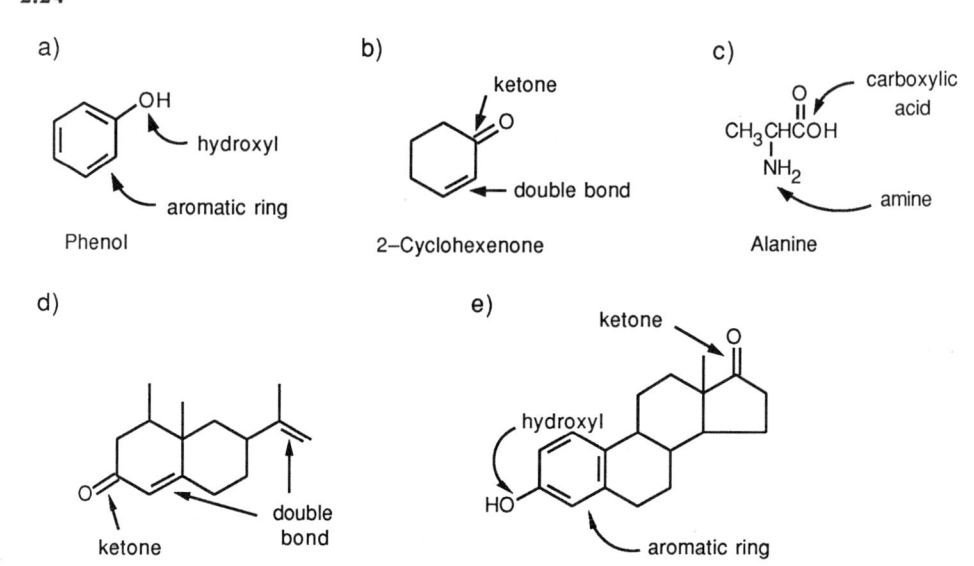

a) Phenol: OH — hydroxyl; aromatic ring

b) 2-Cyclohexenone: ketone; double bond

c) Alanine: CH₃CH(NH₂)COOH — carboxylic acid; amine

d) Nootkatone: ketone; double bond

e) Estrone: ketone; hydroxyl; aromatic ring

2.25 There are many acceptable answers to each part of this problem and of Problem 2.26. Correct answers include:

Shorthand structure

a) CH₃CH₂CH₂CH₂CH=CH₂

b) cyclopentene (H₂C–CH₂–CH₂–CH=CH ring)

c) CH₃CH₂COCH₂CH₃

d) CH₃CH₂CONHCH₃

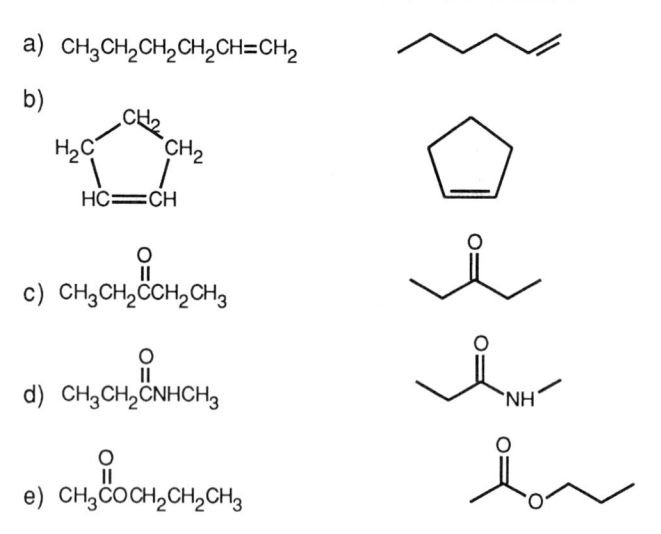

e) CH₃COOCH₂CH₂CH₃

The Nature of Organic Compounds: Alkanes 23

f) [structure: benzene ring with CH₂OH substituent]   [structure: benzene ring with OH substituent via CH₂]

2.26

Formula	Structure	Shorthand structure
a) C_7H_{14}	$CH_3CH_2CH_2CH_2CH_2CH=CH_2$	[zig-zag with terminal double bond]
b) C_3H_4	[cyclopropene structure: HC=CH with CH₂]	[triangle with double bond]
c) C_4H_8O	$CH_3CH_2\overset{O}{\underset{\|}{C}}CH_3$	[butan-2-one skeletal]
d) C_5H_9N	$CH_3CH_2CH_2CH_2C\equiv N$	[zig-zag with CN]
e) C_5H_8	$CH_3CH=CHCH=CH_2$	[pentadiene skeletal]
f) $C_4H_6O_2$	$\overset{O}{\underset{\|}{H\overset{\|}{C}}}CH_2CH_2\overset{O}{\underset{\|}{\overset{\|}{C}}}H$	[succinaldehyde skeletal with two CHO groups]

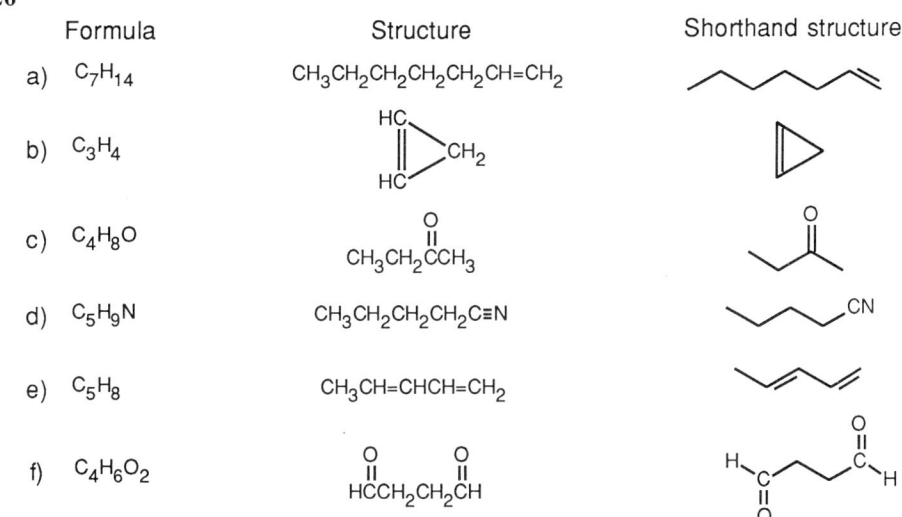

2.27

a) To solve this problem, you must examine all possibilities in a systematic way. The following procedure may be helpful.

1. Draw the simplest straight-chain parent alkane. (Here, it is $CH_3CH_2CH_2CH_3$.)

2. Find the number of different sites to which a functional group may be attached. (Here, CH_3- and CH_2- are the two possible sites for attaching the hydroxyl functional group.)

3. At each different site, replace an –H with an –OH, and draw the structure.

   $CH_3CH_2CH_2CH_2-OH$ and $CH_3CH_2\underset{\underset{OH}{|}}{C}HCH_3$

4. Draw the simplest branched alkane.

   $CH_3\underset{\underset{CH_3}{|}}{C}HCH_3$

5. Locate the number of different sites. (Here, there are two sites.)

Chapter 2

6. Replace an –H with an –OH, and draw the isomer.

$$CH_3CHCH_2-OH \quad \text{and} \quad CH_3CHCH_3$$
$$\quad\quad |\quad\quad\quad\quad\quad\quad\quad\quad\quad |\quad\quad\quad$$
$$\quad\quad CH_3\quad\quad\quad\quad\quad\quad\quad\quad OH\quad CH_3$$

7. Repeat steps 4-6 with the next simplest branched alkane. In this problem, we have already drawn all isomers.

b)
$CH_3CH_2CH_2CH_2CH_2NH_2$

$CH_3CH_2CH_2CH(NH_2)CH_3$

$CH_3CH_2CH(NH_2)CH_2CH_3$

$CH_3CH_2CH(CH_3)CH_2NH_2$

$CH_3CH_2C(NH_2)(CH_3)CH_3$

$CH_3CH(NH_2)CH(CH_3)CH_3$

$H_2NCH_2CH_2CH(CH_3)CH_3$

$CH_3C(CH_3)(CH_3)CH_2NH_2$

$CH_3CH_2CH_2CH_2NHCH_3$

$CH_3CH_2CH_2NHCH_2CH_3$

$CH_3CH_2CH(CH_3)NHCH_3$

$CH_3CHCH_2NHCH_3$ with CH_3 branch

$CH_3CN(CH_3)HCH_3$ with CH_3 branch

$CH_3CH_2NHCH(CH_3)CH_3$

$CH_3CH_2CH_2N(CH_3)CH_3$

$CH_3CH_2N(CH_3)CH_2CH_3$

$CH_3CH(CH_3)N(CH_3)CH_3$

There are 17 isomers of $C_5H_{13}N$. Nitrogen can be bonded to one, two or three alkyl groups.

c)
$CH_3CH_2CH_2COCH_3$

$CH_3CH_2COCH_2CH_3$

$CH_3CH(CH_3)COCH_3$

There are 3 ketone isomers with the formula $C_5H_{10}O$.

d)
$CH_3CH_2CH_2CH_2CHO$

$CH_3CH(CH_3)CH_2CHO$

$CH_3CH_2CH(CH_3)CHO$

$CH_3C(CH_3)(CH_3)CHO$

There are 4 isomeric aldehydes with the formula $C_5H_{10}O$. Remember that the aldehyde functional group can occur only at the end of a chain.

e)
$CH_3CH_2OCH_2CH_3$

$CH_3OCH_2CH_2CH_3$

$CH_3OCH(CH_3)CH_3$

There are 3 ethers with the formula $C_4H_{10}O$.

The Nature of Organic Compounds: Alkanes   25

f)
$$CH_3CH_2\overset{\overset{O}{\|}}{C}OCH_3 \quad CH_3\overset{\overset{O}{\|}}{C}OCH_2CH_3 \quad H\overset{\overset{O}{\|}}{C}OCH_2CH_2CH_3 \quad H\overset{\overset{O}{\|}}{C}O\overset{\overset{CH_3}{|}}{C}HCH_3$$

There are 4 esters with the formula $C_4H_8O_2$.

2.28

$CH_3CH_2CH_2CH_2CH_2Br \quad CH_3CH_2CH_2\overset{\overset{Br}{|}}{C}HCH_3 \quad CH_3CH_2\overset{\overset{Br}{|}}{C}HCH_2CH_3$

2.29

$CH_3\overset{\overset{CH_3}{|}}{C}HCH_2CH_2\overset{\overset{CH_3}{|}}{C}HCH_2Cl \quad CH_3\overset{\overset{CH_3}{|}}{C}HCH_2CH_2\overset{\overset{CH_3}{|}}{C}CH_3 \quad CH_3\overset{\overset{CH_3}{|}}{C}HCH_2\overset{\overset{CH_3}{|}}{C}HCHCH_3$
$\quad\quad\quad\quad\quad\quad\quad\quad\quad\quad\quad\quad\quad\quad\quad\quad\quad\quad\quad\quad\quad\quad Cl \quad\quad\quad\quad\quad\quad\quad\quad\quad\quad\quad\quad\quad\quad\quad Cl$

2.30

$CH_3CH_2CH_2OH \quad CH_3\overset{\overset{}{\underset{\underset{OH}{|}}{}}}{C}HCH_3 \quad CH_3OCH_2CH_3$

Three isomers have the formula C_3H_8O.

2.31

a) $CH_3\overset{\overset{CH_3}{|}}{\underset{\underset{CH_3}{|}}{C}}CH_3$

b) cyclohexane with CH₃ groups at 1,2,4,5 positions

c) benzene with CHCH₃ / CH₃ substituent (isopropylbenzene)

d) cyclohexane with CH₃ at 1,3 positions

e) $CH_3\overset{}{\underset{\underset{NH_2}{|}}{C}}HCH_3$

2.32

a) $\overset{\overset{O}{\|}}{C}$ sp^2

b) —C≡N sp

c) $\overset{}{C}-O-\overset{}{C}$ sp^3

d) $\overset{}{C}-OH$ sp^3

2.33

a) $H-\overset{\overset{H}{|}}{\underset{\underset{H}{|}}{C}}-\overset{\overset{H}{|}}{\underset{\underset{H}{|}}{C}}-\overset{\overset{H}{|}}{\underset{\underset{H}{|}}{C}}-\overset{\overset{H}{|}}{\underset{\underset{H}{|}}{C}}-H$

$H-\overset{\overset{H}{|}}{\underset{\underset{H}{|}}{C}}-\overset{\overset{H-\overset{\overset{H}{|}}{C}-H}{|}}{\underset{\underset{H}{|}}{C}}-\overset{\overset{H}{|}}{\underset{\underset{H}{|}}{C}}-H$

$H-\overset{\overset{H}{|}}{\underset{\underset{H}{|}}{C}}-\overset{\overset{H}{|}}{C}-\overset{\overset{H}{|}}{\underset{\underset{H}{|}}{C}}-H$
$\quad\quad\quad H-\overset{\overset{H}{|}}{\underset{\underset{H}{|}}{C}}-H$

same   same

26   Chapter 2

b)   H H Br H   H Br H H   H—C—H
      H—C—C—C—C—H   H—C—C—C—C—H   H H H
      H H H H   H H H H   H—C—C—C—H
                                            H Br H
      same   same   same

c)   CH₃   CH₃
      CH₃CHBrCHCH₃   CH₃CHCHBrCH₃   (CH₃)₂CHCHBrCH₂CH₃
      same   same

d)   OH   OH   OH
        ⌬—OH   ⌬   ⌬—OH
                           OH
      same   same

2.34

a)   CH₃CH₂CH₂CH₂CH₂CHCH₃   b)   CH₃CH₂CHCH₂CHCH₃
                | | |
               CH₃   CH₃   CH₂CH₃
      2-Methylheptane   4-Ethyl-2-methylhexane

c)   CH₃ CH₃   d)   CH₃ CH₃
      CH₃CH₂CH₂CH₂C—CHCH₂CH₃   CH₃CH₂CH₂CCH₂CHCH₃
                  | |
                CH₂CH₃   CH₃
      4-Ethyl-3,4-dimethyloctane   2,4,4-Trimethylheptane

e)   ⬠—CH₃   f)   CH₃CHCH₃
              CH₃ |
                                          CH₃CH₂CH₂CHCHCH₂CH₃
                                                     |
                                                   CH₃
      1,1-Dimethylcyclopentane   4-Isopropyl-3-methylheptane

2.35

a)   CH₃   b)   CH₃
      CH₃CH₂CH₂CHCHCH₃   CH₃CH₂CH₂CHCHCH₃
                | |
               CH₃   CH₂CH₂CH₂CH₃
      2,3-Dimethylhexane   4-Isopropyloctane

c) 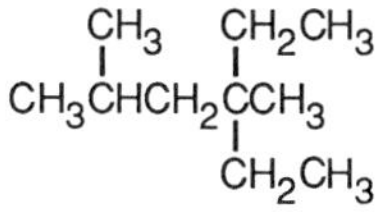 $\begin{array}{c}\text{CH}_3\ \ \text{CH}_2\text{CH}_3\\ |\ \ \ \ \ \ |\\ \text{CH}_3\text{CHCH}_2\text{CCH}_3\\ |\\ \text{CH}_2\text{CH}_3\end{array}$

4-Ethyl-2,4-dimethylhexane

d) 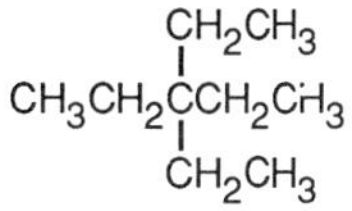 $\begin{array}{c}\text{CH}_2\text{CH}_3\\ |\\ \text{CH}_3\text{CH}_2\text{CCH}_2\text{CH}_3\\ |\\ \text{CH}_2\text{CH}_3\end{array}$

3,3-Diethylpentane

2.36

(a) 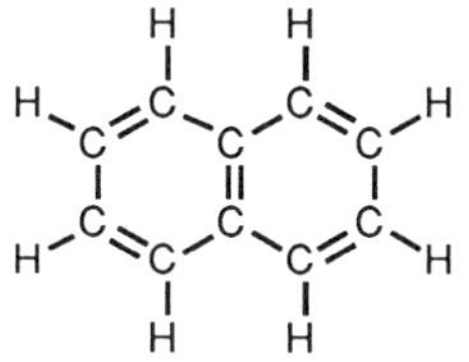 [structure of naphthalene shown with explicit H atoms, and line-drawing below]

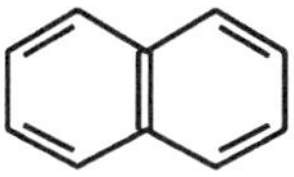

(b) 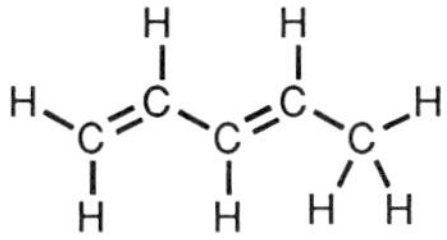 [structure of 1,3-pentadiene shown with explicit H atoms, and line-drawing below]

2.37

a) 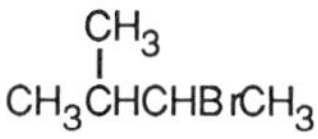 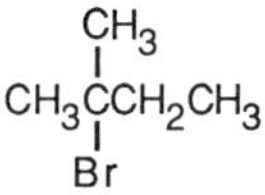 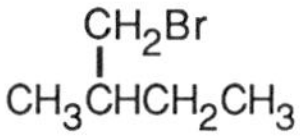 $\text{CH}_3\underset{|}{\overset{\text{CH}_3}{\text{C}}}\text{HCHBrCH}_3$ $\text{CH}_3\underset{\text{Br}}{\overset{\text{CH}_3}{\underset{|}{\overset{|}{\text{C}}}}}\text{CH}_2\text{CH}_3$ $\text{CH}_3\underset{}{\overset{\text{CH}_2\text{Br}}{\underset{|}{\text{CHCH}_2\text{CH}_3}}}$ plus other structures

b) 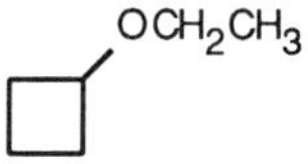 [cyclobutane with OCH₂CH₃ substituent]   Many other structures can be drawn.

c) 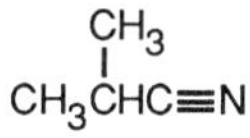 $\text{CH}_3\overset{\text{CH}_3}{\underset{|}{\text{CHC}}}\equiv\text{N}$

d) 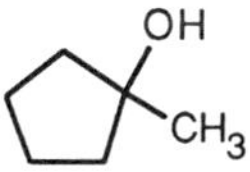 [cyclopentane with OH and CH₃ on same carbon]

A very large number of cyclic alcohols can be drawn.

e) No aldehyde isomer of this structure is possible.

f) 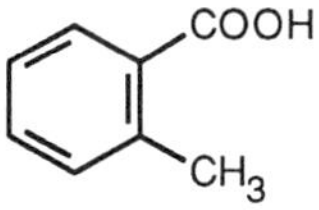 [benzene ring with COOH and CH₃ ortho substituents]

28 Chapter 2

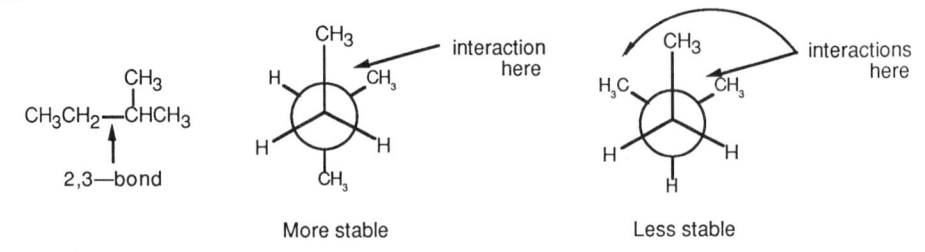

2.38

CH₃CH₂—CHCH₃ (with CH₃ branch)

2,3—bond

Newman projection (More stable): CH₃ groups with "interaction here" arrow

Newman projection (Less stable): with "interactions here" arrow

In the less stable isomer, there are two interactions between methyl groups that are 60° apart. Only one of these interactions occurs in the more stable isomer.

2.39

Lower energy Newman projection

Higher energy Newman projection — "strain worse here"

In the higher energy isomer, the two methyl groups that are eclipsed produce more strain than occurs in the other isomer.

2.40 Since *cis*–1–*tert*-butyl–4–methylcyclohexane exists in the conformation shown, a *tert*-butyl group must be much larger than a methyl group.

2.41

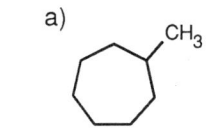

a) Methylcycloheptane

b) *cis*–1,3–Dimethylcyclopentane

c) *trans*–1,2–Dimethylcyclohexane

d) *trans*–1–Isopropyl–2–methylcyclobutane

e) 1,1,4–Trimethylcyclohexane

2.42

CH₃CH₂CH₂CH₂CH₂CH₃
Hexane

CH₃CH₂CH₂CHCH₃
          |
         CH₃
2-Methylpentane

CH₃CH₂CHCH₂CH₃
      |
     CH₃
3-Methylpentane

         CH₃
          |
CH₃CH₂CCH₃
          |
         CH₃
2,2-Dimethylbutane

       CH₃
        |
CH₃CHCHCH₃
          |
         CH₃
2,3-Dimethylbutane

2.43

CH₃CH₂CH₂CH₂CH₂CH₂CH₃
Heptane

              CH₃
               |
CH₃CH₂CH₂CH₂CHCH₃
2-Methylhexane

           CH₃
            |
CH₃CH₂CH₂CHCH₂CH₃
3-Methylhexane

         CH₃
          |
CH₃CH₂CH₂CCH₃
          |
         CH₃
2,2-Dimethylpentane

       CH₃
        |
CH₃CH₂CHCHCH₃
          |
         CH₃
2,3-Dimethylpentane

       CH₃
        |
CH₃CHCH₂CHCH₃
                |
              CH₃
2,4-Dimethylpentane

         CH₃
          |
CH₃CH₂CCH₂CH₃
          |
         CH₃
3,3-Dimethylpentane

CH₃CH₂CHCH₂CH₃
       |
      CH₂CH₃
3-Ethylpentane

   CH₃ CH₃
    | |
CH₃CH—CCH₃
           |
          CH₃
2,2,3-Trimethylbutane

2.44

a)
                        CH₃
                         |
CH₃CH₂CH₂CH₂CH₂CH₂CCH₃
                         |
                        CH₃
2,2-Dimethyloctane

b)
   CH₃ CH₂CH₃
    | |
CH₃CCH₂CCH₂CH₃
    | |
   CH₃ CH₂CH₃
4,4-Diethyl-2,2-dimethylhexane

c)
       CH₃
        \
         CH₃
        /
       CH₃
1,1,2-Trimethylcyclohexane

2.45

a) CH₃CH(CH₂CH₃)CH₂CH₂C(CH₃)₂CH₃   The longest chain is an octane.

correct name: 2,2,6–Trimethyloctane

b) CH₃CH(CH₃)CH(CH₂CH₃)CH₂CH₃   The longest chain is a hexane; numbering should start from the other end.

correct name: 3-Ethyl-2-methylhexane

c) CH₃CH₂C(CH₃)(CH₃)CH(CH₂CH₃)CH₂CH₃   Numbering should start from the other end.

correct name: 4-Ethyl-3,3-dimethylhexane

d) CH₃CH₂CH(CH₃)C(CH₃)(CH₃)CH₂CH₂CH₂CH₃   Numbering should start from the other end.

correct name: 3,4,4–Trimethyloctane

2.46 The strain energy of the highest energy conformation of bromoethane is 3.6 kcal/mol. Since this includes two H–H eclipsing interactions of 0.9 kcal/mol each, the value of an H–Br interaction is 3.6 – 2(0.9) = 1.8 kcal/mol.

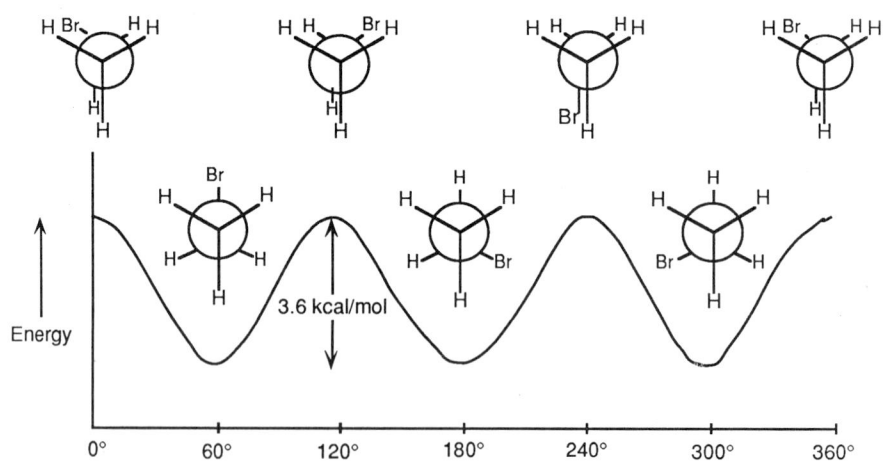

2.47

2.48 a) Because malic acid has 2 –COOH groups, the formula for the rest of the molecule is C_2H_4O. Possible structures for malic acid are:

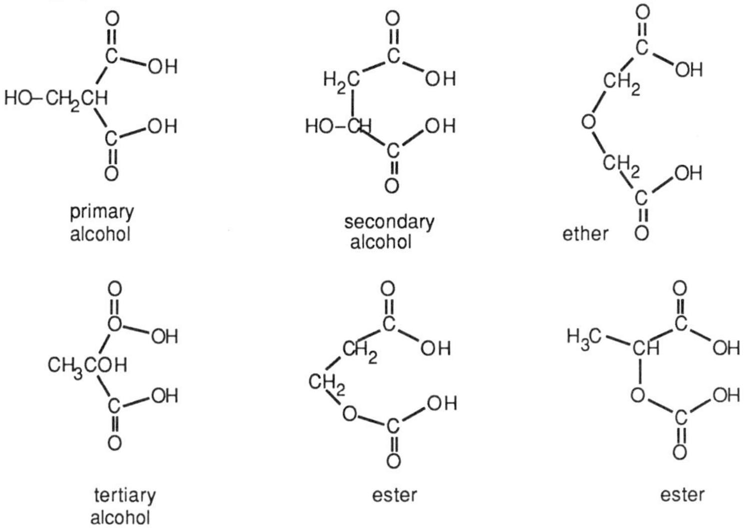

[Structures shown:
- primary alcohol: HO–CH₂CH(COOH)(COOH)
- secondary alcohol: H₂C(COOH)–CH(OH)(COOH)
- ether: (COOH)–CH₂–O–CH₂–COOH
- tertiary alcohol: CH₃C(OH)(COOH)(COOH)
- ester: CH₂(COOH)–CH₂–O–C(O)–OH
- ester: H₃C–CH(COOH)–O–C(O)–OH]

b) Because only one of these compounds (the second one) is also a secondary alcohol, it must be malic acid.

2.49

a)  $CH_2(CH_2Br)(CH_2Br)$ $\xrightarrow{2\ Na}$ cyclopropane ring (CH₂ with two CH₂) + 2 NaBr

b)  $C(CH_2Br)_4$ $\xrightarrow{4\ Na}$ spiro bicyclic structure (two cyclopropane-like rings sharing central C) + 4 NaBr

The two rings are perpendicular in order to keep the geometry of the central carbon as close to tetrahedral as possible.

2.50

a) 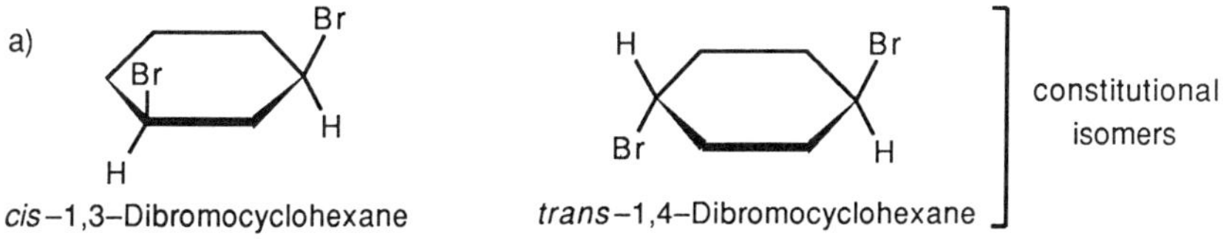 cis-1,3-Dibromocyclohexane and trans-1,4-Dibromocyclohexane — constitutional isomers

32 Chapter 2

b)

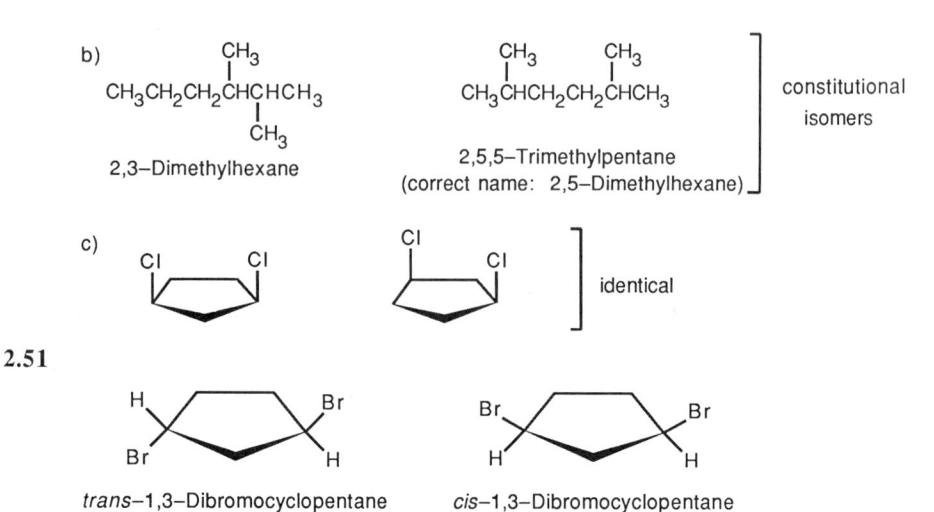

$$\underset{\text{2,3–Dimethylhexane}}{\text{CH}_3\text{CH}_2\text{CH}_2\overset{\overset{\text{CH}_3}{|}}{\underset{\underset{\text{CH}_3}{|}}{\text{CH}}}\text{CHCH}_3}$$

$$\underset{\substack{\text{2,5,5–Trimethylpentane}\\(\text{correct name: 2,5–Dimethylhexane})}}{\text{CH}_3\overset{\overset{\text{CH}_3}{|}}{\text{CH}}\text{CH}_2\text{CH}_2\overset{\overset{\text{CH}_3}{|}}{\text{CH}}\text{CH}_3}$$

] constitutional isomers

c) [cyclopentane with two Cl] [cyclopentane with two Cl]] identical

2.51

trans–1,3–Dibromocyclopentane   cis–1,3–Dibromocyclopentane

Many other constitutional isomers can also be drawn.

2.52

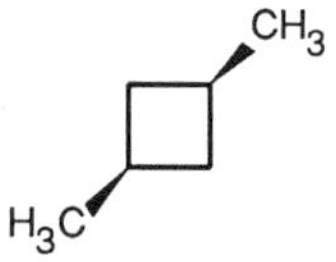

cis–1,3–Dimethylcyclobutane

2.53

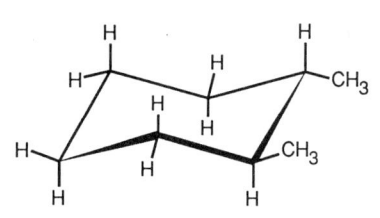

The methyl groups are equatorial in the more stable chair conformation of trans–1,2–dimethylcyclohexane.

2.54

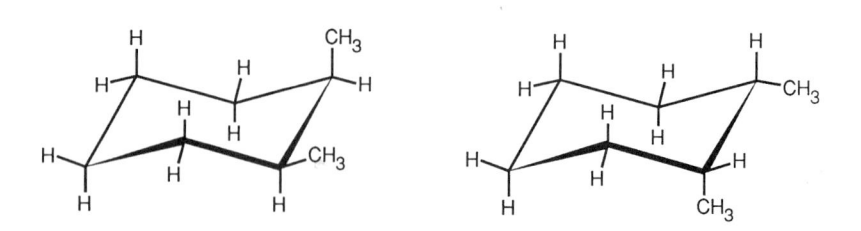

The Nature of Organic Compounds: Alkanes   33

There are two chair conformations of *cis*–1,2–dimethylcyclohexane that are of equal stability. In each conformation, one methyl group is axial and one is equatorial. Both *cis* conformations are less stable than the more stable conformation of *trans*–1,2–dimethylcyclohexane because of steric strain caused by the axial methyl group.

2.55

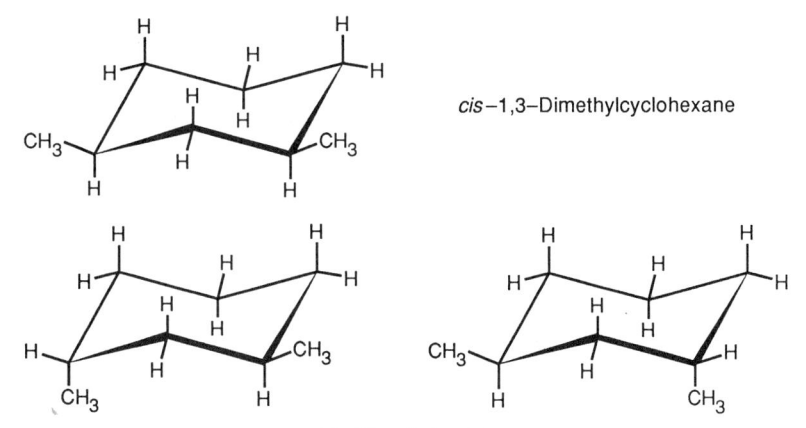

cis-1,3-Dimethylcyclohexane

trans-1,3-Dimethylcyclohexane

The lowest energy conformations of both dimethylcyclohexanes are drawn. *Cis*–1,3–Dimethylhexane is the more stable isomer because both methyl groups are equatorial in the most stable conformation. For *trans*–1,3–dimethylhexane, one methyl group must always be in the higher energy axial orientation. (A high energy diaxial conformation of *cis*–1,3–dimethylcyclohexane can also be drawn.)

2.56 Since the methyl group of *N*–methylpiperidine prefers an equatorial conformation, the steric requirements of a methyl group must be greater than those of an electron pair.

2.57

Glucose

2.58

A   B

Two cis–trans isomers of 1,3,5–trimethylcyclohexane are possible. In one isomer (A), all methyl groups are cis; in B, one methyl group is trans to the other two.

2.59

The two *trans*–1,2–dimethylcyclopentanes are mirror images.

Chapter Outline

I. Functional groups (Section 2.1)
II. Alkanes and alkyl groups (Sections 2.2 – 2.5, 2.7 – 2.8)
   A. Introduction to alkanes (Section 2.2)
      1. Aliphatic compounds
      2. Straight–chain alkanes
      3. Branched–chain alkanes
      4. Constitutional isomers
      5. Alkyl groups
   B. Naming branch–chain alkanes (Section 2.3)
   C. Properties of alkanes (Section 2.4)
   D. Conformations of ethane (Section 2.5)
      1. Conformers
      2. Sawhorse representations of alkanes
      3. Newman projections of alkanes
      4. Staggered conformation
      5. Eclipsed conformation
      6. Skew conformation
      7. Torsional strain
III. Drawing chemical structures (Section 2.6)
IV. Cycloalkanes (Sections 2.7 – 2.11)
   A. Naming cycloalkanes (Section 2.7)
   B. *Cis-trans* isomerism in cycloalkanes (Section 2.8)
   C. Conformations of cycloalkanes (Sections 2.9 – 2.11)
      1. Chair conformation of cyclohexane (Section 2.9)
      2. Axial and equatorial bonds in cyclohexane (Section 2.10)
      3. Conformational mobility of cyclohexane (Section 2.11)
         a. Ring-flips
         b. 1,3–Diaxial interactions
         c. Steric strain

Study Skills for Chapter 2

After studying this chapter, you should be able to:

1. Locate and identify functional groups in organic molecules (Problems 2.1, 2.2, 2.24, 2.25)
2. Draw all isomers of a given formula (Problems 2.3, 2.4, 2.5, 2.24, 2.26, 2.27, 2.28, 2.29, 2.30, 2.42, 2.43)
3. Name alkanes (Problems 2.8, 2.35, 2.44, 2.45)
4. Draw the structures of alkanes corresponding to a given name (Problems 2.9, 2.34, 2.42)
5. Identify carbon atoms as primary, secondary, tertiary, or quaternary (Problems 2.6, 2.7, 2.31)
6. Draw conformations of alkanes and assign energy values to them (Problems 2.10, 2.11, 2.12, 2.13, 2.38, 2.39, 2.46, 2.47)
7. Convert line structures to molecular formulas and vice versa (Problems 2.14, 2.15, 2.36)
8. Name cycloalkanes (Problems 2.16, 2.41)
9. Draw cycloalkane structures corresponding to a given name (Problems 2.17, 2.18, 2.19, 2.20, 2.53)
10. Draw the isomer of a given alkane or cycloalkane (Problems 2.37, 2.50, 2.51, 2.52, 2.58, 2.59)
11. Explain the stability of substituted cyclohexanes (Problems 2.21, 2.22, 2.23, 2.40, 2.54, 2.55, 2.56, 2.57)

Chapter 3 – Alkenes: The Nature of Organic Reactions

3.1

a) $\underset{1\ 2\ 3\ 4\ 5}{H_2C=CHCH_2\overset{\overset{\displaystyle CH_3}{|}}{C}HCH_3}$

1. Find the longest carbon chain containing the double bond, and name the parent compound. Here, the longest chain contains five carbons, and the compound is a *pentene*.

2. Number the carbon atoms, giving to the double bond the lowest possible number.

3. Identify the substituents. Here, there is a methyl group at C4.

4. Name the compound. Here, the name is 4–methyl–1–pentene.

b) $CH_3CH_2CH=CHCH_2CH_2CH_3$
3–Heptene

c) $H_2C=CHCH_2CH_2CH=CH_2CH_3$
1,5–Heptadiene

d) $CH_3CH_2CH=CHCH(CH_3)_2$
2–Methyl–3–hexene

3.2

a) 1,2–Dimethylcyclohexene

b) 4,4–Dimethylcycloheptene

c) 3–Isopropylcyclopentene

3.3

a) $CH_3CH_2CH_2CH_2\overset{\overset{\displaystyle CH_3}{|}}{C}=CH_2$
2–Methyl–1–hexene

b) $(CH_3)_3CCH=CHCH_3$
4,4–Dimethyl–2–pentene

c) $H_2C=CHCH_2CH_2\overset{\overset{\displaystyle CH_3}{|}}{C}=CH_2$
2–Methyl–1,5–hexadiene

d) $CH_3CH_2CH_2CH=\overset{\overset{\displaystyle CH_2CH_3}{|}}{C}(CH_3)_3$
3–Ethyl–2,2–dimethyl–3–heptene

3.4 Compounds (c) and (d) can exist as pairs of *cis–trans* isomers.

c) cis: Cl,Cl on same side / trans: Cl,H

d) cis: CH₃CH₂, CH₃ on same side / trans

Compounds (e) and (f) can exist as pairs of isomers that are better described by the *E, Z* nomenclature system, which is explained in Section 3.4.

Alkenes: The Nature of Organic Reactions 37

e) $\underset{H}{\overset{CH_3CH_2}{\diagup}}C=C\underset{Br}{\overset{CH_3}{\diagup}}$ and $\underset{H}{\overset{CH_3CH_2}{\diagup}}C=C\underset{CH_3}{\overset{Br}{\diagup}}$

f) $\underset{H}{\overset{CH_3CH_2CH_2}{\diagup}}C=C\underset{CH_3}{\overset{CH_2CH_3}{\diagup}}$ and $\underset{H}{\overset{CH_3CH_2CH_2}{\diagup}}C=C\underset{CH_2CH_3}{\overset{CH_3}{\diagup}}$

3.5

cis–2–Methyl–3–hexene   trans–2–Methyl–3–hexene

The *trans* isomer is more stable. The *cis* isomer has two large substituents, which cause steric strain, on the same side of the double bond.

3.6 A model of cyclohexene shows that a six-membered ring is too small to contain a *trans* double bond without causing severe strain to the ring.

3.7 Review the sequence rules of Section 3.4. In summary:

Rule 1: A high-atomic-weight atom has priority over a low-atomic-weight atom.

Rule 2: If a decision can't be reached by Rule 1, look at the second, third or fourth atom out until a decision can be made.

Rule 3: Multiple-bonded atoms are considered to be equivalent to the same number of single bonded atoms.

	High	Low	Rule
a)	–Br	–H	1
b)	–Br	–Cl	1
c)	–CH₂CH₃	–CH₃	2
d)	–OH	–NH₂	1
e)	–CH₂OH	–CH₃	2
f)	–CH=O	–CH₂OH	3

3.8

High	Low	Rule
$\underset{1\ \ 2\ \ 3}{-\overset{\overset{O}{\|}}{C}-O-CH_3}$	$\underset{1\ \ 2\ \ 3}{-\overset{\overset{O}{\|}}{C}-O-H}$	2

In this problem, the decision is made by considering the *third* atom away from the substituent. Here, –CH₃ is of higher priority than –H.

38   Chapter 3

3.9

High	Low	Rule
$\overset{2}{C}H_3$ $-\underset{\underset{2}{H}}{\overset{\overset{2}{C}H_3}{C}}-CH_3$	$\overset{2}{H}$ $-\underset{\underset{2}{H}}{\overset{\overset{2}{H}}{C}}-CH_2(CH_2)_5CH_3$	2

The "second" atoms in an isopropyl group are –C, –C, –H, which are of higher priority than –C, –H, –H of an n-octyl group.

3.10

a)   (h) CH$_3$O   Cl (h)
              \\   /
               C=C   Z
              /   \\
         (l) H   CH$_3$ (l)

First, consider substituents on the left-hand carbon. CH$_3$O– ranks higher than H– by Sequence Rule 1. On the right side, –Cl ranks higher than –CH$_3$. The isomer has Z configuration because the higher ranking substituents are on the same side of the double bond.

b)
     (h) H$_3$C   C–OCH$_3$ (l)
              \\   /   E
               C=C
              /   \\
         (l) H   OCH$_3$ (h)

3.11

a) CH$_3$Br + KOH $\longrightarrow$ CH$_3$OH + KBr   substitution reaction

b) CH$_3$CH$_2$OH $\longrightarrow$ H$_2$C=CH$_2$ + H$_2$O   elimination reaction

c) H$_2$C=CH$_2$ + H$_2$ $\longrightarrow$ CH$_3$CH$_3$ addition reaction

3.12 Use Figure 1.17 to identify the more electronegative element, and draw a δ^- over it. Draw a δ^+ over the more electropositive element.

a) $\overset{\delta^+}{\underset{/}{\diagdown}}C=\overset{\delta^-}{O}$ b) $-\overset{\delta^+}{\underset{|}{C}}-\overset{\delta^-}{Cl}$ c) $-\overset{\delta^+}{\underset{|}{C}}-\overset{\delta^-}{O}H$ d) $-\overset{\delta^-}{\underset{|}{C}}-\overset{\delta^+}{Li}$

   Ketone   Alkyl chloride   Alcohol   Alkyllithium

3.13

a) $CH_3\overset{\overset{\overset{\delta^-}{O}}{\|}}{\underset{\delta^+}{C}}CH_3$ b) $\overset{\delta^+}{CH_3CH_2}-\overset{\delta^-}{Cl}$ c) $\overset{\delta^+}{CH_3}-\overset{\delta^-}{S}H$ d) $CH_3\overset{\delta^-}{CH_2}-\overset{\overset{\overset{\delta^-}{CH_2CH_3}}{|\delta^+}}{\underset{\underset{\delta^-}{CH_2CH_3}}{\underset{|}{Pb}}}-\overset{\delta^-}{CH_2CH_3}$

   ketone   alkyl halide   thiol   organometallic

Alkenes: The Nature of Organic Reactions   39

3.14 *Electrophile or nucleophile?*   *Reason*
a) H⁺ is an electrophile.   Cations (electron-poor) are electrophiles.
b) HO:⁻ is a nucleophile.   Anions (electron-rich) are nucleophiles.
c) Br⁺ is an electrophile.   Cations are electrophiles.
d) :NH₃ is a nucleophile.   Compounds with lone pair electrons are usually nucleophiles.
e) HC≡CH is a nucleophile.   The electron-rich double bond is nucleophilic.
f) CO₂ is an electrophile.   The carbon is positively polarized and electron poor.

3.15

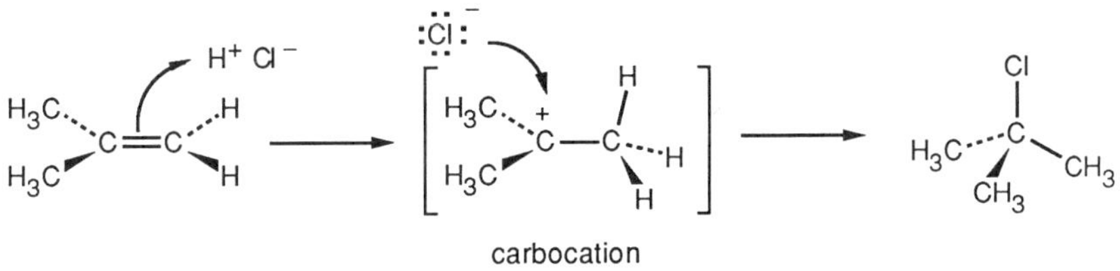

[reaction scheme showing alkene + H⁺Cl⁻ → carbocation intermediate → chlorinated product]

3.16

$CH_3CH_2CH=CHCH_3 \xrightarrow{H^+}$ [$CH_3CH_2CH_2\overset{+}{C}HCH_3$ and $CH_3CH_2\overset{+}{C}HCH_2CH_3$] $\xrightarrow{:\ddot{C}l:^-}$ $CH_3CH_2CH_2\underset{Cl}{C}HCH_3$ and $CH_3CH_2\underset{Cl}{C}HCH_2CH_3$

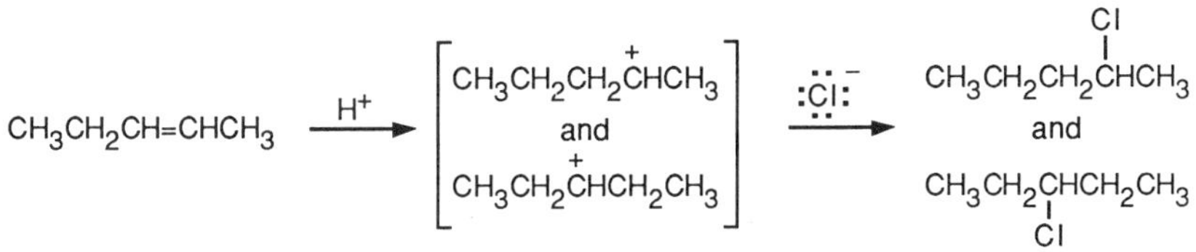

3.17 A reaction with $\Delta H = -10$ kcal/mol is more exothermic because its ΔH is negative.

3.18 A reaction with $K_{eq} = 1000$ more exothermic than one with $K_{eq} = 0.001$.

3.19 A reaction with $E_{act} = 15$ kcal/mol is faster than a reaction with $E_{act} = 20$ kcal/mol. It is not possible to predict the *size* of K_{eq} from E_{act}, because E_{act} measures the energy difference between reactant and transition state, not the energy difference between reactant and product, which is described by K_{eq}.

3.20 a) Exothermic one-step reaction: (b) Endothermic one-step reaction:

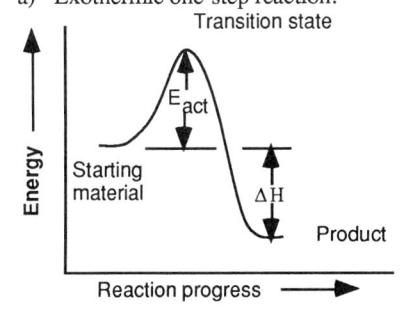

[energy diagrams showing (a) exothermic reaction with starting material higher than product, and (b) endothermic reaction with product higher than starting material; both labeled with Transition state, E_act, and ΔH]

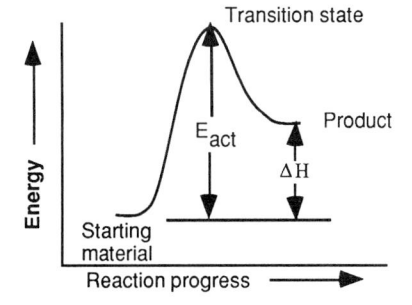

3.21

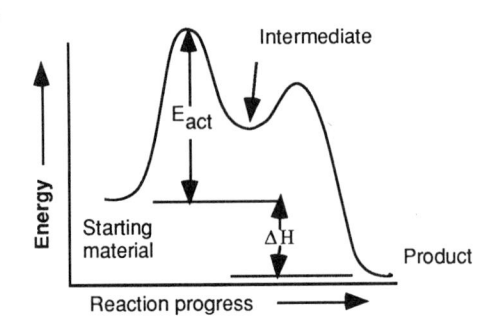

[Energy diagram showing a reaction progress with Starting material, an Intermediate between two peaks, E_act marking activation energy to the first (higher) peak, ΔH marking the enthalpy change from Starting material down to Product.]

3.22-3.23

a) $\overset{\delta+}{CH_3CH_2}\overset{\delta-}{C}\equiv N$ — nitrile

b) cyclopentyl–$\overset{\delta+}{O}$–$\overset{\delta+}{CH_3}$ with $\delta-$ on O — ether

c) $CH_3\overset{\delta-}{\underset{\delta+}{\overset{O}{\|}}}CCH_2\overset{\delta-}{\underset{\delta+}{\overset{O}{\|}}}C-\overset{\delta-}{O}-\overset{\delta+}{CH_3}$ — ketone, ester

d) para-benzoquinone with ketone C=O ($\delta+$ on C, $\delta-$ on O) and double bond labeled

e) CH₃–CH=CH–C(=O)–NH₂ with $\delta-$ on O, $\delta+$ on C, $\delta-$ on N — amide, double bond

f) benzaldehyde Ph–CHO with $\delta-$ on O, $\delta+$ on H — aldehyde, aromatic ring

3.24

a) Amphetamine: Ph–CH₂CH(NH₂)CH₃ — aromatic ring, amine (NH₂)

b) Thiamine — amine (NH₂ on pyrimidine), amine (ring N⁺), thiol (S in thiazolium), alcohol (CH₂CH₂OH), Cl⁻

Alkenes: The Nature of Organic Reactions   41

3.25   An *addition reaction* takes place when two reactants form a single product.

An *elimination reaction* takes place when one reactant splits apart to give two products.

A *substitution reaction* occurs when two reactants exchange parts to yield two different products.

A *rearrangement reaction* occurs when a reactant undergoes a reorganization of bonds to give a different product.

3.26   Nucleophiles: Cl⁻, CH₃NH₂, CN⁻

Electrophiles: Mg²⁺, CH₃⁺

3.27

a) 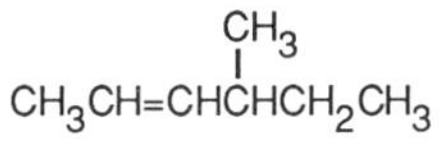   CH₃CH=CHCH(CH₃)CH₂CH₃
    4-Methyl-2-hexene

b) 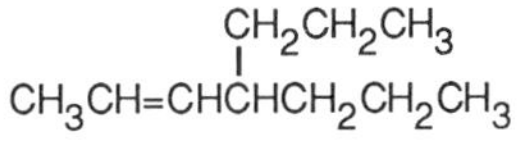   CH₃CH=CHCH(CH₂CH₂CH₃)CH₂CH₃
    4-Propyl-2-heptene

c) 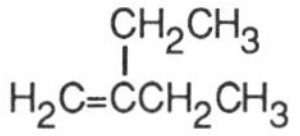   H₂C=C(CH₂CH₃)CH₂CH₃
    2-Ethyl-1-butene

d)   H₂C=C=CHCH₃
    1,2-Butadiene

3.28

a) 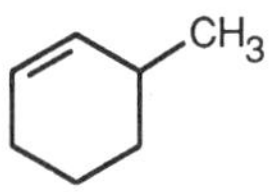   3-Methylcyclohexene

b) 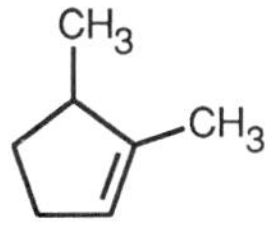   2,3-Dimethylcyclopentene

c) 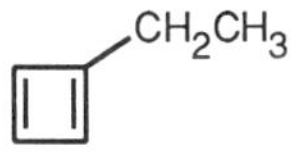   Ethyl-1,3-cyclobutadiene

d) 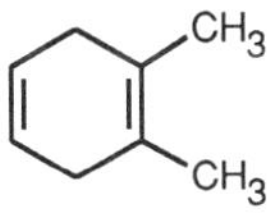   1,2-Dimethyl-1,4-cyclohexadiene

3.29

a) 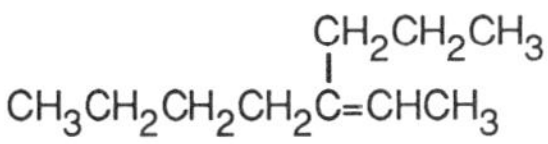   CH₃CH₂CH₂CH₂C(CH₂CH₂CH₃)=CHCH₃
    3-Propyl-2-heptene

b) 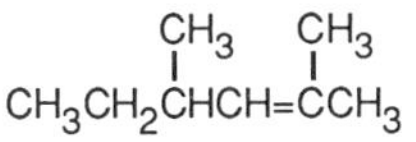   CH₃CH₂CH(CH₃)CH=C(CH₃)CH₃
    2,4-Dimethyl-2-hexene

c)   CH₃CH₂CH=CHCH₂CH₂CH=CH₂
    1,5-Octadiene

d) 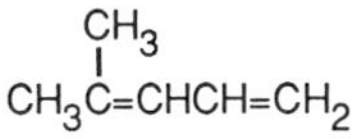   CH₃C(CH₃)=CHCH=CH₂
    4-Methyl-1,3-pentadiene

42 Chapter 3

e) cis-4,4-Dimethyl-2-hexene

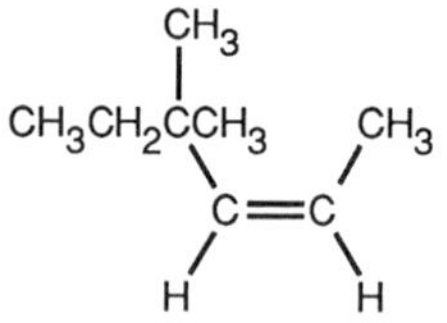

f) (E)-3-Methyl-3-heptene

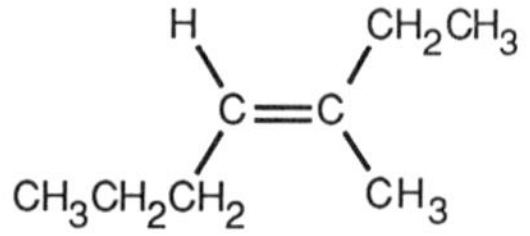

3.30

a) cis-4,5-Dimethylcyclohexene

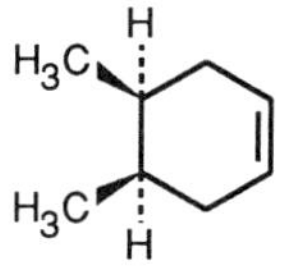

b) 3,3,4,4-Tetramethylcyclobutene

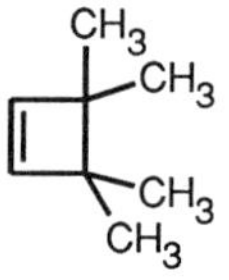

3.31

a) 3-Methylcyclopentene

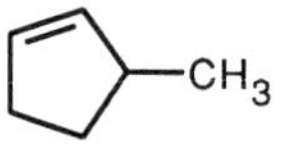

(The double bond of a cycloalkene occurs between C1 and C2.)

b) CH₃CH₂CH₂CH=CHCH₃

2-Hexene

(The methyl group is part of the carbon chain.)

c) 4-Ethylcycloheptene

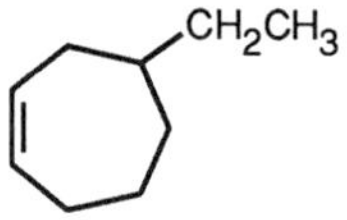

(The ethyl group receives the lowest possible number.)

d) 2-Ethyl-3-methylcyclohexene

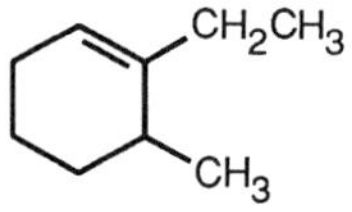

(Substituents should be listed alphabetically.)

3.32 a) No cis-trans isomerism.

b)

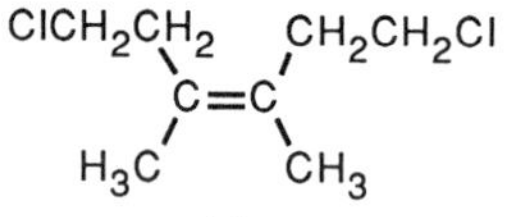

ClCH₂CH₂   CH₂CH₂Cl
     \\C=C/
     /   \\
   H₃C   CH₃
        Z

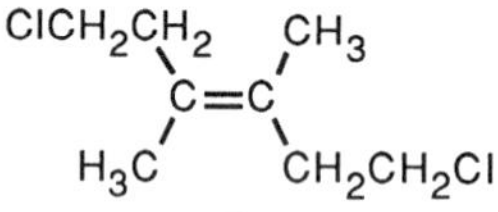

ClCH₂CH₂   CH₃
     \\C=C/
     /   \\
   H₃C   CH₂CH₂Cl
        E

c) Z, 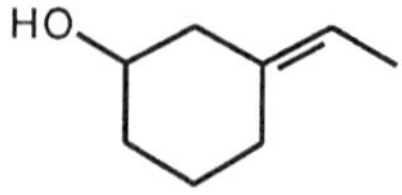 E

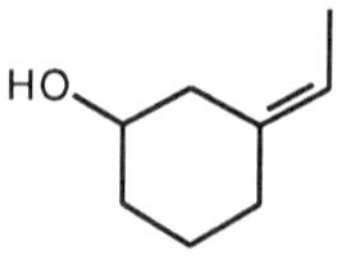

3.33

a)

CH₃CH₂CH₂CH₂CH=CH₂
1-Hexene

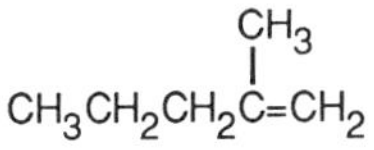

CH₃CH₂CH₂C(CH₃)=CH₂
2-Methyl-1-pentene

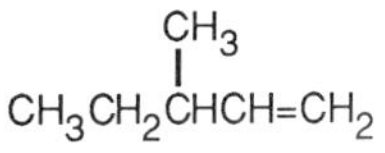

CH₃CH₂CH(CH₃)CH=CH₂
3-Methyl-1-pentene

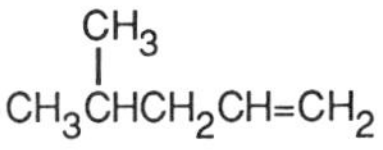

CH₃CH(CH₃)CH₂CH=CH₂
4-Methyl-1-pentene

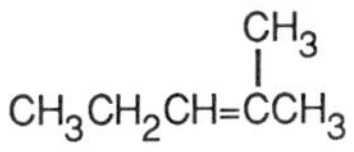

CH₃CH₂C(CH₃)=CHCH₃
2-Methyl-2-pentene

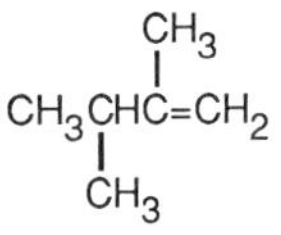

(CH₃)₂CHC(CH₃)=CH₂
2,3-Dimethyl-1-butene

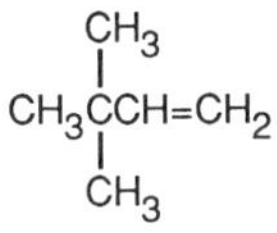

(CH₃)₃CCH=CH₂
3,3-Dimethyl-1-butene

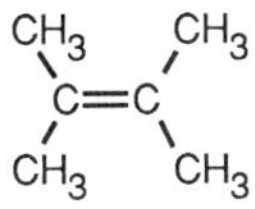

(CH₃)₂C=C(CH₃)₂
2,3-Dimethyl-2-butene

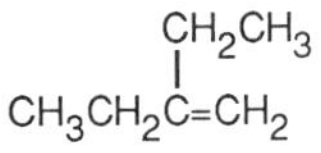

CH₃CH₂C(CH₂CH₃)=CH₂
2-Ethyl-1-butene

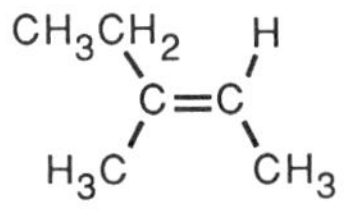

b) E-3-Methyl-2-pentene — CH₃CH₂ and H on one carbon (cis to each other on the left), H₃C and CH₃ on the other.

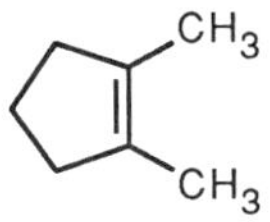

c) 1,2-Dimethylcyclopentene

3.34

CH₃CH₂CH=CH₂   CH₃CH=CHCH₃   CH₃C(CH₃)=CH₂
1-Butene   2-Butene   2-Methylpropene

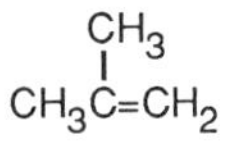

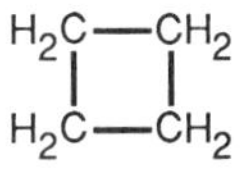

Cyclobutane

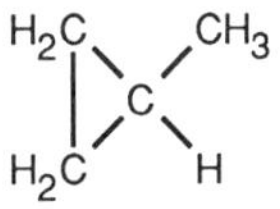

Methylcyclopropane

3.35 Of the above structures, only 2–butene shows *cis–trans* isomerism.

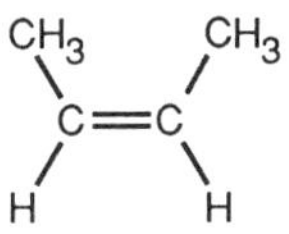

cis-2-Butene (both CH₃ on same side)

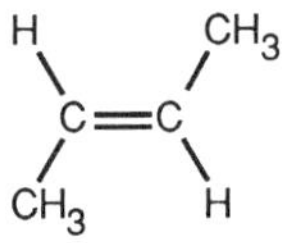

trans-2-Butene (CH₃ groups on opposite sides)

3.36

a) C₆H₁₀

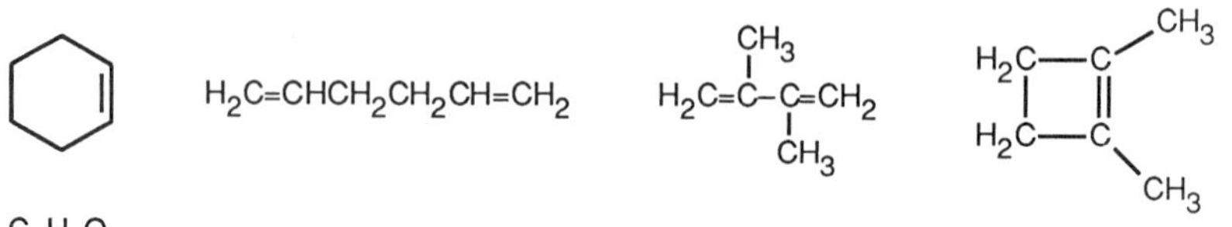

[cyclohexene]   H₂C=CHCH₂CH₂CH=CH₂   H₂C=C(CH₃)-C(CH₃)=CH₂   [1,2-dimethylcyclobutene]

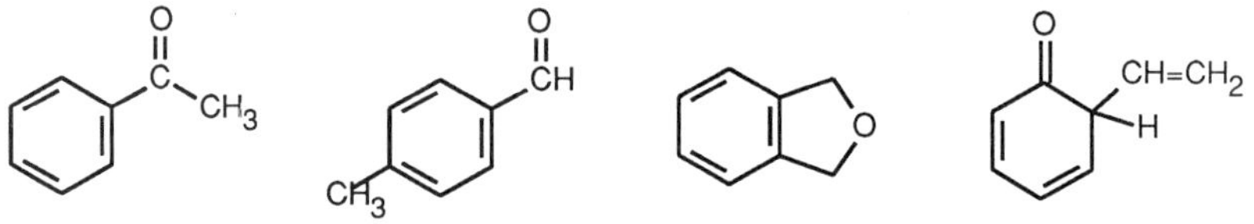

b) C₈H₈O

[PhC(=O)CH₃]   [4-methylbenzaldehyde]   [phthalan/isobenzofuran]   [cyclohexadienone with CH=CH₂]

c) C₇H₁₀Cl₂

[3,3-dichloro-1-methylcyclohexene]   ClCH=CHC(CH₃)(CH₃)CH=CHCl   [dichlorocyclopentane with =CHCH₃]   [dichloronorbornane]

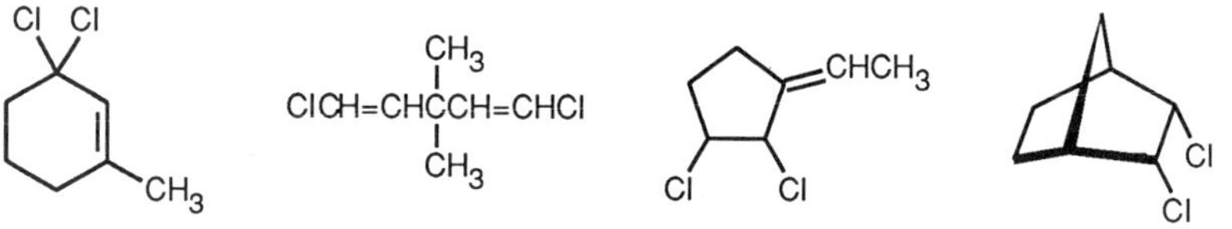

Many other structures corresponding to these formulas can be drawn.

3.37 As was explained in Problem 3.6, a six membered ring is too small to contain a *trans* double bond without causing severe ring strain. A model shows that a ten-membered ring is flexible enough to include either a *cis* or a *trans* double bond, although the *cis* isomer has less ring strain than the *trans* isomer.

3.38 Highest priority ———> Lowest priority

a) –I, –Br, –CH₃, –H
b) –OCH₃, –OH, –COOH, –H
c) –COOH, –CHO, –CH₂OH, –CH₃
d) –CH=CH₂, –CH(CH₃)₂, –CH₂CH₃, –CH₃

3.39

a) 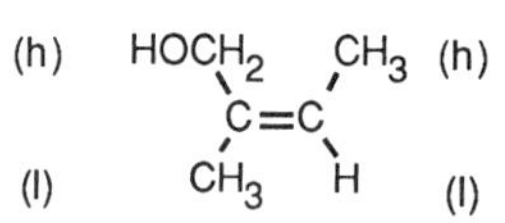 (h) HOCH₂   CH₃ (h)   Z
       C=C
   (l)   CH₃   H (l)

b) 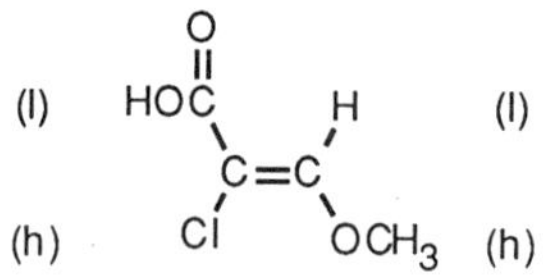 (l) HOOC   H (l)   Z
       C=C
   (h) Cl   OCH₃ (h)

Alkenes: The Nature of Organic Reactions 45

3.40

CH$_3$CH$_2$CH$_2$CH=CH$_2$
1-Pentene

CH$_3$CH$_2$CH=CHCH$_3$
2-Pentene

CH$_3$CH$_2$C(CH$_3$)=CH$_2$
2-Methyl-1-butene

CH$_3$CH(CH$_3$)CH=CH$_2$
3-Methyl-1-butene

CH$_3$CH=C(CH$_3$)CH$_3$
2-Methyl-2-butene

3.41

Menthene (1-isopropyl-4-methylcyclohexene)

3.42

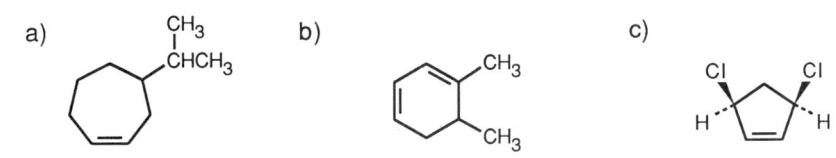

a) 4-Isopropylcycloheptene

b) 1,6-Dimethyl-1,3-cyclohexadiene

c) cis-3,5-Dichlorocyclopentene

3.43

Electrophile: Zn^{2+}

Nucleophiles: CH$_3$NH$_2$, CH$_3$C(=O)O:$^-$, HS:$^-$

3.44-3.45

α-Farnesene

(3E,6E)-3,7,11-Trimethyl-1,3,6,10-dodecatetraene

46   Chapter 3

3.46 (a) A *polar reaction* is a process that involves electronically unsymmetrical bond breaking and bond formation. In a polar reaction, electron-rich sites in the functional groups of one molecule react with electron-poor sites in the functional groups of another molecule.

(b) A *radical reaction* is a reaction in which odd-electron species are produced or consumed.

(c) A *functional group* is a group of atoms within a molecule that has a characteristic reactivity.

(d) A *reaction intermediate* is a structure, often quite reactive, that is formed during the course of a multi-step reaction and that lies on an energy minimum between two transition states.

3.47

a) H^+ — proton

b) $:\ddot{Br}:^-$ — bromide anion

c) an aldehyde (H–C(=O)–)

3.48

$$K_{eq} = \frac{[\text{Products}]}{[\text{Reactants}]} \qquad \text{Here, } K_{eq} = 0.001$$

In this reaction, 0.001 moles of product are formed for every mole of reactants. It is apparent that formation of product is not favored and that the reaction is endothermic.

3.49 A reaction with $E_{act} = 5$ kcal/mol at room temperature is likely to be fast since 5 kcal/mol is a small value for E_{act}, and a small E_{act} indicates a fast reaction.

3.50 A reaction with $\Delta H = 12$ kcal/mol at room temperature is endothermic, since a positive ΔH indicates an endothermic reaction. It is not possible to predict if the reaction is fast or slow; reaction rate is determined by E_{act}, not ΔH.

3.51

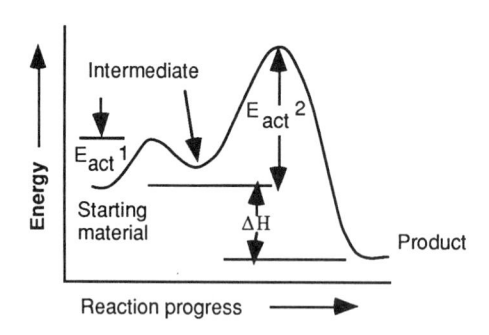

The first step is faster because $E_{act\ 1}$ is smaller than $E_{act\ 2}$.

3.52

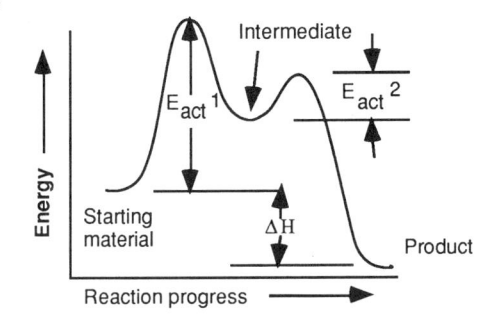

In this reaction, the second step is faster because $E_{act\ 2}$ is smaller than $E_{act\ 1}$.

3.53

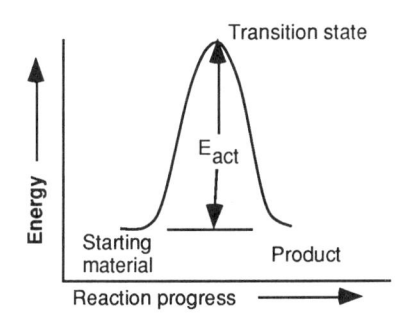

A reaction with $K_{eq} = 1$ has $\Delta H = 0$ (when ΔS also $= 0$).

3.54 Transition states and intermediates are both relatively unstable species that are produced during a reaction. A transition state represents a structure occurring at an energy maximum. An intermediate occurs at an energy minimum between two transition states. Even though an intermediate may be of such high energy that it cannot be isolated, it is still of lower energy than a transition state.

3.55

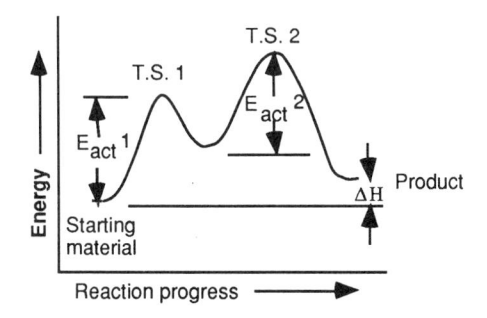

There are two steps in the reaction. The second step is faster because $E_{act\ 2}$ is smaller than $E_{act\ 1}$. There are two transition states.

48   Chapter 3

3.56 a) The reaction is a polar rearrangement.

b) [reaction scheme]

c) $K_{eq} = \dfrac{[\text{Products}]}{[\text{Reactants}]} = \dfrac{.70}{.30} = 2.3$

Chapter Outline

I. Alkenes (Sections 3.1 - 3.4)
   A. Naming of Alkenes (Section 3.1)
   B. Electronic structure of alkenes (Section 3.2)
   C. Isomerism in alkenes (Sections 3.3 - 3.4)
      1. cis - trans isomerism (Section 3.3)
      2. E, Z isomerism (Section 3.4)
II. Organic reactions (Sections 3.5 - 3.11)
   A. Kinds of reactions (Section 3.5)
      1. Addition reactions
      2. Elimination reactions
      3. Substitution reactions
      4. Rearrangement reactions
   B. How reactions occur: mechanisms (Sections 3.6 - 3.8)
      1. Radical reactions -- homolytic processes (Section 3.6)
      2. Polar reactions -- heterolytic processes;
         a. General features -- nucleophiles and electrophiles
         b. An example: addition of HCl to ethylene (Section 3.7)
         c. The mechanism of addition: carbocation intermediates (Section 3.8)
   C. Describing a reaction (Sections 3.9 - 3.11)
      1. Rates and equilibria (Section 3.9)
         a. Equilibrium constant, K_{eq}
         b. Heat of reaction, ΔH
      2. Reaction energy diagrams (Section 3.10)
         a. Transition state
         b. Energy of activation, E_{act}
      3. Intermediates (Section 3.11)

Study Skills for Chapter 3

After studying this chapter, you should be able to:

1. Name alkenes and cycloalkenes (Problems 3.1, 3.2, 3.27, 3.28, 3.31, 3.33, 3.34, 3.40, 3.42, 3.44).
2. Draw the structures of alkenes and cycloalkenes corresponding to given names (Problems 3.3, 3.29, 3.30, 3.33, 3.34, 3.40, 3.41).
3. Identify double-bond substituents as *cis* or *trans* (Problems 3.4, 3.5, 3.6, 3.32, 3.35, 3.37).
4. Assign priorities to double-bond substituents according to the sequence rules (Problems 3.7, 3.8, 3.9, 3.38).
5. Assign *E, Z* configurations to double bonds (Problems 3.10, 3.39, 3.45).
6. Classify organic reactions (Problems 3.11, 3.56).
7. Indicate the polarity of bonds in functional groups (Problems 3.12, 3.13, 3.23).
8. Identify chemical species as nucleophiles or electrophiles (Problems 3.14, 3.26, 3.43).
9. Formulate the mechanism of electrophilic addition reactions (Problems 3.14, 3.16, 3.56).
10. Use ΔH and E_{act} to predict the favorability and rate of reactions (Problems 3.17, 3.18, 3.19, 3.48, 3.49, 3.50).
11. Draw and label reaction energy diagrams (Problems 3.20, 3.21, 3.51, 3.52, 3.53, 3.55).
12. Define the important terms in this chapter (Problems 3.25, 3.46, 3.47, 3.54).

Chapter 4 – Alkenes and Alkynes

4.1

a)
    one alkyl group   no alkyl groups

$$CH_3CH_2CH=CH_2 + HCl \longrightarrow CH_3CH_2CHClCH_3$$

H^+ adds to the carbon with fewer alkyl groups.

b) $(CH_3)_2C=CHCH_2CH_3 + HI \longrightarrow (CH_3)_2CICH_2CH_2CH_3$

c)  cyclohexene + HCl ⟶ chlorocyclohexane

4.2

a)  Cyclopentene + HBr ⟶ Bromocyclopentane

b) $CH_3CH_2CH=CHCH_2CH_3 + HBr \longrightarrow CH_3CH_2CHBrCH_2CH_2CH_3$
    3–Hexene

c) 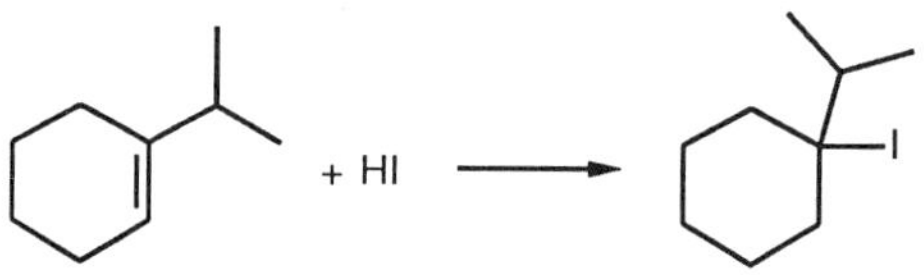 1-Isopropylcyclohexene + HI ⟶ 1-Iodo-1-isopropylcyclohexane

d) 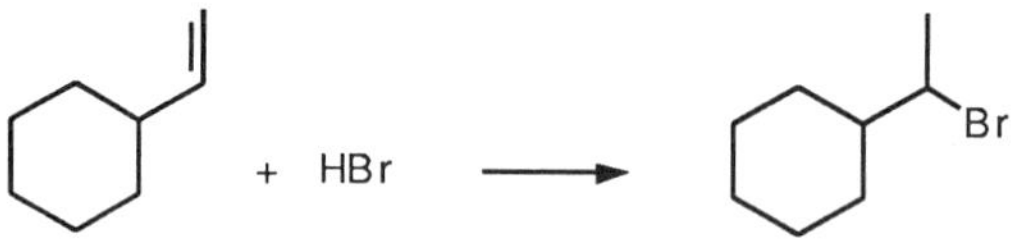 (methylenecyclohexane) + HBr ⟶ (1-bromoethylcyclohexane)

Only this alkene starting material will give the desired product.

4.3

a)
$$CH_3CH_2\underset{CH_3}{\underset{|}{C}}=CH\underset{CH_3}{\underset{|}{C}}HCH_3 + H^+ \longrightarrow \left[CH_3CH_2\underset{CH_3}{\underset{|}{\overset{+}{C}}}-CH_2\underset{CH_3}{\underset{|}{C}}HCH_3 \right] \xrightarrow{Br^-} CH_3CH_2\underset{Br}{\underset{|}{\overset{CH_3}{\underset{|}{C}}}}-CH_2\underset{CH_3}{\underset{|}{C}}HCH_3$$

tertiary carbocation

b) cyclopentylidene=CHCH₃ + H⁺ ⟶ [cyclopentyl⁺–CH₂CH₃] (tertiary carbocation) —I⁻→ cyclopentane with CH₂CH₃ and I substituents

4.4

a) $CH_3CH_2\underset{CH_3}{\underset{|}{C}}=CHCH_2CH_3 + H_2O \xrightarrow[\text{catalyst}]{H^+} CH_3CH_2\underset{CH_3}{\underset{|}{\overset{OH}{\underset{|}{C}}}}CH_2CH_2CH_3$

b) cyclopentene-CH₃ + H₂O → cyclopentane with CH₃ and OH (H⁺ catalyst)

c) $CH_3CH_2\underset{CH_3}{\underset{|}{C}}HCH_2CH=\underset{CH_3}{\underset{|}{C}}-CH_3 + H_2O \xrightarrow[\text{catalyst}]{H^+} CH_3CH_2\underset{CH_3}{\underset{|}{C}}HCH_2CH_2\underset{OH}{\underset{|}{\overset{CH_3}{\underset{|}{C}}}}-CH_3$

4.5

a) $CH_3CH=CHCH_3$
   or
   $CH_3CH_2CH=CH_2$ + H₂O $\xrightarrow[\text{catalyst}]{H^+}$ $CH_3CH_2\underset{OH}{\underset{|}{C}}HCH_3$

b) $CH_3CH_2\underset{CH_3}{\underset{|}{C}}=CHCH_3$
   or
   $CH_3CH_2\underset{\underset{CH_2}{\|}}{C}CH_2CH_3$ + H₂O $\xrightarrow[\text{catalyst}]{H^+}$ $CH_3CH_2\underset{CH_3}{\underset{|}{\overset{OH}{\underset{|}{C}}}}CH_2CH_3$

Chapter 4

c) [structure: 1,2-dimethylcyclohexene]

or

[structure: 2,3-dimethylcyclohexene] + H₂O —H⁺ catalyst→ [structure: 1,2-dimethylcyclohexanol]

or

[structure: methylene-methylcyclohexane]

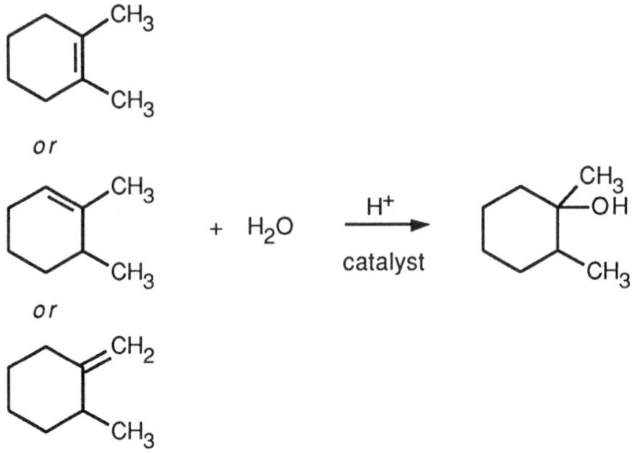

4.6-4.7

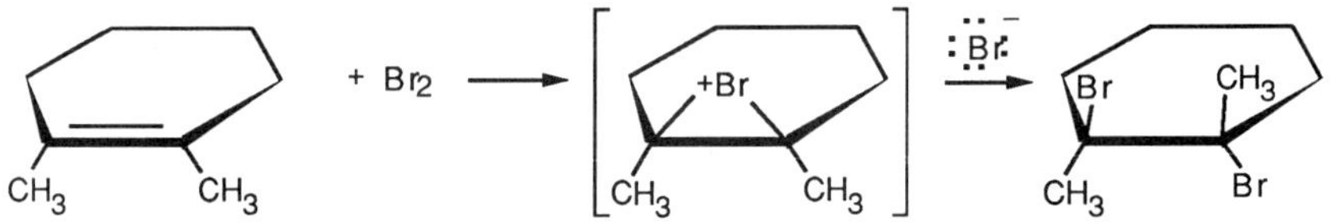

[1,2-Dimethylcyclohexene] + Br₂ ⟶ [Bromonium ion intermediate] ⟶ [trans-1,2-Dibromo-1,2-dimethylcyclohexane]

4.8

a) $(CH_3)_2C=CHCH_2CH_3$ —H₂, Pd catalyst→ $(CH_3)_2CHCH_2CH_2CH_3$

b) [1,1-dimethylcyclopentene structure] —H₂, Pd catalyst→ [1,1-dimethylcyclopentane structure]

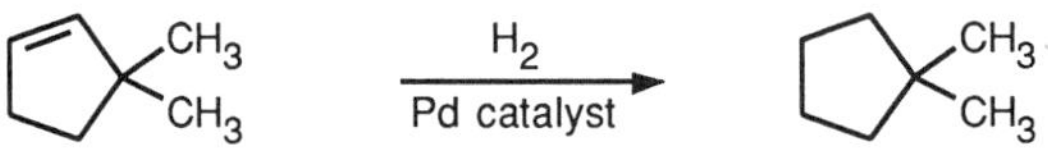

4.9

In both of these reactions, the double bond is cleaved. The product contains an oxygen atom double-bonded to the carbon at each end of the original double bond.

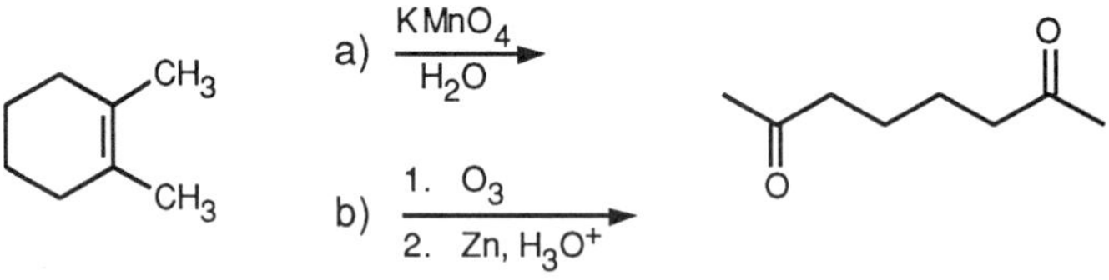

[1,2-dimethylcyclohexene]

a) KMnO₄ / H₂O

b) 1. O₃ 2. Zn, H₃O⁺

→ [diketone product: 2,7-octanedione structure]

4.10

To draw the correct alkene, remove oxygen and connect the remaining fragments by drawing a double bond between them.

a) $(CH_3)_2C=CH_2$ —1. O₃; 2. Zn, H₃O⁺→ $(CH_3)_2C=O$ + $O=CH_2$

Alkenes and Alkynes 53

b) $CH_3CH_2CH=CHCH_2CH_3$ $\xrightarrow{\text{1. }O_3}{\text{2. Zn, }H_3O^+}$ $CH_3CH_2CH=O + O=CHCH_2CH_3$

c) [cyclohexane ring]=CHCH_3 $\xrightarrow{\text{1. }O_3}{\text{2. Zn, }H_3O^+}$ [cyclohexanone] + O=CHCH_3

4.11

$$CH_2=\underset{CH_3}{\overset{|}{CH}} \longrightarrow \left[CH_2\underset{CH_3}{\overset{|}{CH}}CH_2\underset{CH_3}{\overset{|}{CH}}CH_2\underset{CH_3}{\overset{|}{CH}} \right]$$

Propylene → Polypropylene

4.12

$$CH_3CH_2\underset{Br}{\overset{CH_3}{\underset{|}{\overset{|}{C}}}}CH_3 \xrightarrow{KOH} CH_3CH=\underset{CH_3}{\overset{|}{C}}CH_3 + CH_3CH_2\underset{CH_3}{\overset{|}{C}}=CH_2$$

2-Bromo-2-methylbutane

2-Methyl-2-butene major
(more substituted double bond)

2-Methyl-1-butene minor

4.13

a) $CH_3\underset{|}{\overset{CH_3}{CH}}CH_2CH_2\underset{|}{\overset{CH_3}{CH}}CH_2CH_2Br + KOH \xrightarrow{CH_3CH_2OH} CH_3\underset{|}{\overset{CH_3}{CH}}CH_2CH_2\underset{|}{\overset{CH_3}{CH}}CH=CH_2$

Only this alkyl halide gives the desired alkene as the major product.

b) [cyclopentane with Br, two CH_3 groups] + KOH $\xrightarrow{CH_3CH_2OH}$ [cyclopentene with two CH_3 groups]

4.14

a) $CH_3CH_2CH_2\underset{Br}{\overset{CH_3}{\underset{|}{\overset{|}{C}}}}CH_3 \xrightarrow{KOH}{C_2H_5OH} CH_3CH_2CH=\underset{CH_3}{\overset{|}{C}}CH_3 + CH_3CH_2CH_2\underset{CH_3}{\overset{|}{C}}=CH_2$

2-Bromo-2-methylpentane

2-Methyl-2-pentene major

2-Methyl-1-pentene minor

54   Chapter 4

b)

$$\underset{\underset{CH_3}{|}}{CH_3CH-\overset{\overset{H_3C}{|}}{\underset{|}{C}}-CH_2CH_3} \xrightarrow{H_2SO_4} \underset{H_3C}{\overset{H_3C}{>}}C=C\underset{CH_3}{\overset{CH_2CH_3}{<}} + \underset{H_3C}{\overset{CH_3CH(CH_3)}{>}}C=C\underset{H}{\overset{CH_3}{<}} +$$

major   minor

$$\underset{H_3C}{\overset{CH_3CH(CH_3)}{>}}C=C\underset{CH_3}{\overset{H}{<}} + \underset{CH_3CH_2}{\overset{CH_3CH(CH_3)}{>}}C=CH_2$$

minor   minor

4.15

a) [cyclohexane with two CH₃ groups and OH] $\xrightarrow{H_2SO_4}$ [cyclohexene with two CH₃, major] + [cyclohexene with two CH₃, minor]

major   minor

b) $\underset{\underset{}{\overset{OH}{|}}}{CH_3CH_2CH_2CHCH_2CH_2CH_3} \xrightarrow{H_2SO_4} CH_3CH_2CH=CHCH_2CH_2CH_3$

Whenever possible, use the alcohol that gives the desired product as the only product.

4.16

$$CH_2=CH-CH=CH_2$$
$$\downarrow Br_2$$

$$\left[\overset{+}{C}H_2-CH=CH-\underset{\underset{}{\overset{Br}{|}}}{C}H_2 \longleftrightarrow CH_2=CH-\underset{\underset{}{\overset{Br}{|}}}{\overset{+}{C}H}-CH_2 \right]$$

$$\downarrow :\ddot{B}\ddot{r}:^-$$

$$\underset{\underset{}{\overset{Br}{|}}}{CH_2}-CH=CH-\underset{\underset{}{\overset{Br}{|}}}{CH_2} \quad + \quad CH_2=CH-\underset{\underset{}{\overset{Br}{|}}}{CH}-\underset{\underset{}{\overset{Br}{|}}}{CH_2}$$

1,4–addition   1,2–addition

4.17

CH$_3$CH=CHCH=CH$_2$ 1,3-Pentadiene

	Product	Name	Results from:
(1)	CH$_3$CHClCH=CHCH$_3$	4-Chloro-2-pentene	1,2 addition 1,4 addition
(2)	CH$_3$CH$_2$CHClCH=CH$_2$	3-Chloro-1-pentene	1,2 addition
(3)	CH$_3$CHClCH$_2$CH=CH$_2$	4-Chloro-1-pentene	1,2 addition
(4)	CH$_3$CH$_2$CH=CHCH$_2$Cl	1-Chloro-2-pentene	1,4 addition

4.18

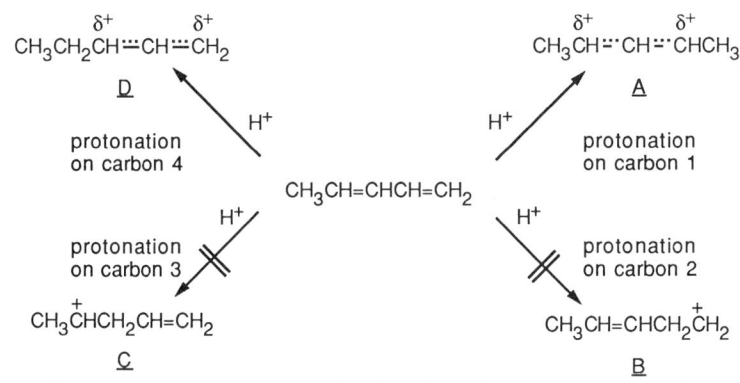

$$\underset{D}{CH_3CH_2\overset{\delta+}{CH}\!\!=\!\!\!=\!\!\overset{\delta+}{CH}\!\!=\!\!\!=\!\!CH_2} \quad\quad \underset{A}{CH_3\overset{\delta+}{CH}\!\!=\!\!\!=\!\!CH\!\!=\!\!\!=\!\!\overset{\delta+}{CH}CH_3}$$

protonation on carbon 4 ← H$^+$ — CH$_3$CH=CHCH=CH$_2$ — H$^+$ → protonation on carbon 1

protonation on carbon 3 ↙ H$^+$ H$^+$ ↘ protonation on carbon 2

$$\underset{C}{CH_3\overset{+}{C}HCH_2CH=CH_2} \quad\quad \underset{B}{CH_3CH=CHCH_2\overset{+}{C}H_2}$$

The positive charge of allylic carbocation <u>A</u> is delocalized over two secondary carbons; the positive charge of carbocation <u>D</u> is delocalized over one secondary carbon and one primary carbon; the positive charge of carbocations <u>B</u> and <u>C</u> is not delocalized. We therefore predict that carbocation <u>A</u> is the major intermediate formed, and that product (1) in Problem 4.17 will predominate. Note that product (1) results from both 1,2– and 1,4– addition.

4.19

a) [cyclohexenyl cation with +CH$_2$ ↔ methylenecyclohexane cation]

b) CH$_3$–C(=O:)–CH$_2$ ↔ CH$_3$–C(–Ö:⁻)=CH$_2$

c) [benzene cation resonance structures]

4.20

a) CH₃CH₂C≡CCH₂CH(CH₃)₂
6-Methyl-3-heptyne

b) HC≡CC(CH₃)₃
3,3-Dimethyl-1-butyne

c) CH₃CH(CH₃)CH₂C≡CCH₃
5-Methyl-2-hexyne

d) CH₃CH=CHCH₂C≡CCH₃
2-Hepten-5-yne

4.21

a) CH₃CH₂CH₂C≡CH + 1 equiv Cl₂ ⟶
   1-Pentyne

$$\underset{\text{E-1,2-dichloro-1-pentene}}{\overset{CH_3CH_2CH_2}{\underset{Cl}{>}}C=C\overset{Cl}{\underset{H}{<}}}$$

b) CH₃CH₂CH₂C≡CCH₂CH₃ + 1 equiv HBr ⟶
   3-Heptyne

$$\underset{\text{Z-3-Bromo-3-heptene}}{\overset{CH_3CH_2CH_2}{\underset{H}{>}}C=C\overset{Br}{\underset{CH_2CH_3}{<}}}$$

+

$$\underset{\text{Z-4-Bromo-3-heptene}}{\overset{CH_3CH_2CH_2}{\underset{Br}{>}}C=C\overset{H}{\underset{CH_2CH_3}{<}}}$$

c) CH₃CHCH₂C≡CCH₂CH₃ + H₂ $\xrightarrow{\text{Lindlar catalyst}}$
   |
   CH₃
   6-Methyl-3-heptyne

$$\underset{\text{cis-6-Methyl-3-heptene}}{\overset{CH_3CHCH_2}{\underset{H}{>}}C=C\overset{CH_2CH_3}{\underset{H}{<}}}$$
with CH₃ branch

4.22

CH₃CH₂CH₂C≡CCH₂CH₂CH₃ + H₂O $\xrightarrow[\text{HgSO}_4]{\text{H}_2\text{SO}_4}$ [CH₃CH₂CH₂CH=C(OH)CH₂CH₂CH₃]
4-Octyne

⟶ CH₃CH₂CH₂CH₂C(=O)CH₂CH₂CH₃

4.23

a) CH₃CH₂CH₂C≡CH $\xrightarrow[\text{HgSO}_4]{\text{H}_2\text{O, H}_2\text{SO}_4}$ CH₃CH₂CH₂C(=O)CH₃

b) CH₃CH₂C≡CCH₂CH₃ $\xrightarrow[\text{HgSO}_4]{\text{H}_2\text{O, H}_2\text{SO}_4}$ CH₃CH₂CH₂C(=O)CH₂CH₃

Alkenes and Alkynes 57

4.24

a) HC≡CH + Na⁺NH₂⁻ ⟶ HC≡C⁻Na⁺ + NH₃

(CH₃)₂CHCH₂CH₂Br + HC≡C⁻Na⁺ ⟶ (CH₃)₂CHCH₂CH₂C≡CH
                                                                      5-Methyl-1-hexyne

b) CH₃C≡CH + Na⁺NH₂⁻ ⟶ CH₃C≡C⁻Na⁺ + NH₃

CH₃CH₂CH₂Br + CH₃C≡C⁻Na⁺ ⟶ CH₃CH₂CH₂C≡CCH₃
                                                                       2-Hexyne

or

CH₃CH₂CH₂C≡CH + Na⁺NH₂⁻ ⟶ CH₃CH₂CH₂C≡C⁻Na⁺ + NH₃

CH₃CH₂CH₂C≡C⁻Na⁺ + CH₃Br ⟶ CH₃CH₂CH₂C≡CCH₃
                                                                     2-Hexyne

c) (CH₃)₂CHC≡CH + Na⁺NH₂⁻ ⟶ (CH₃)₂CHC≡C⁻Na⁺ + NH₃

(CH₃)₂CHC≡C⁻Na⁺ + CH₃Br ⟶ (CH₃)₂CHC≡CCH₃
                                                               4-Methyl-2-pentyne

Note: The reaction of CH₃C≡C⁻Na⁺ with (CH₃)₂CHBr will not yield the desired product because (CH₃)₂CHBr is not a primary alkyl halide.

4.25

a) CH₃CH=CHC(CH₃)=CHCH₃
   3-Methyl-2,4-hexadiene

b) CH₃CH=CHCH(CH₂CH₂CH₃)CH₂C≡CH
   4-Propyl-5-hepten-1-yne

c) CH₂=C=C(CH₃)₂
   3-Methyl-1,2-butadiene

d) HC≡CCH₂C≡CCH(CH₃)₂
   6-Methyl-1,4-heptadiyne

4.26

a) CH₃CH₂CH₂CH₂CH(CH₂CH₃)C≡CH
   3-Ethyl-1-heptyne

b) CH₃C(CH₃)=CHCH(CH₃)C≡CH
   3,5-Dimethyl-4-hexen-1-yne

c) CH₃C≡CCH₂CH₂C≡CH
   1,5-Heptadiyne

d) [cyclopentadiene ring]—CH₃
   1-Methyl-1,3-cyclopentadiene

4.27

a) C_6H_8

(CH_3)_2C=CHC≡CH   [cyclohexa-1,3-diene structure]   H_2C=CHCH=CHCH=CH_2

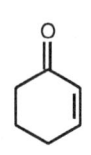

b) C_6H_8O

[cyclohex-2-enone structure]   CH_3CH_2C≡CCH_2CHO   H_2C=CHCH(CH_2OH)C≡CH

4.28

a) CH_3CH_2C≡CCH_2CH_2CH_3
3-Heptyne

b) CH_3CH_2C≡CC(CH_3)_3
2,2-Dimethyl-3-hexyne

c) CH_3C≡CCH_2C≡CCH_2CH_3
2,5-Octadiyne

d) H_2C=CHCH=CHC≡CH
1,3-Hexadien-5-yne

4.29

a) CH_3CH_2C≡CCH_2CH_2CH_3
3-Heptyne

b) CH_3CH_2CH_2C≡CC(CH_3)_2CH_2CH_3 with CH_3 branch
3,3-Dimethyl-4-octyne

c) [cyclodecyne ring structure with two methyls]
3,4-Dimethylcyclodecyne

d) (CH_3)_3CC≡CC(CH_3)_3
2,2,4,4-Tetramethyl-3-hexyne

4.30

CH_3CH_2CH_2C≡CH
1-Pentyne

CH_3CH_2C≡CCH_3
2-Pentyne

CH_3CH(CH_3)C≡CH
3-Methyl-1-butyne

4.31

Isomer	Name	Conjugated?
CH₃CH₂CH=C=CH₂	1,2-Pentadiene	no
CH₃CH=CHCH=CH₂	1,3-Pentadiene	yes
H₂C=CHCH₂CH=CH₂	1,4-Pentadiene	no
CH₃CH=C=CHCH₃	2,3-Pentadiene	no
H₂C=CHC(CH₃)=CH₂	2-Methyl-1,3-butadiene	yes
CH₃C(CH₃)=C=CH₂	3-Methyl-1,2-butadiene	no

4.32 a) CH₃CH=CHC≡CC≡CCH=CHCH=CHCH=CH₂
1,3,5,11–Tridecatetraen–7,9–diyne

Using *E–Z* notation: (3*E*,5*E*,11*E*)–1,3,5,11–Tridecatetraen–7,9–diyne
The parent alkane of this hydrocarbon is tridecane.

b) CH₃C≡CC≡CC≡CC≡CCH=CH₂
1–Tridecen–3,5,7,9,11–pentayne

This hydrocarbon is also of the tridecane family.

4.33

a) PhCH=CH₂ + H₂ (Pd catalyst) → PhCH₂CH₃

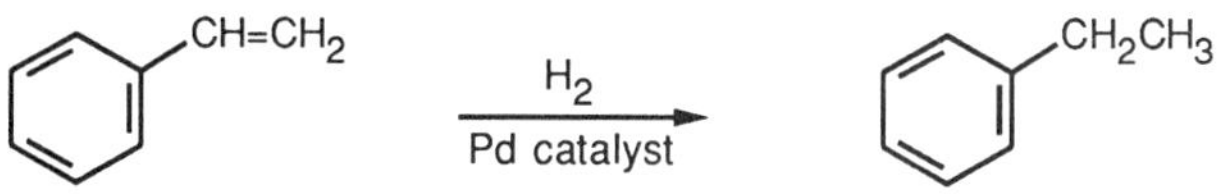

b) PhCH=CH₂ + Br₂ → PhCHBrCH₂Br

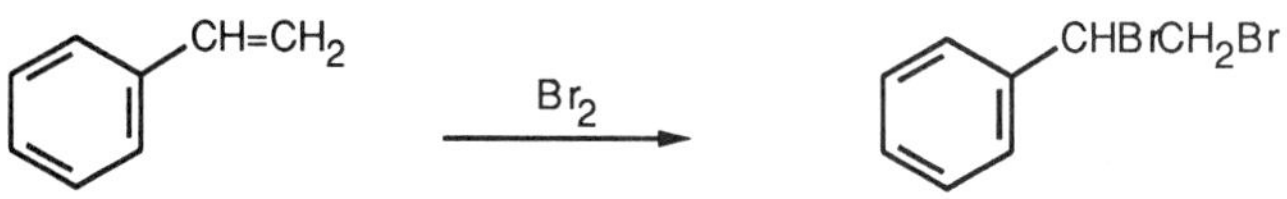

c) PhCH=CH₂ + HBr → PhCHBrCH₃

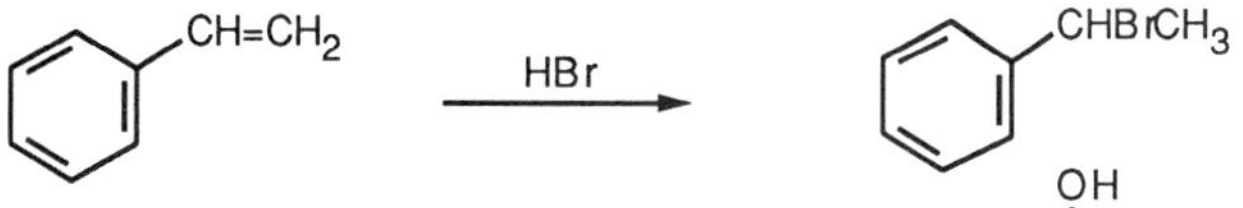

d) PhCH=CH₂ + KMnO₄ / NaOH → PhCH(OH)CH₂OH

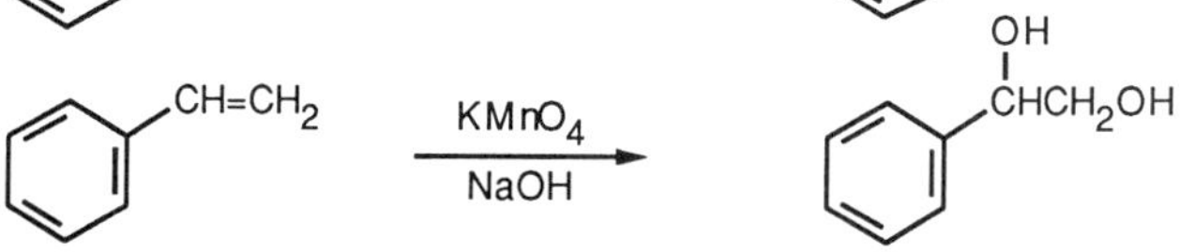

4.34

a)

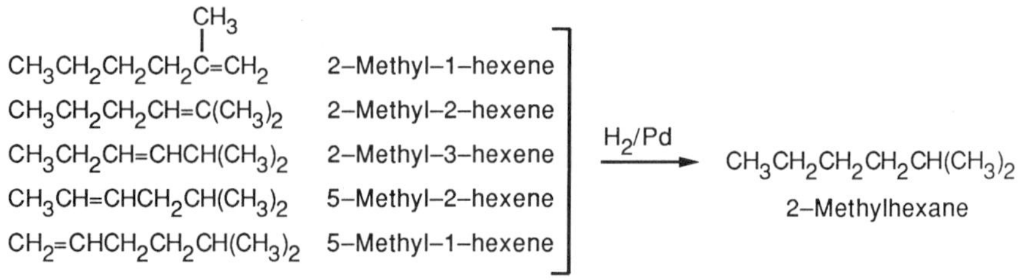

CH₃CH₂CH₂CH₂C(CH₃)=CH₂   2-Methyl-1-hexene
CH₃CH₂CH₂CH=C(CH₃)₂   2-Methyl-2-hexene
CH₃CH₂CH=CHCH(CH₃)₂   2-Methyl-3-hexene
CH₃CH=CHCH₂CH(CH₃)₂   5-Methyl-2-hexene
CH₂=CHCH₂CH₂CH(CH₃)₂   5-Methyl-1-hexene

$\xrightarrow{H_2/Pd}$ CH₃CH₂CH₂CH₂CH(CH₃)₂   2-Methylhexane

b) CH₃CH=CHCH₂CH(CH₃)₂ $\xrightarrow{Br_2/CCl_4}$ CH₃CHBrCHBrCH₂CH(CH₃)₂

5-Methyl-2-hexene   2,3-Dibromo-5-methylhexane

c) CH₃CH₂CH₂CH₂CH(CH₃)CH=CH₂ $\xrightarrow{HBr}$ CH₃CH₂CH₂CH₂CH(CH₃)CH(Br)CH₃

3-Methyl-1-heptene   2-Bromo-3-methylheptane

d) CH₃CH(CH₃)CH₂CH=CHCH₂CH₃ $\xrightarrow[OH^-]{KMnO_4}$ CH₃CH(CH₃)CH₂CH(OH)–CH(OH)CH₂CH₃

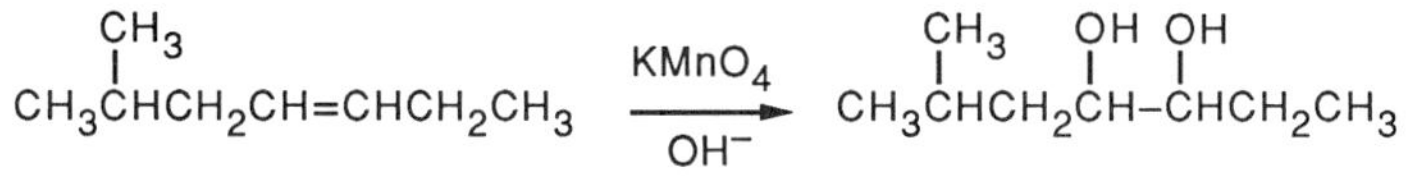

6-Methyl-3-heptene

4.35

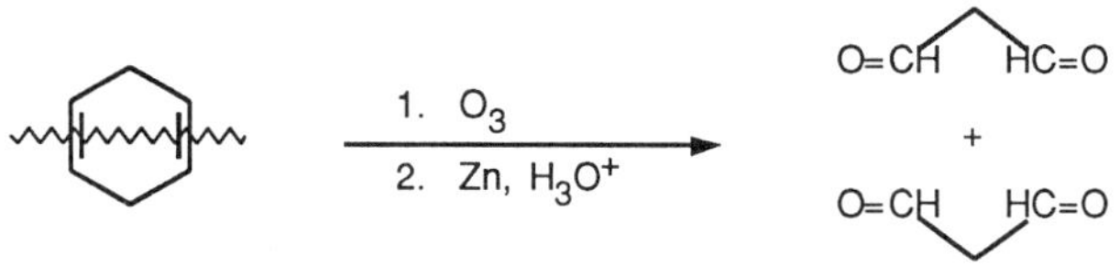

[cyclohexadiene] $\xrightarrow[\text{2. Zn, H}_3\text{O}^+]{\text{1. O}_3}$ O=CH–HC=O   +   O=CH–CH₂–HC=O

This cyclohexadiene yields only one product.

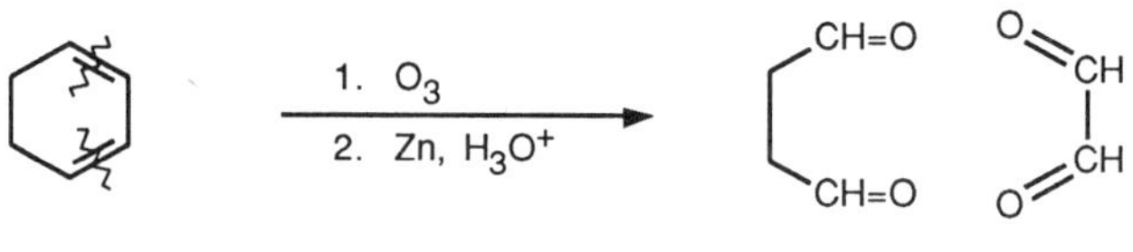

[cyclohexadiene] $\xrightarrow[\text{2. Zn, H}_3\text{O}^+]{\text{1. O}_3}$ CH=O, CH=O (dialdehyde) + O=CH–CH=O

Two different products result from oxidative cleavage of this diene.

4.36

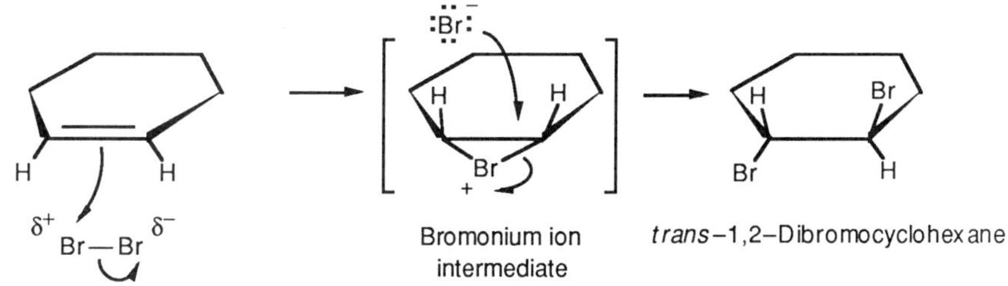

Bromonium ion intermediate

trans-1,2-Dibromocyclohexane

4.37

a) 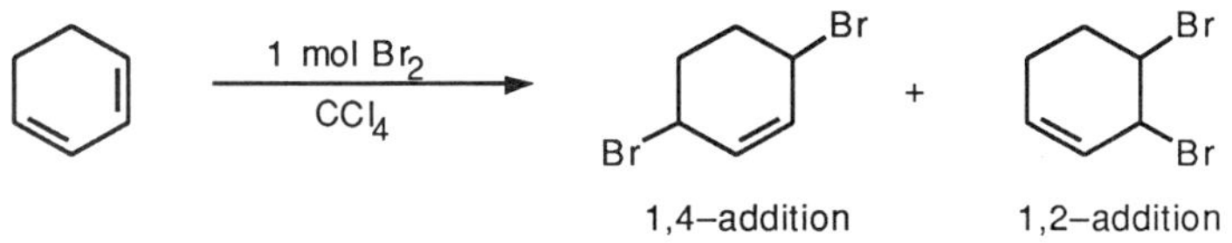 cyclohexadiene + 1 mol Br$_2$ / CCl$_4$ → 1,4-addition product + 1,2-addition product

b) 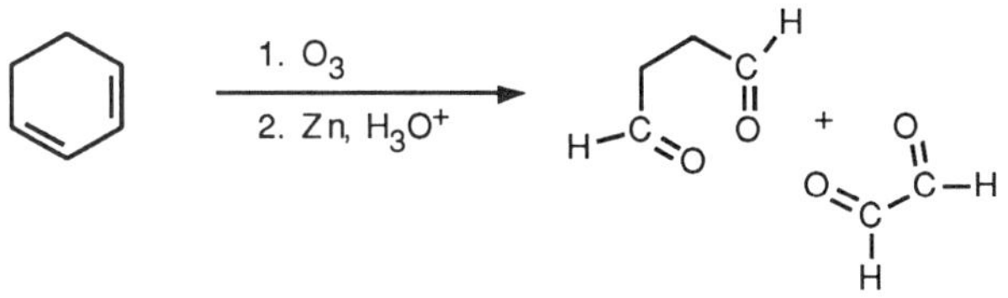 cyclohexadiene + 1. O$_3$; 2. Zn, H$_3$O$^+$ → OHC-CH$_2$-CH$_2$-CHO + OHC-CHO

c) 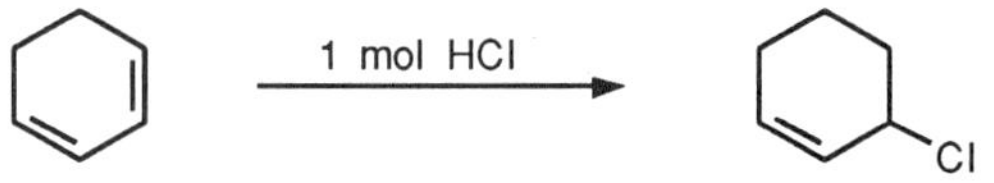 cyclohexadiene + 1 mol HCl → 3-chlorocyclohexene

d) 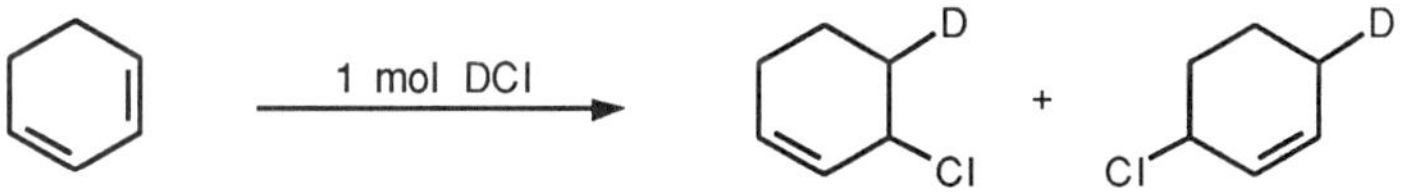 cyclohexadiene + 1 mol DCl → 3-chloro-6-deuteriocyclohexene + 3-chloro-4-deuteriocyclohexene

e) 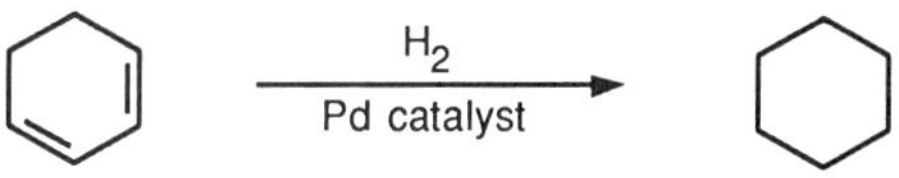 cyclohexadiene + H$_2$ / Pd catalyst → cyclohexane

4.38

$$\text{CH}_3\text{CH}_2\text{CH}_2\text{CH}_2\text{CH=CHCH}_2\text{CH}_2\text{CH}_2\text{CH}_3 \text{ (cis)}$$

or

$$\text{CH}_3\text{CH}_2\text{CH}_2\text{CH}_2\text{CH=CHCH}_2\text{CH}_2\text{CH}_2\text{CH}_3 \text{ (trans)}$$

1 mol H$_2$ / Pd catalyst → CH$_3$(CH$_2$)$_8$CH$_3$

1. O$_3$; 2. Zn, H$_3$O$^+$ → CH$_3$CH$_2$CH$_2$CH$_2$CHO

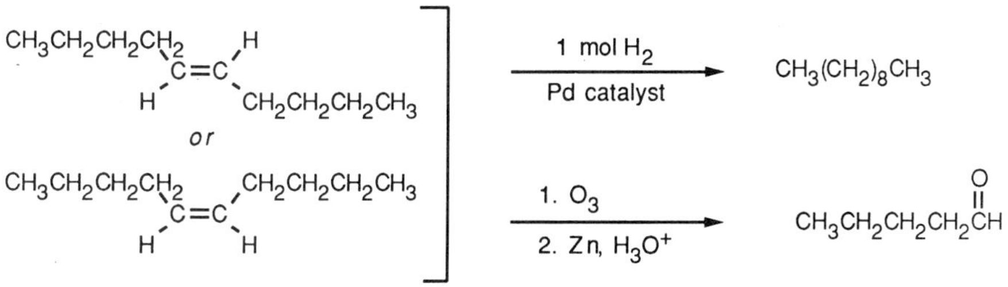

4.39

[cyclohexene with CH₃ substituent] $\xrightarrow[\text{2. Zn, H}_3\text{O}^+]{\text{1. O}_3}$ $\text{H}\overset{\text{O}}{\overset{\|}{\text{C}}}\text{CH}_2\text{CH}_2\text{CH}_2\text{CH}_2\overset{\text{O}}{\overset{\|}{\text{C}}}\text{CH}_3$

4.40

a) $\text{CH}_3\text{CH=CHCH}_3$
   or
   $\text{CH}_3\text{CH}_2\text{CH=CH}_2$
   $+ \text{ H}_2\text{O} \xrightarrow[\text{catalyst}]{\text{H}^+}$ $\text{CH}_3\text{CH}_2\overset{\text{OH}}{\overset{|}{\text{C}}}\text{HCH}_3$

b) [cyclohexene] $+ \text{ H}_2\text{O} \xrightarrow[\text{catalyst}]{\text{H}^+}$ [cyclohexanol]

c) [cyclohexyl-CH=CH₂] $+ \text{ H}_2\text{O} \xrightarrow[\text{catalyst}]{\text{H}^+}$ [cyclohexyl-CH(OH)-CH₃]

4.41

a) $\text{CH}_3\overset{\text{CH}_3}{\overset{|}{\text{C}}}\text{HCH}_2\text{C}\equiv\text{CH} + \text{H}_2\text{O} \xrightarrow[\text{HgSO}_4]{\text{H}_2\text{SO}_4}$ $\text{CH}_3\overset{\text{CH}_3}{\overset{|}{\text{C}}}\text{HCH}_2\overset{\text{O}}{\overset{\|}{\text{C}}}\text{CH}_3$

b) [PhC≡CH] $+ \text{ H}_2\text{O} \xrightarrow[\text{HgSO}_4]{\text{H}_2\text{SO}_4}$ [PhC(=O)CH₃]

4.42

[1,5-cyclooctadiene] $\xrightarrow[\text{Pd catalyst}]{2\text{ H}_2}$ [cyclooctane]

1,5–Cyclooctadiene

$\xrightarrow[\text{2. Zn, H}_3\text{O}^+]{\text{1. O}_3}$ $\text{H}-\overset{\text{O}}{\overset{\|}{\text{C}}}\overset{\text{CH}_2-\text{CH}_2}{}\overset{\text{O}}{\overset{\|}{\text{C}}}-\text{H}$

$+$

$\text{H}-\overset{\text{O}}{\overset{\|}{\text{C}}}\overset{}{}\overset{\text{O}}{\overset{\|}{\text{C}}}-\text{H}$
$\phantom{\text{H}-\text{C}}\text{CH}_2-\text{CH}_2$

4.43

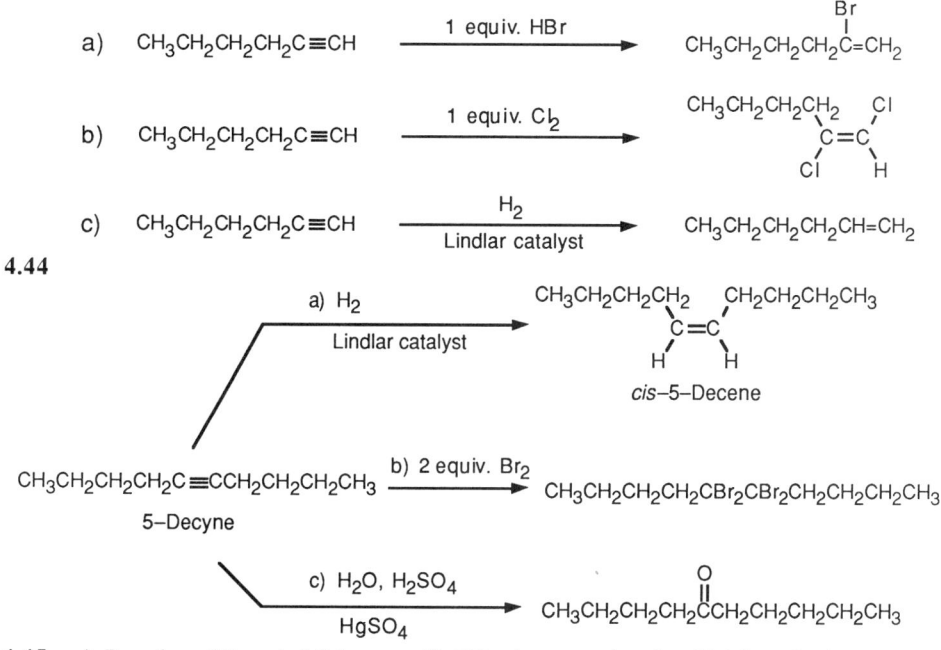

a) $CH_3CH_2CH_2CH_2C\equiv CH \xrightarrow{\text{1 equiv. HBr}} CH_3CH_2CH_2CH_2\underset{\underset{Br}{|}}{C}=CH_2$

b) $CH_3CH_2CH_2CH_2C\equiv CH \xrightarrow{\text{1 equiv. Cl}_2} \begin{array}{c} CH_3CH_2CH_2CH_2 \\ \diagdown \\ \end{array}\!\!\!\!C=C\!\!\!\!\begin{array}{c} Cl \\ \diagup \\ \end{array}$
$ Cl H$

c) $CH_3CH_2CH_2CH_2C\equiv CH \xrightarrow[\text{Lindlar catalyst}]{H_2} CH_3CH_2CH_2CH_2CH=CH_2$

4.44

$CH_3CH_2CH_2CH_2C\equiv CCH_2CH_2CH_2CH_3$
5–Decyne

a) H_2, Lindlar catalyst → $\begin{array}{c}CH_3CH_2CH_2CH_2\\ \diagdown \\ \end{array}C=C\begin{array}{c}CH_2CH_2CH_2CH_3\\ \diagup\\ \end{array}$
$ H H$
cis–5–Decene

b) 2 equiv. Br$_2$ → $CH_3CH_2CH_2CH_2CBr_2CBr_2CH_2CH_2CH_2CH_3$

c) H$_2$O, H$_2$SO$_4$, HgSO$_4$ → $CH_3CH_2CH_2CH_2\overset{\overset{O}{\|}}{C}CH_2CH_2CH_2CH_2CH_3$

4.45 a) Reaction of 2-methyl-2-butene with HBr gives a product in which bromine is bonded to the *more* substituted carbon.

b) Hydroxylation of double bonds produces cis, not trans, diols.

c) Ozonolysis of a double bond yields aldehyde and ketone products. To obtain the products shown, use KMnO$_4$ in neutral or acidic solution.

4.46

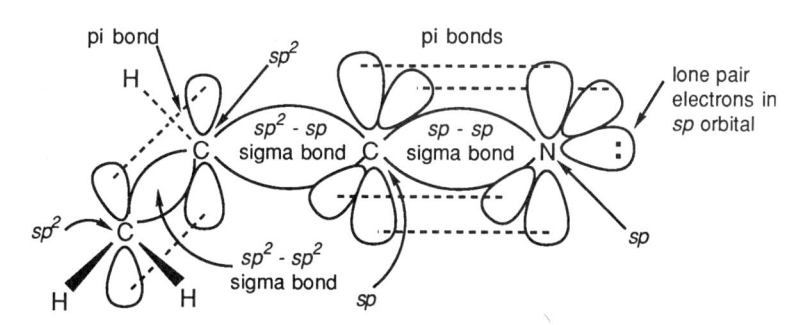

Each of the two pi bonds between carbon and nitrogen is formed by overlap of one *p* orbital of carbon with one *p* orbital of nitrogen. Acrylonitrile is conjugated.

4.47

$$CH_3C\equiv CH \xrightarrow[\text{2. } CH_3Br]{\text{1. } NaNH_2} CH_3C\equiv CCH_3 \xrightarrow[\text{catalyst}]{H_2 \text{ Lindlar}} \underset{\text{cis-2-Butene}}{\overset{HH}{\underset{H_3CCH_3}{C=C}}}$$

4.48

a) $CH_3CH_2C\equiv CH$ (1-Butyne) $\xrightarrow[\text{Pd catalyst}]{2 \text{ equiv. } H_2}$ $CH_3CH_2CH_2CH_3$ (Butane)

b) $CH_3CH_2C\equiv CH \xrightarrow[CCl_4]{2 \text{ } Cl_2}$ $CH_3CH_2\underset{Cl}{\overset{Cl}{C}}CHCl_2$ (1,1,2,2-Tetrachlorobutane)

c) $CH_3CH_2C\equiv CH \xrightarrow[\text{Lindlar Catalyst}]{H_2} CH_3CH_2CH=CH_2 \xrightarrow{HBr} CH_3CH_2\underset{}{\overset{Br}{C}}HCH_3$ (2-Bromobutane)

or $\xrightarrow{HBr} CH_3CH_2\overset{Br}{C}=CH_2 \xrightarrow[\text{Pd catalyst}]{H_2} CH_3CH_2\overset{Br}{C}HCH_3$

d) $CH_3CH_2C\equiv CH \xrightarrow[HgSO_4]{H_2O, H_2SO_4} CH_3CH_2\overset{O}{\overset{\|}{C}}CH_3$ (2-Butanone)

4.49

$\underset{\text{2,3-Dimethyl-2-butene}}{\overset{CH_3CH_3}{\underset{CH_3CH_3}{C=C}}} \xrightarrow[\text{2. } Zn, H_3O^+]{\text{1. } O_3} \overset{CH_3}{\underset{CH_3}{C}}=O + O=\overset{CH_3}{\underset{CH_3}{C}}$

4.50

Styrene (Ph-CH=CH₂)

$\xrightarrow[\text{Pd catalyst}]{H_2}$ Ph-CH₂CH₃ **A**

$\xrightarrow{4 \text{ equiv } H_2}$ cyclohexyl-CH₂CH₃ **B**

$\xrightarrow[H_2O]{KMnO_4}$ Ph-CO₂H + CO₂ **C**

4.51

$$CH_3(CH_2)_{12}CH=CH(CH_2)_7CH_3 \xrightarrow[H_3O^+]{KMnO_4} CH_3(CH_2)_{12}COOH + CH_3(CH_2)_7COOH$$

Muscalure

4.52

$$HC\equiv CH \xrightarrow[2.\ CH_3(CH_2)_{11}CH_2Br]{1.\ NaNH_2} CH_3(CH_2)_{12}C\equiv CH$$

$$\downarrow \begin{array}{l} 1.\ NaNH_2 \\ 2.\ CH_3(CH_2)_6CH_2Br \end{array}$$

$$\begin{array}{c} CH_3(CH_2)_{11}CH_2 \quad CH_2(CH_2)_7CH_3 \\ C=C \\ H \quad\quad H \end{array} \xleftarrow[\text{Lindlar catalyst}]{H_2} CH_3(CH_2)_{12}C\equiv C(CH_2)_7CH_3$$

Muscalure

4.53

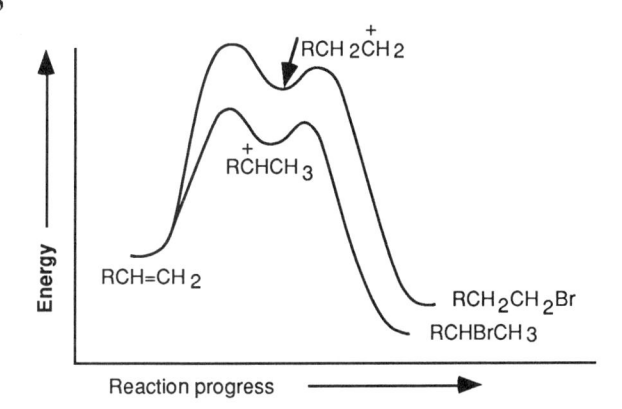

[Energy diagram: Reaction progress vs Energy showing two curves from RCH=CH₂ through transition states $RCH_2\overset{+}{C}H_2$ and $R\overset{+}{C}HCH_3$ to products RCH_2CH_2Br and $RCHBrCH_3$]

4.54

$CH_3CH_2CH_2CH=CH_2$

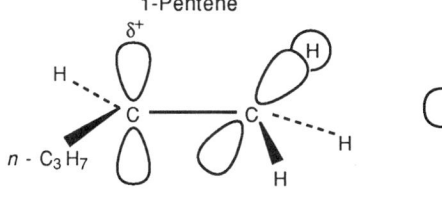

1-Pentene

[Transition State 1: orbital diagram with δ+, H, n-C₃H₇ substituents]

Transition State 1

[Transition State 2: orbital diagram with CH₃, H, n-C₃H₇ and :Br⁻]

Transition State 2

4.55

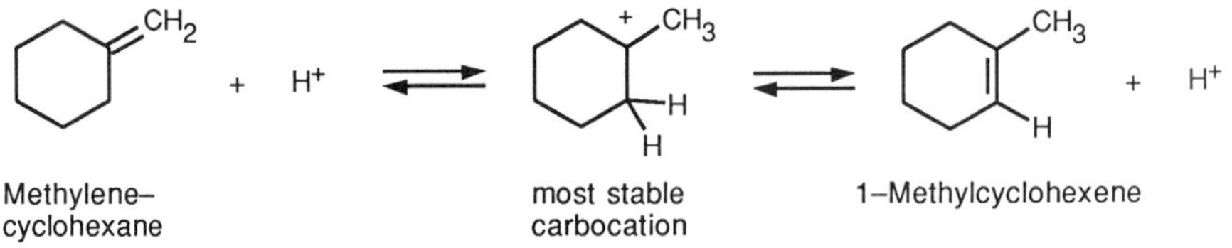

Methylene-cyclohexane + H⁺ ⇌ most stable carbocation ⇌ 1-Methylcyclohexene + H⁺

Protonation occurs to produce the most stable carbocation, which can then lose H⁺ to form either of two alkenes. Because 1–methylcyclohexene is the major product of this equilibrium, it must be the more stable product.

4.56

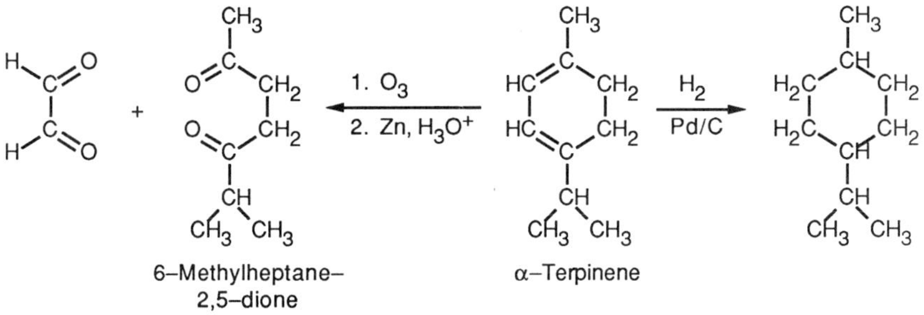

[Reaction scheme: aldehyde + aldehyde → 6-Methylheptane-2,5-dione ←(1. O₃; 2. Zn, H₃O⁺)— α-Terpinene —(H₂, Pd/C)→ 1-methyl-4-isopropylcyclohexane]

Chapter Outline

I. Alkenes (Sections 4.1 - 4.9)
   A. Reactions of alkenes (Sections 4.1 - 4.8)
      1. Addition of HX (Sections 4.1 - 4.3)
         a. Orientation of addition - Markovnikov's rule (Section 4.2)
         b. Carbocation stability (Section 4.3)
      2. Hydration of alkenes (Section 4.4)
      3. Addition of halogens to alkenes (Section 4.5)
         a. Stereochemistry of addition
         b. Mechanism of addition
      4. Hydrogenation of alkenes (Section 4.6)
      5. Oxidation of alkenes (Section 4.7)
      6. Alkene polymers (Section 4.8)
   B. Preparation of alkenes (Section 4.9)
      1. Elimination reactions
      2. Dehydrohalogenation of alkyl halides
      3. Dehydration of alcohols

Alkenes and Alkynes   67

II. Conjugated dienes (Sections 4.10 - 4.12)
   A. Electrophilic addition to conjugated dienes (Section 4.10 - 4.11)
      1. 1,4-Addition products (Section 4.10)
      2. Mechanism of 1,4-addition: allylic cations (Section 4.11)
   B. Drawing resonance structures (Section 4.12)
III. Alkynes (Sections 4.13 - 4.16)
   A. Naming of alkynes (Section 4.13)
   B. Reactions of alkynes (Sections 4.14 - 4.15)
      1. Addition of HX and X_2 (Section 4.14)
      2. Hydrogenation of alkynes (Section 4.14)
      3. Hydration of alkynes (Section 4.15)
      4. Formation of acetylide anions (Section 4.16)

Study Skills for Chapter 4

After studying this chapter, you should be able to:

1. Predict the products of reaction of alkenes (Problems 4.1, 4.2, 4.4, 4.5, 4.6, 4.8, 4.9, 4.10, 4.33, 4.34, 4.40).
2. Predict the products of reaction of conjugated dienes (Problems 4.16, 4.17, 4.37).
3. Predict the products of reaction of alkynes (Problems 4.21, 4.22, 4.23, 4.24, 4.41, 4.43, 4.44, 4.48).
4. Choose the correct starting material to prepare alkenes (Problems 4.12, 4.13, 4.14, 4.15, 4.47, 4.52).
5. Draw resonance forms of simple molecules (Problem 4.19).
6. Name alkynes and dienes (Problems 4.20, 4.25, 4.28, 4.32).
7. Draw structures of alkynes and dienes corresponding to a given name (Problems 4.26, 4.29).
8. Draw isomers of alkenes, alkynes, and dienes corresponding to a given structural formula (Problems 4.27, 4.30, 4.31).
9. Use the reactions in this chapter to determine the structure of an unknown compound (Problems 4.35, 4.38, 4.39, 4.42, 4.49, 4.50, 4.51, 4.56).
10. Formulate mechanisms of electrophilic addition reactions and draw carbocation intermediates (Problems 4.3, 4.7, 4.18, 4.36, 4.53, 4.54, 4.55).

Chapter 5 – Aromatic Compounds

5.1 According to Kekulé, four dibromobenzenes are possible.

[Four dibromobenzene structures: 1,4- (para), 1,3- (meta), and two 1,2- (ortho) forms]

Kekulé would say that the two isomers on the right interconvert rapidly, and only one isomer can be isolated.

5.2 According to resonance theory, *o*–dibromobenzene is not described properly by either of the two structures shown on the right in Problem 5.1, but is a resonance hybrid of the two.

5.3 Number the positions on the ring. Give number 1 to one substituent and the lowest possible number to the other substituent. Substituents at positions 1 and 2 have an *ortho* relationship; substituents at positions 1 and 3 have a *meta* relationship; substituents at positions 1 and 4 have a *para* relationship.

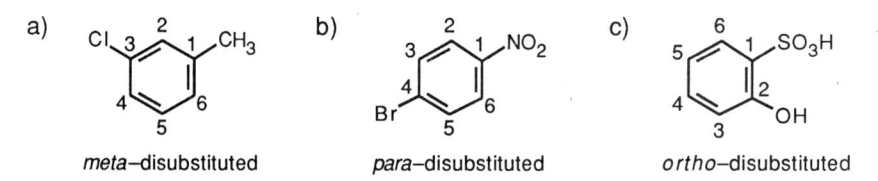

a) Cl at 3, CH₃ at 1
meta–disubstituted

b) NO₂ at 1, Br at 4
para–disubstituted

c) SO₃H at 1, OH at 2
ortho–disubstituted

5.4 For compounds with two substituents, determine the *ortho*, *meta*, *para* relationship of the substituents, and cite them in alphabetical order.

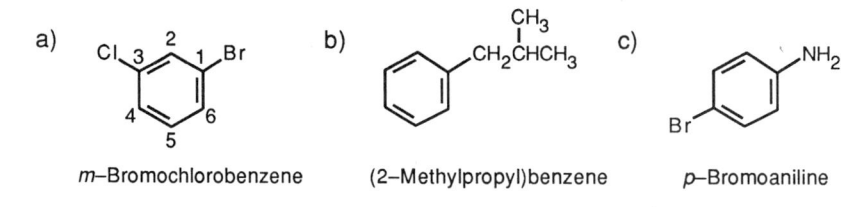

a) Cl at 3, Br at 1
m–Bromochlorobenzene

b) CH₂CH(CH₃)CH₃ on benzene
(2–Methylpropyl)benzene

c) Br and NH₂ para
p–Bromoaniline

5.5

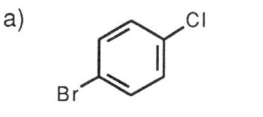

a) p-Bromochlorobenzene

b) p-Bromotoluene

c) m-Chloroaniline

d) 1-Chloro-3,5-dimethylbenzene

5.6

Toluene $\xrightarrow{Br_2, FeBr_3}$ o-Bromotoluene + m-Bromotoluene + p-Bromotoluene

5.7 The mechanism of nitration is the same as the mechanism of other electrophilic aromatic substitution mechanisms we have studied.

benzene + $^+NO_2$ → [carbocation intermediate] → nitrobenzene + H^+

5.8

o-Xylene $\xrightarrow{Cl_2, FeCl_3}$ A + B

Chlorination at position "a" of o-xylene yields product A; chlorination at position "b" yields product B.

p-Xylene $\xrightarrow{Cl_2, FeCl_3}$ 2-chloro-1,4-dimethylbenzene

Only one product results from chlorination of p-xylene because all sites on the ring are equivalent.

5.9

m-Xylene → A + B + C

(A: 1-Cl-2-CH₃-4-CH₃ benzene; B: 1,3-Cl with 3,5-CH₃; C: chlorine between the two methyls)

Three products might form on chlorination of *m*-xylene. Product C is unlikely to form because substitution rarely occurs between two *meta* substituents.

5.10

Benzene + D⁺ → [carbocation intermediate] → deuterated benzene + H⁺

Benzene can be protonated by strong acids. The resulting intermediate can lose either deuterium or hydrogen. If H⁺ is lost, deuterated benzene is produced. Attack by D⁺ can occur at all positions of the ring and leads to eventual replacement of all hydrogens by deuterium.

5.11

a) Benzene + CH₃CH₂Cl —AlCl₃→ Ethylbenzene

b) *p*-Xylene + CH₃CH₂Cl —AlCl₃→ 2-Ethyl-1,4-dimethylbenzene

5.12

a) Benzene + (CH₃)₃CCl —AlCl₃→ *tert*-butylbenzene (C(CH₃)₃)

b) Benzene + CH₃CH₂COCl —AlCl₃→ phenyl ethyl ketone (C(=O)CH₂CH₃)

5.13 Figure 5.9 lists groups in order of the strength of their activating or deactivating effects.

   Least reactive ———> Most reactive
   a) Nitrobenzene, toluene, phenol
   b) Benzoic acid, chlorobenzene, benzene, phenol
   c) Benzaldehyde, bromobenzene, benzene, aniline

5.14

a) Nitrobenzene + HNO$_3$ →(H$_2$SO$_4$) m-Dinitrobenzene

b) Bromobenzene + HNO$_3$ →(H$_2$SO$_4$) o-Bromonitrobenzene + p-Bromonitrobenzene

c) Toluene + HNO$_3$ →(H$_2$SO$_4$) o-Nitrotoluene + p-Nitrotoluene

d) Benzoic acid + HNO$_3$ →(H$_2$SO$_4$) m-Nitrobenzoic acid

e) p-Xylene + HNO$_3$ →(H$_2$SO$_4$) 1,4-Dimethyl-2-nitrobenzene

5.15

a) *Para* attack:

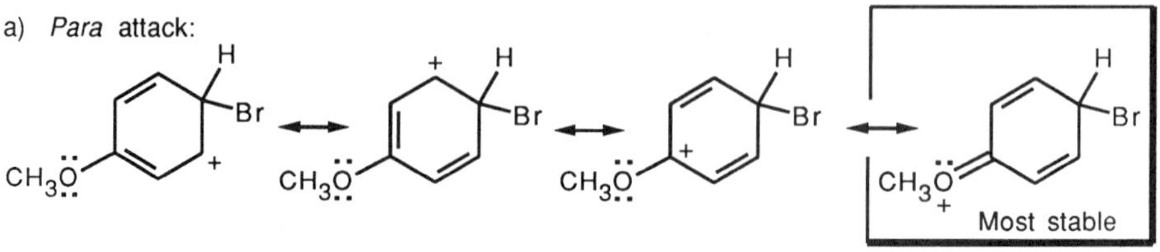
Most stable

b) *Meta* attack:

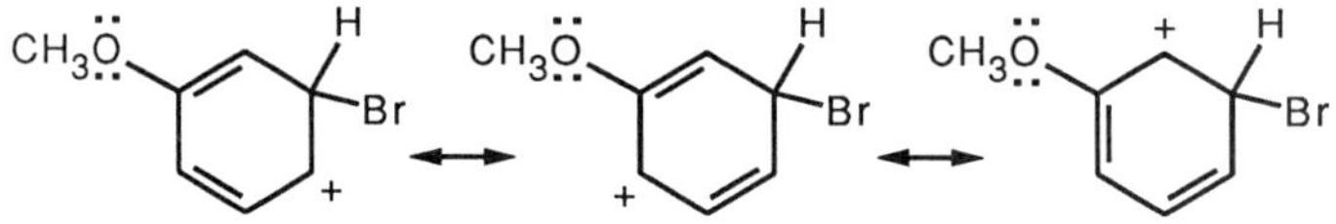

c) *Ortho* attack:

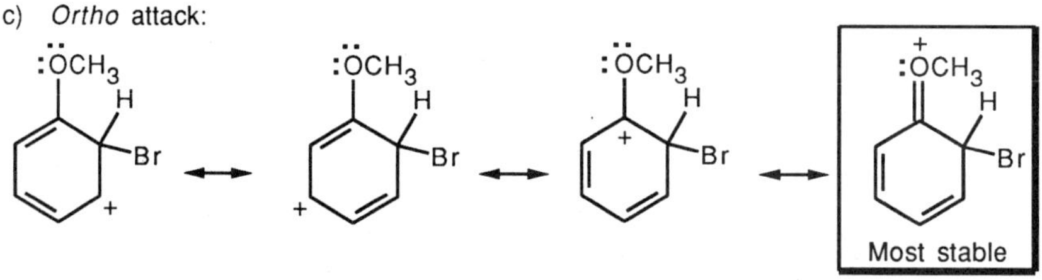
Most stable

The carbocation intermediates in ortho-para substitution can be stabilized by the oxygen atom of the $-OCH_3$ substituent. Thus, ortho-para substitution is favored.

5.16

Ortho attack:

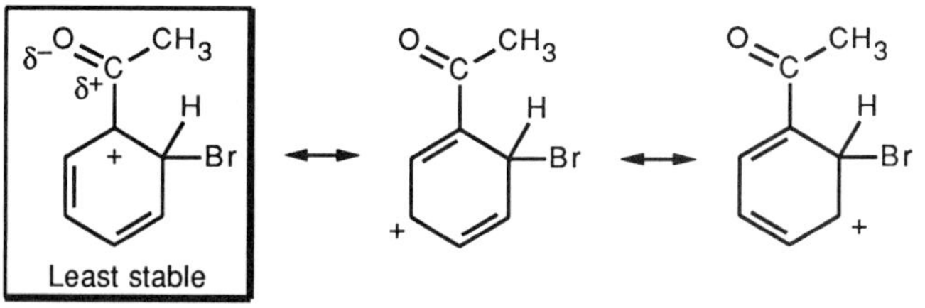
Least stable

Meta attack:

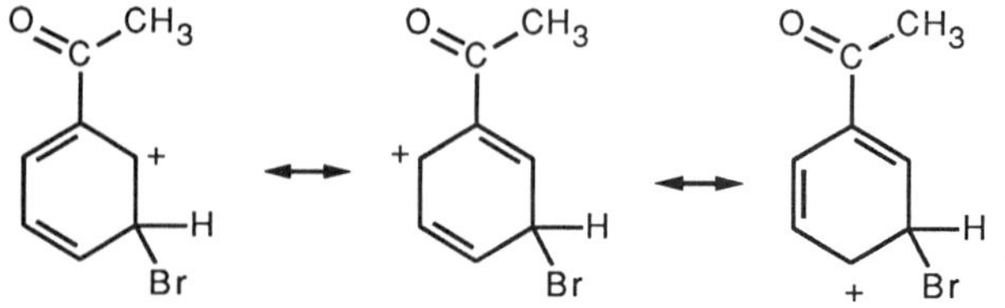

Aromatic Compounds 73

Para attack:

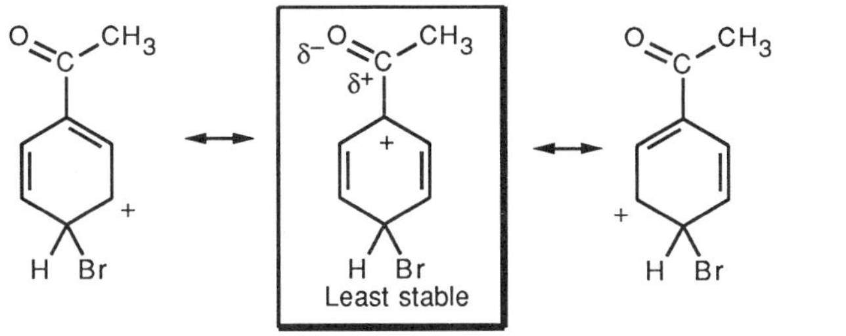

[structures showing para attack resonance forms with carbocation intermediates, middle one boxed as "Least stable"]

The indicated carbocation intermediates of ortho and para attack are least stable because they place a positive charge next to a positively polarized carbon atom. Thus, meta substitution is favored.

5.17

a)

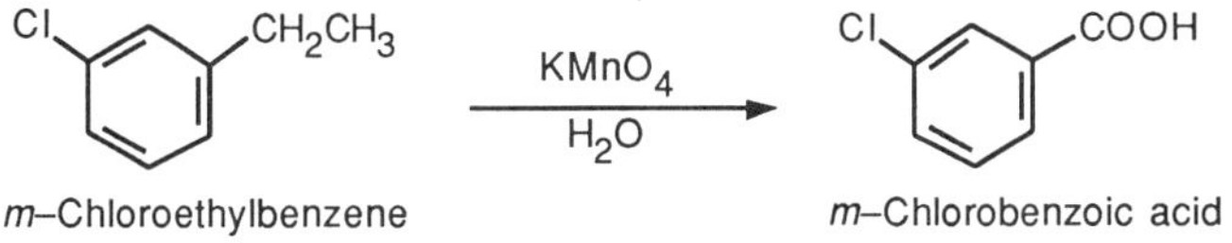

m-Chloroethylbenzene — KMnO$_4$, H$_2$O → m-Chlorobenzoic acid

b)

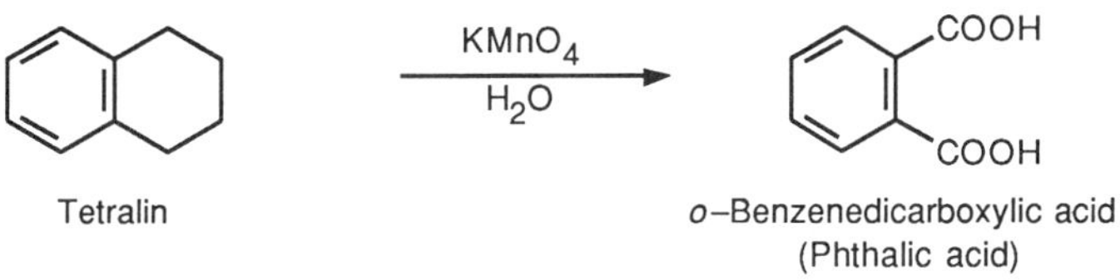

Tetralin — KMnO$_4$, H$_2$O → o-Benzenedicarboxylic acid (Phthalic acid)

5.18

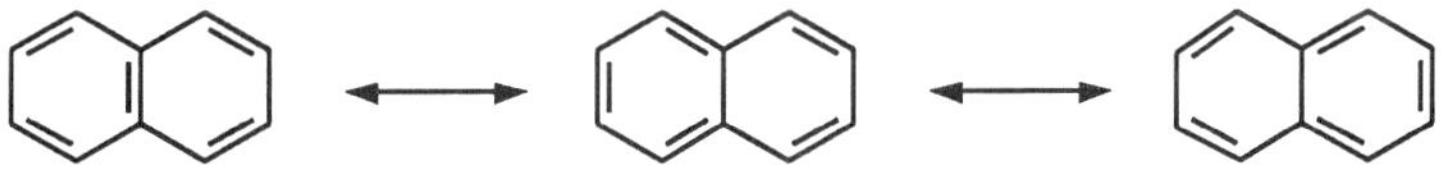

[three resonance structures of naphthalene]

5.19 In order to obtain the desired product, you must perform the reactions in the correct order. Assume that you can separate *ortho* and *para* isomers.

a) benzene — CH$_3$Cl / AlCl$_3$ → toluene — CH$_3$COCl / AlCl$_3$ → p-Methylacetophenone

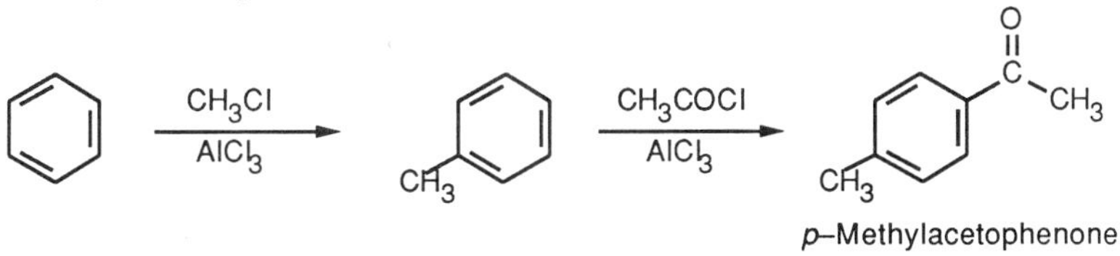

b) benzene — Cl$_2$ / FeCl$_3$ → chlorobenzene — HNO$_3$ / H$_2$SO$_4$ → p-Chloronitrobenzene

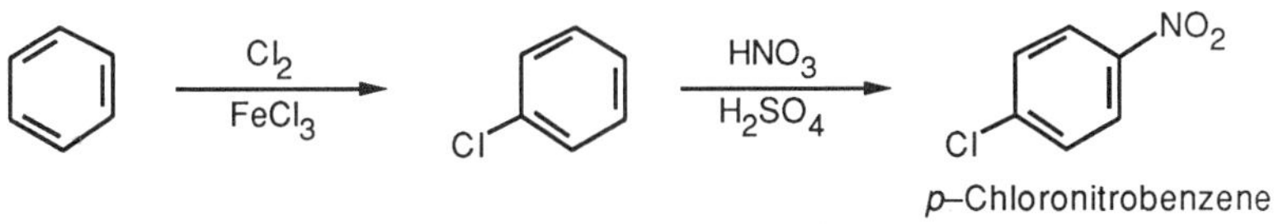

5.20 a) Two routes can be used to synthesize *o*–bromotoluene.

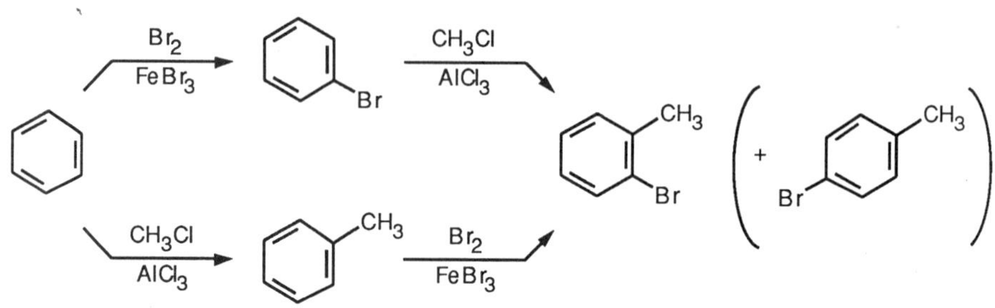

b)

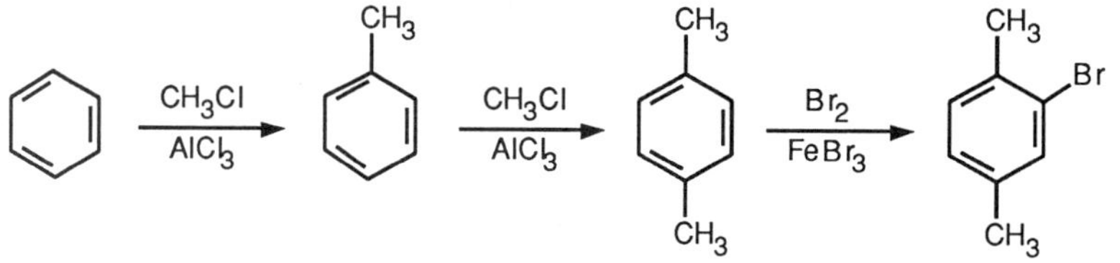

2-Bromo-1,4-dimethylbenzene (plus *ortho* isomer formed in previous step)

5.21

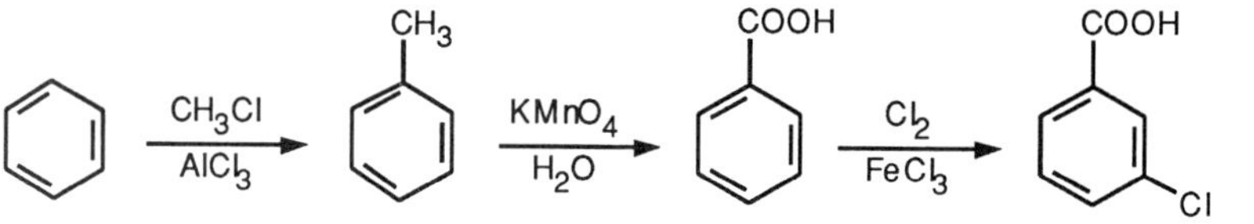

m–Chlorobenzoic acid

5.22

a) (4-Methylpentyl)benzene or 4-Methyl-1-phenylpentane

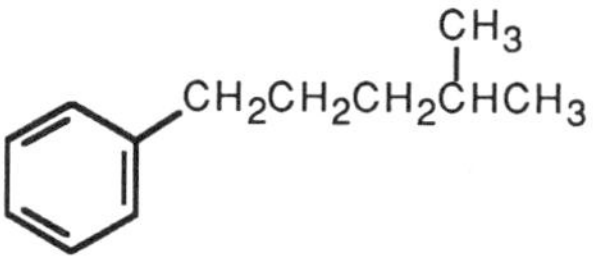

b) *m*-Bromobenzoic acid

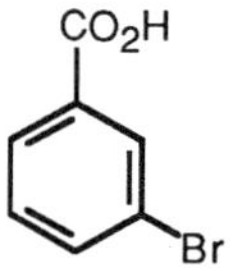

c) 1-Bromo-3,5-dimethylbenzene

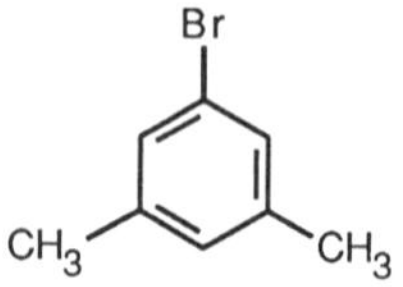

d) *o*-Bromopropylbenzene

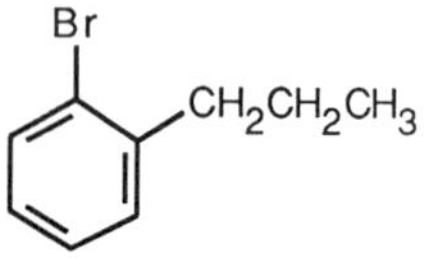

Aromatic Compounds 75

5.23

a) m-Bromophenol

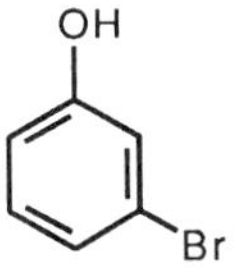

b) 1,3,5-Benzenetriol

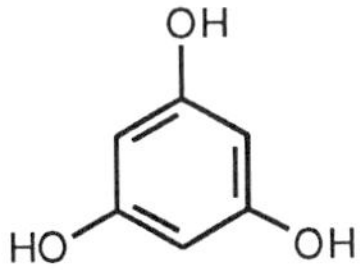

c) p-Iodonitrobenzene

d) 2,4,6-Trinitrotoluene (TNT)

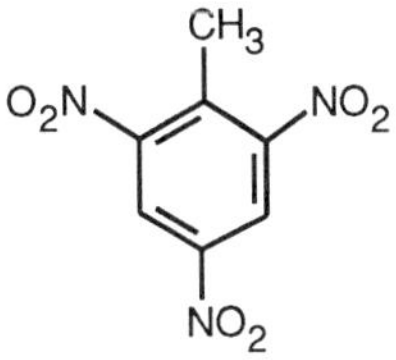

e) o-Aminobenzoic acid

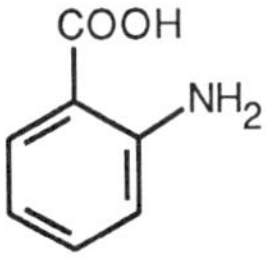

f) 3-Methyl-2-phenylhexane

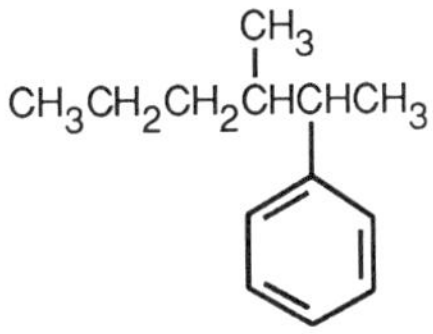

5.24

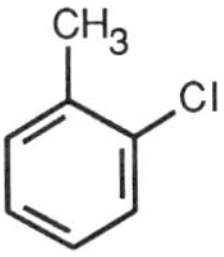

o-Chlorotoluene 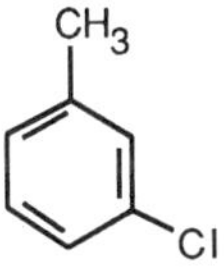   m-Chlorotoluene 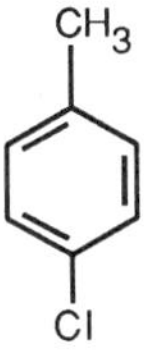   p-Chlorotoluene 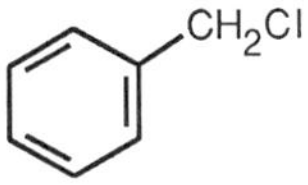   Benzyl chloride

5.25

a) o-Dinitrobenzene 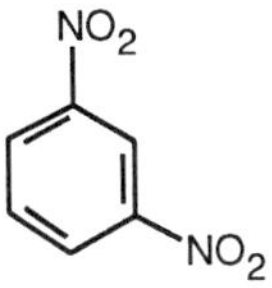   m-Dinitrobenzene    p-Dinitrobenzene

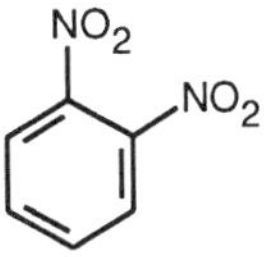

b)

1-Bromo-2,3-dimethylbenzene

4-Bromo-1,2-dimethylbenzene

2-Bromo-1,3-dimethylbenzene

1-Bromo-2,4-dimethylbenzene

1-Bromo-3,5-dimethylbenzene

2-Bromo-1,4-dimethylbenzene

5.26
a) 1,3,5-Trimethylbenzene $\xrightarrow{Br_2, FeBr_3}$ 2-Bromo-1,3,5-trimethylbenzene

b) m-Dimethylbenzene (m-Xylene) $\xrightarrow{Cl_2, AlCl_3}$ 2-Chloro-1,3-dimethylbenzene + 1-Chloro-2,4-dimethylbenzene + 1-Chloro-3,5-dimethylbenzene

Ethylbenzene can also give three products on aromatic chlorination.

c) o-Diethylbenzene $\xrightarrow{HNO_3, H_2SO_4}$ 1,2-Diethyl-3-nitrobenzene + 1,2-Diethyl-4-nitrobenzene

Aromatic Compounds   77

or

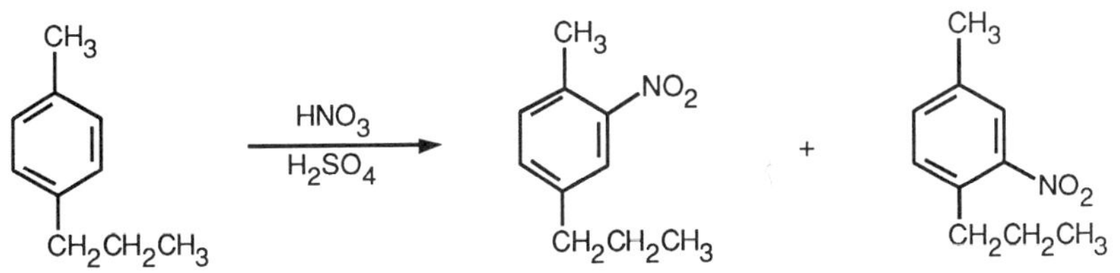

1-Methyl-4-propylbenzene + HNO$_3$/H$_2$SO$_4$ → 1-Methyl-2-nitro-4-propylbenzene + 4-Methyl-2-nitro-1-propylbenzene

or

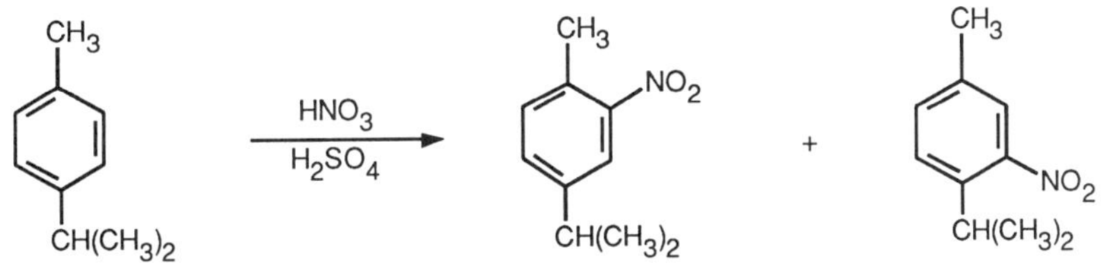

1-Isopropyl-4-methylbenzene + HNO$_3$/H$_2$SO$_4$ → 4-Isopropyl-1-methyl-2-nitrobenzene + 1-Isopropyl-4-methyl-2-nitrobenzene

5.27

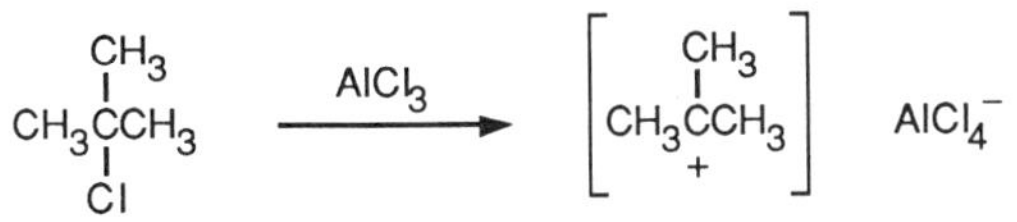

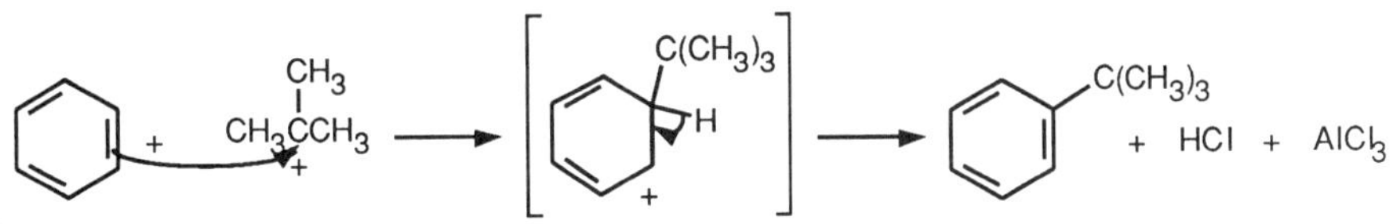

$CH_3CCH_3(CH_3)(Cl)$ + AlCl$_3$ → [$CH_3CCH_3(CH_3)^+$] AlCl$_4^-$

benzene + $CH_3CCH_3(CH_3)^+$ → [cyclohexadienyl cation with C(CH$_3$)$_3$ and H] → *tert*-Butylbenzene (C(CH$_3$)$_3$) + HCl + AlCl$_3$

5.28

a) Bromobenzene + HNO$_3$ →(H$_2$SO$_4$)→ 2-bromonitrobenzene + 4-bromonitrobenzene

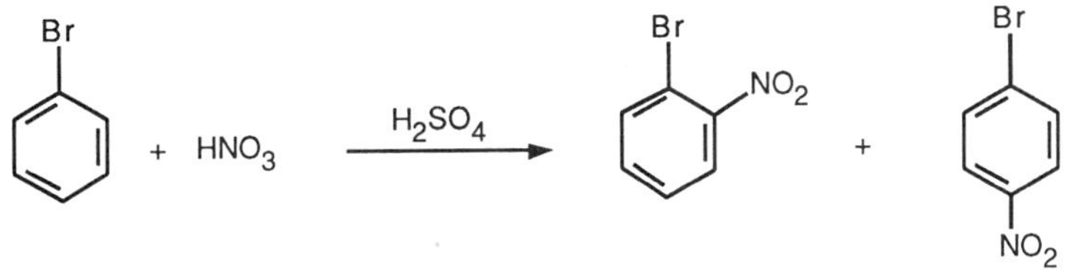

Chapter 5

b) [benzonitrile] + HNO₃ →(H₂SO₄) 3-nitrobenzonitrile

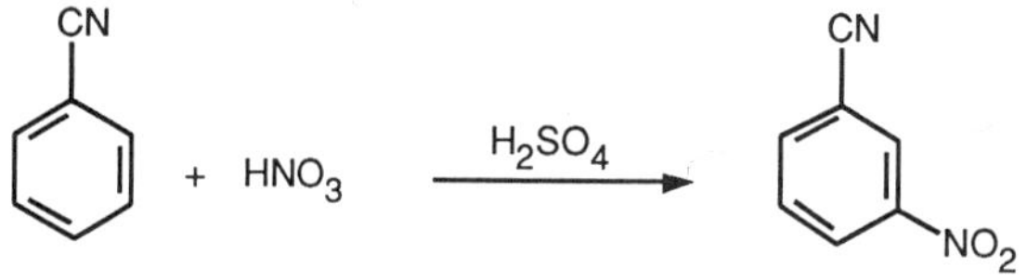

c) [benzoic acid] + HNO₃ →(H₂SO₄) 3-nitrobenzoic acid

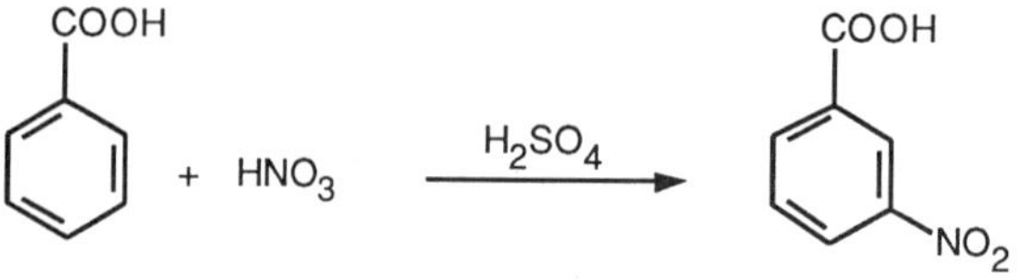

d) [nitrobenzene] + HNO₃ →(H₂SO₄) 1,3-dinitrobenzene

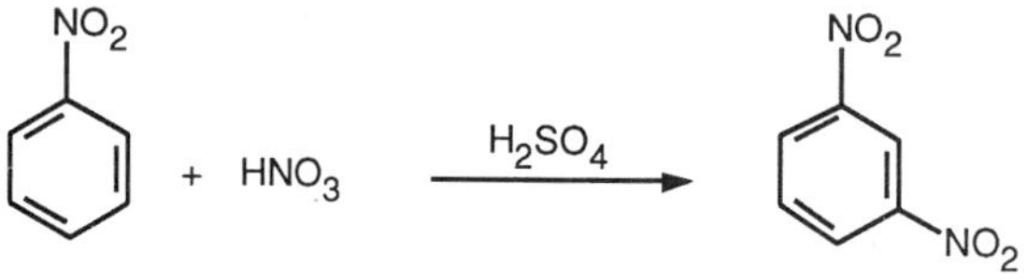

e) [phenol] + HNO₃ →(H₂SO₄) o-nitrophenol + p-nitrophenol

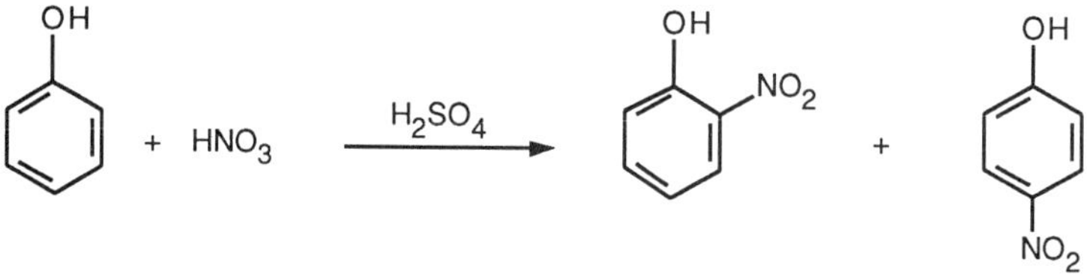

f) [benzaldehyde] + HNO₃ →(H₂SO₄) 3-nitrobenzaldehyde

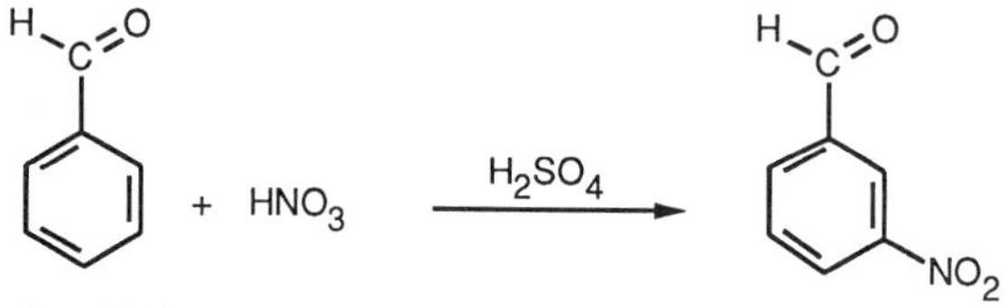

5.29 Only phenol (e) reacts faster than benzene.

5.30

Most reactive ———→ *Least reactive*

a) Benzene > chlorobenzene > o–dichlorobenzene
b) Phenol > nitrobenzene > p–bromonitrobenzene
c) o–Dimethylbenzene > fluorobenzene > benzaldehyde

5.31

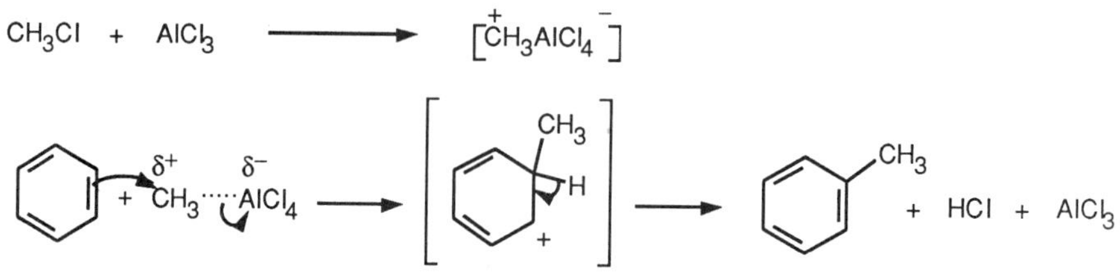

$CH_3Cl + AlCl_3 \longrightarrow [\overset{+}{C}H_3AlCl_4^-]$

[benzene + $\overset{\delta+}{CH_3}\cdots\overset{\delta-}{AlCl_4}$] ⟶ [arenium ion intermediate with CH₃ and H] ⟶ toluene + HCl + AlCl₃

5.32

a) *m*-Nitrophenol + Cl₂ →(FeCl₃) 2-Chloro-5-nitrophenol + 4-Chloro-3-nitrophenol

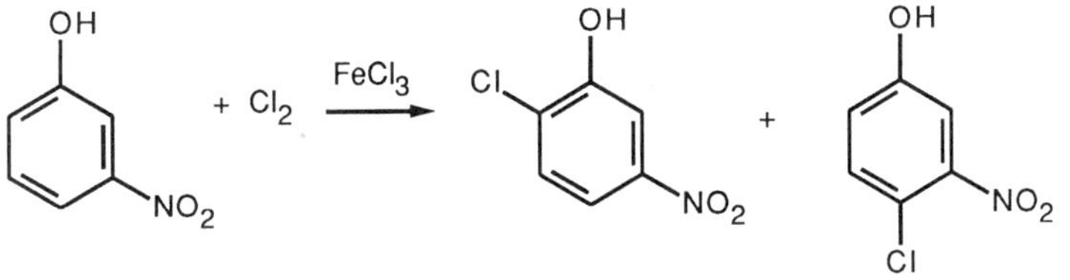

b) *o*-Methylphenol + Cl₂ →(FeCl₃) 4-Chloro-2-methylphenol + 2-Chloro-6-methylphenol

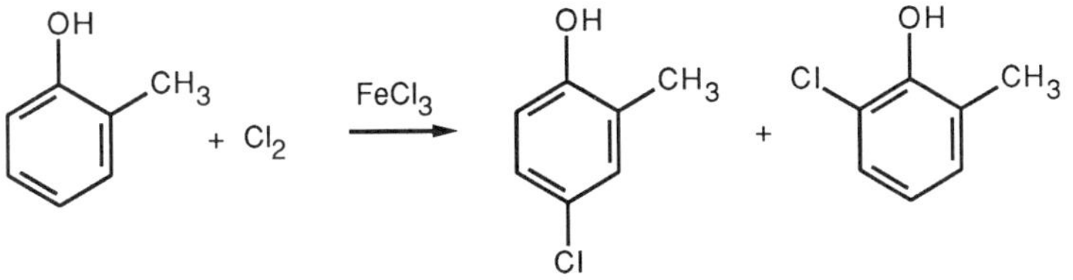

c) *p*-Chloronitrobenzene + Cl₂ →(FeCl₃) 3,4-Dichloronitrobenzene

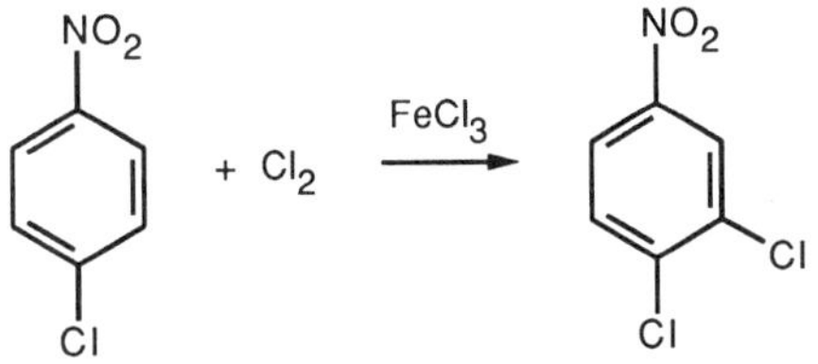

5.33

a) Bromobenzene + SO₃ →(H₂SO₄) 2-bromobenzenesulfonic acid + 4-bromobenzenesulfonic acid

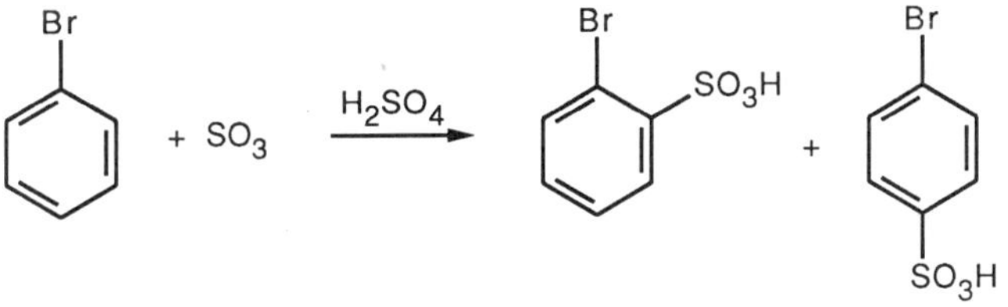

b) 3-Bromophenol + SO₃ →(H₂SO₄) 2-hydroxy-5-bromobenzenesulfonic acid + 4-hydroxy-3-bromobenzenesulfonic acid

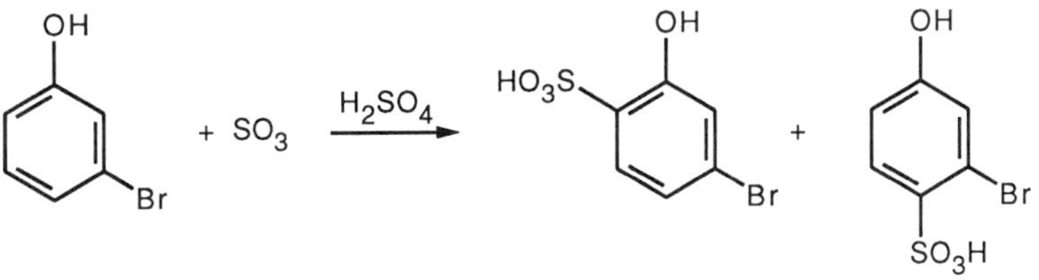

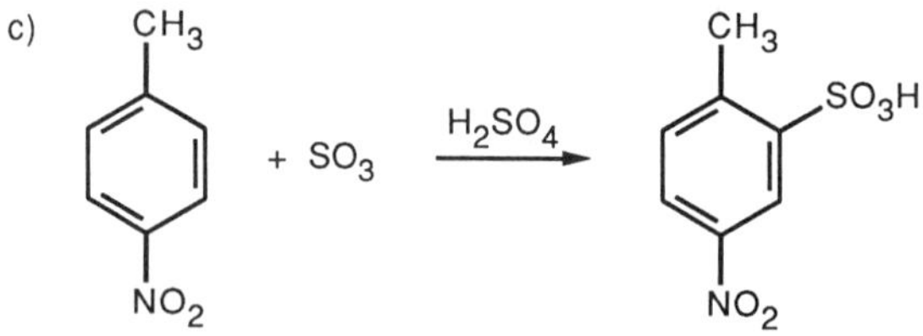

80 Chapter 5

c)

$$\text{4-methyl-nitrobenzene} + SO_3 \xrightarrow{H_2SO_4} \text{2-methyl-5-nitrobenzenesulfonic acid}$$

5.34 Most reactive ─────────────→ Least reactive

Anisole > Toluene > *p*-Bromotoluene > Bromobenzene

Nitrobenzene does not undergo Friedel-Crafts alkylation.

5.35

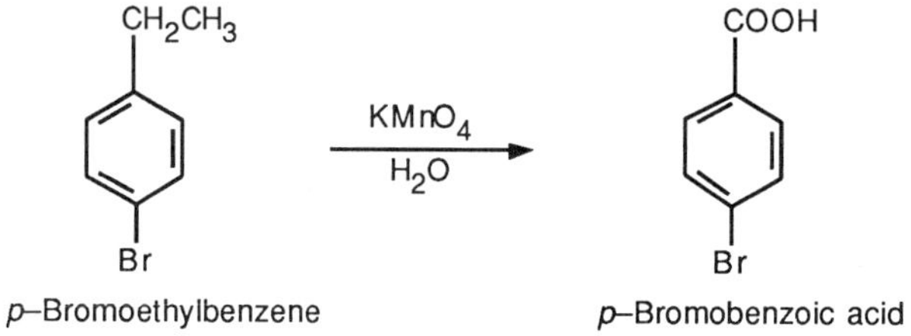

p-Bromoethylbenzene $\xrightarrow{KMnO_4, H_2O}$ *p*-Bromobenzoic acid

5.36

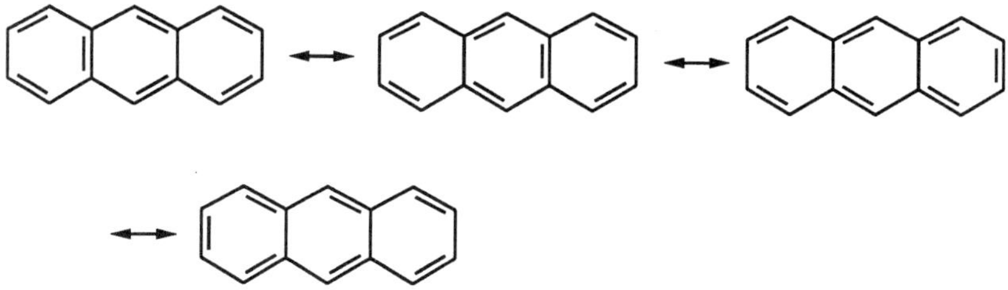

[Anthracene resonance structures]

5.37

[Phenanthrene resonance structures]

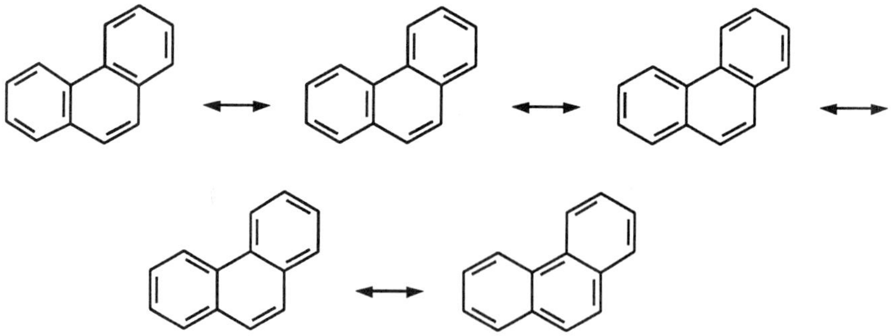

5.38

The positively charged carbocation intermediate formed from *ortho* or *para* attack of bromine can be stabilized by resonance contributions from the second ring. This stabilization is not possible for *meta* attack, and thus bromination occurs at the *ortho* and *para* positions.

5.39 Attack occurs on the unsubstituted ring because bromine is a deactivating substituent. Attack occurs at the *ortho* and *para* positions of the ring because the positively charged intermediate can be stabilized by resonance contributions from bromine and from the second ring (see Problem 5.38).

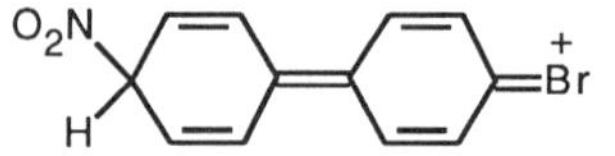

O₂N—C₆H₅—C₆H₄=Br⁺   etc.

5.40

a) Benzene →(SO₃/H₂SO₄)→ benzenesulfonic acid →(Br₂/FeBr₃)→ *m*-Bromobenzenesulfonic acid

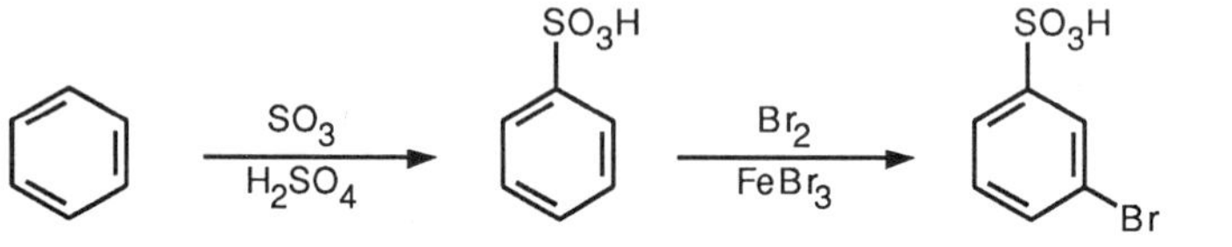

b) Benzene →(Cl₂/FeCl₃)→ chlorobenzene →(SO₃/H₂SO₄)→ *o*-Chlorobenzenesulfonic acid

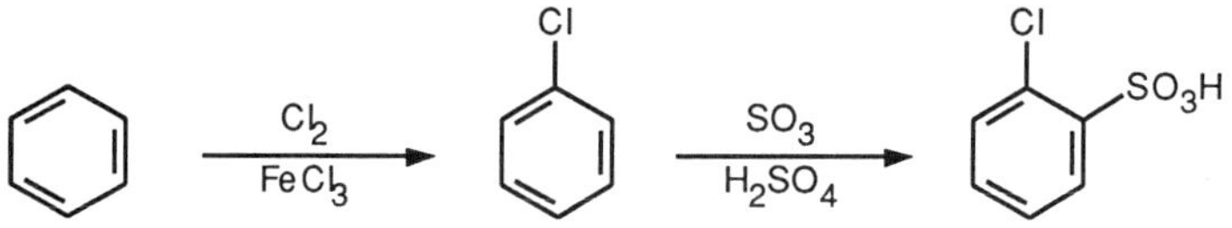

c) Benzene →(Cl₂/FeCl₃)→ chlorobenzene →(CH₃Cl/AlCl₃)→ *p*-chlorotoluene

or Benzene →(CH₃Cl/AlCl₃)→ toluene →(Cl₂/FeCl₃)→ *p*-chlorotoluene

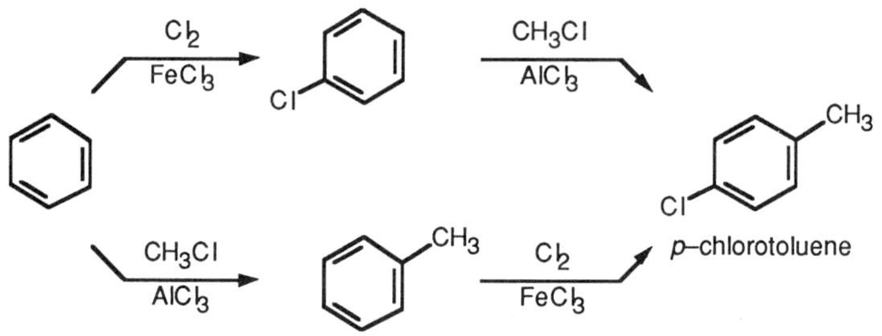

5.41

a) Benzene →(CH₃Cl/AlCl₃)→ toluene →(HNO₃/H₂SO₄)→ o-nitrotoluene (plus para isomer) →(KMnO₄/H₂O)→ o-nitrobenzoic acid

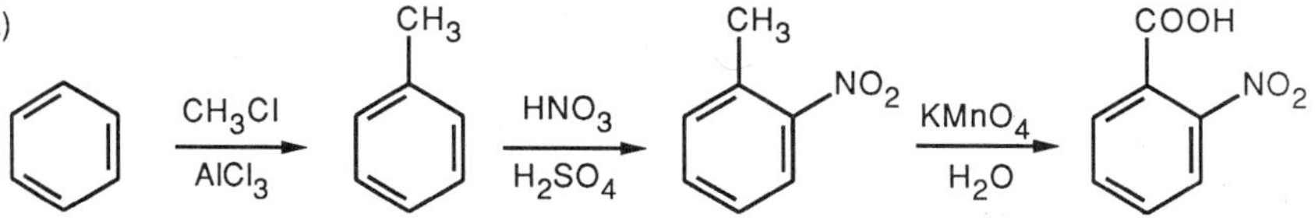

b) Toluene →((CH₃)₃CCl/AlCl₃)→ p-tert-butyltoluene →(KMnO₄/H₂O)→ *p-tert*-Butylbenzoic acid

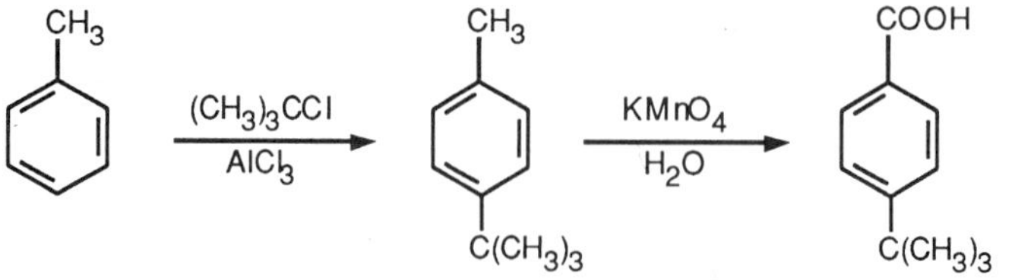

5.42 Resonance forms for the intermediate from attack at C–1:

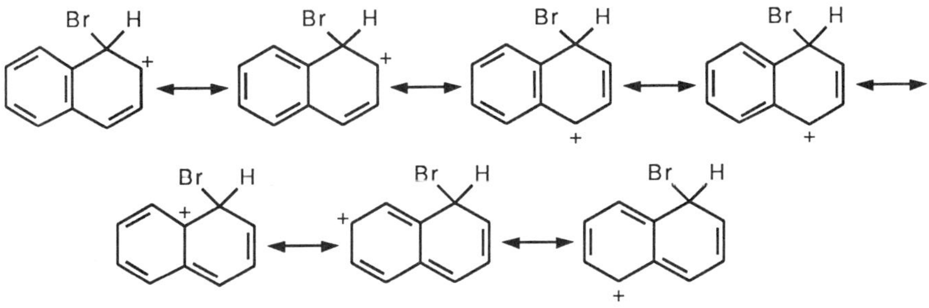

Resonance forms for the intermediate from attack at C2:

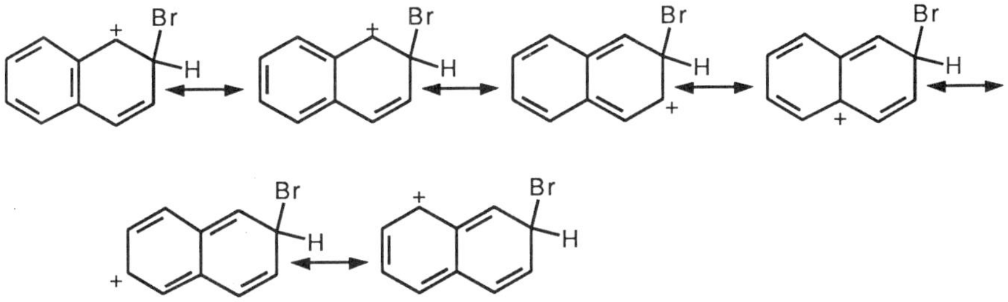

There are seven resonance forms for attack at C–1 and six for attack at C2. For C–1 attack, the second ring is fully aromatic in four of the resonance forms. In the other three forms, the positive charge has been delocalized into the second ring, destroying the ring's aromaticity. For C2 attack, the second ring is fully aromatic in only the first two forms. Since stabilization is lost when aromaticity is disturbed, the intermediate from C2 attack is less stable than the intermediate from C–1 attack, and C–1 attack is favored.

5.43

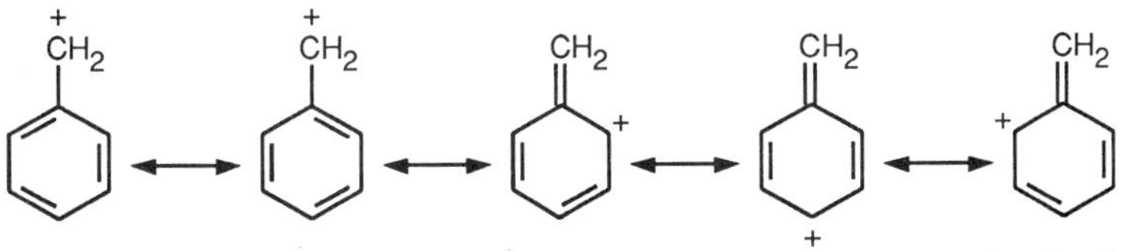

A benzylic carbocation is stabilized because its positive charge can be delocalized over the pi system of the aromatic ring.

5.44

[Reaction scheme: 1-phenylpropene + H–Br → resonance-stabilized carbocation intermediate [PhCHCH₂CH₃]⁺ with :Br:⁻ → PhCHBrCH₂CH₃]

Protonation of the double bond at carbon 2 of 1–phenylpropene leads to an intermediate that can be stabilized by resonance involving the phenyl ring.

5.45 a) Chlorination of toluene occurs at the *ortho* and *para* positions. To synthesize the given product, first oxidize toluene to benzoic acid and then chlorinate.

b) A *tert*-butyl group can't be oxidized by KMnO₄ to a –COOH group because it has no benzylic hydrogens. To obtain the desired compound, alkylate chlorobenzene with CH₃Cl and AlCl₃ and then oxidize.

5.46

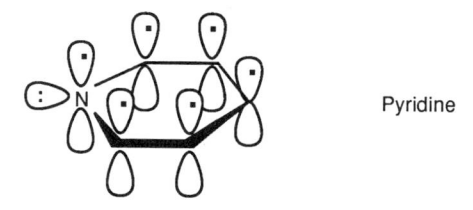

[Orbital diagram of pyridine showing sp² nitrogen lone pair and p orbitals of the ring]

Pyridine

The electronic descriptions of pyridine and benzene are very similar. The pyridine ring is formed by the sigma overlap of carbon and nitrogen sp^2 orbitals. In addition, six *p* orbitals, perpendicular to the plane of the ring, hold six electrons. These six *p* orbitals allow electrons to be delocalized over the pi system of the pyridine ring. The lone pair of nitrogen electrons occupies an sp^2 orbital that lies in the plane of the ring.

5.47

[Structure: phenyl–N⁺(CH₃)₃ Br⁻]

The trimethylammonium group is deactivating because it is positively charged and because it has no lone-pair electrons to donate to the aromatic ring.

5.48

toluene (PhCH₃) —KMnO₄/H₂O→ benzoic acid (PhCOOH) —HNO₃/H₂SO₄→ *m*-nitrobenzoic acid

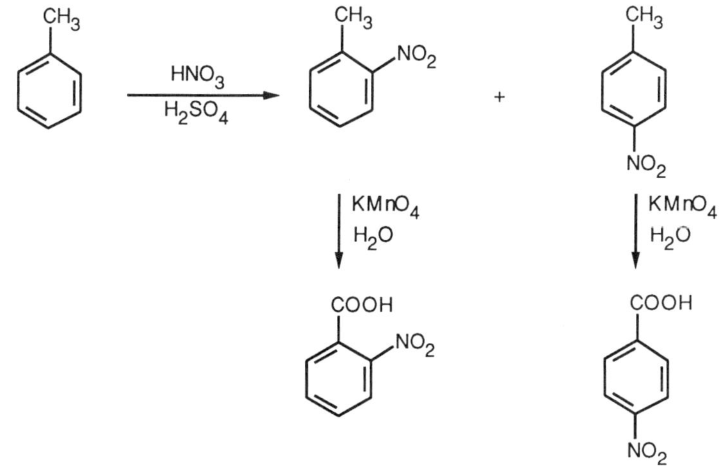

Aromatic Compounds 85

5.49

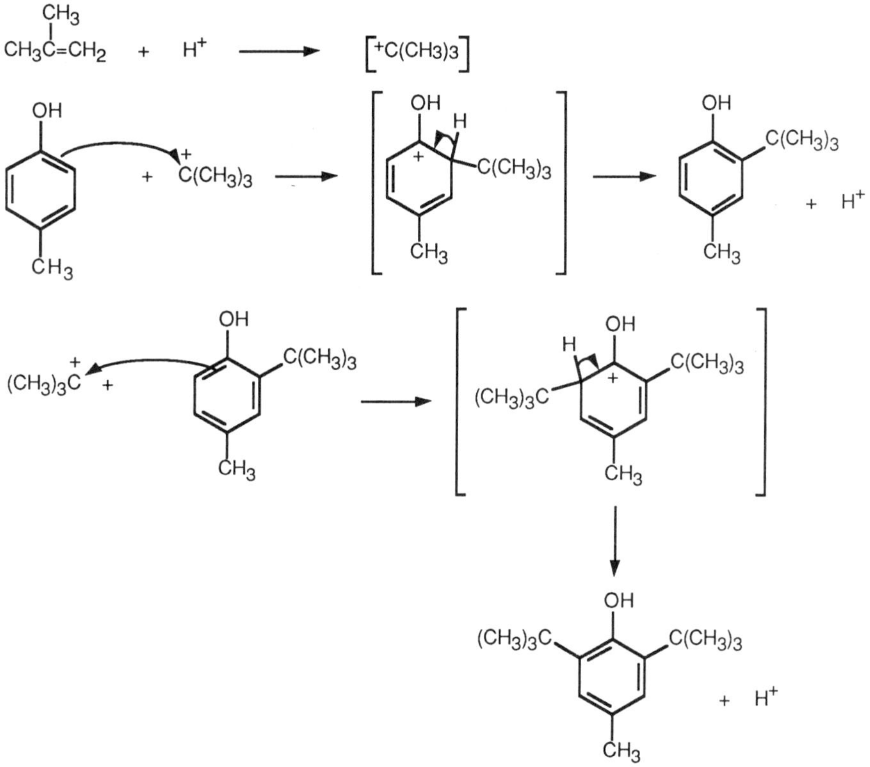

5.50 Problem 5.44 shows the mechanism of the addition of HBr to 1–phenylpropene and shows how the aromatic ring stabilizes the carbocation intermediate. Similar resonance forms can be drawn for the intermediates of reaction of the substituted styrenes with HBr. For the methoxy-substituted styrene, an additional form can be drawn. For the nitro-substituted styrene, no additional form is possible. In addition, one other resonance form is not important because it places two positive charges next to each other.

Thus, the intermediate resulting from addition of HBr to the methoxy substituted styrene is more stable, and reaction of *p*–methoxystyrene is faster.

Chapter Outline

I. Aromatic compounds (Sections 5.1 - 5.5)
   A. Naming aromatic compounds (Section 5.1)
   B. Benzene (Sections 5.2 - 5.4)
      1. Structure of benzene (Section 5.2)
      2. Stability of benzene (Section 5.3)
      3. Structure of benzene - resonance (Section 5.4)
II. Electrophilic aromatic substitution of benzene (Sections 5.5 - 5.8)
   A. Bromination (Section 5.6)
   B. Other electrophilic aromatic substitutions (Section 5.7)
      1. Chlorination
      2. Iodination
      3. Nitration
      4. Sulfonation
   C. Friedel-Crafts alkylation and acylation (Section 5.8)
III. Substituent effects in electrophilic aromatic substitution (Sections 5.9 - 5.11)
   A. Reactivity (Section 5.9)
      1. Activating substituents
      2. Deactivating substituents
   B. Orientation of reactions (Section 5.10)
      1. Ortho, para-directing activators
      2. Meta-directing deactivators
      3. Ortho, para-directing deactivators

IV. Oxidation and reduction of aromatic compounds (Section 5.12)
   A. Side-chain oxidation
   B. Ring reduction
V. Polycyclic aromatic hydrocarbons (Section 5.13)
VI. Organic synthesis (Section 5.14)

Study Skills for Chapter 5

After studying this chapter, you should be able to:

1. Name simple aromatic compounds (Problems 5.2, 5.26, 5.28, 5.29).
2. Draw structures corresponding to given names (Problems 5.3, 5.27, 5.28, 5.29, 5.30).
3. Formulate the mechanisms of electrophilic aromatic substitution reactions (Problems 5.7, 5.10, 5.11, 5.12, 5.31, 5.35, 5.33).
4. Predict the products of electrophilic aromatic substitution reactions (Problems 5.6, 5.8, 5.9, 5.16, 5.17, 5.32, 5.36, 5.39).
5. Predict the reactivity and orientation of aromatic ring and substituents in electrophilic aromatic substitution reactions (Problems 5.13, 5.14, 5.15, 5.18, 5.19, 5.20, 5.33, 5.34, 5.42, 5.51).
6. Draw resonance structures for polycyclic aromatic compounds (Problems 5.22, 5.39, 5.40, 5.41, 5.46, 5.47, 5.48).
7. Synthesize substituted benzenes (Problems 5.23, 5.24, 5.43, 5.44, 5.45, 5.49, 5.42).

Chapter 6 – Stereochemistry

6.1 Chiral: bean stalk, screw, shoe
Not chiral: screwdriver

6.2 Draw each compound, and identify all carbons that are *not* stereogenic. Nonstereogenic carbons include:

CH_3-, CH_2-, CX_2-, $-\underset{|}{C}=\underset{|}{C}-$, $-C\equiv C-$, $-\underset{|}{C}=O$, all benzene ring carbons.

Cross out these carbons. If all carbons are crossed out, the compound is achiral. If any carbons remain, they should be bonded to four different groups, and the compound is chiral. (Stereogenic carbons are starred.)

a) 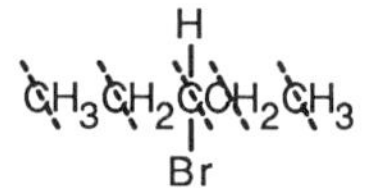
$$\overset{H}{\underset{Br}{C\!H_3\,C\!H_2\,C\!O\!H_2\,C\!H_3}}$$
achiral

b) 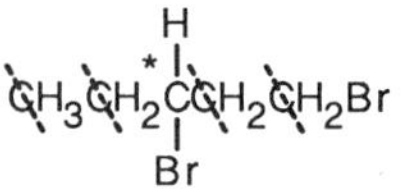
$$\overset{H}{\underset{Br}{C\!H_3\,C\!H_2\,\overset{*}{C}C\!H_2\,C\!H_2Br}}$$
chiral

Carbon 3 is bonded to four different groups: CH_3CH_2-, $BrCH_2CH_2-$, $Br-$, $H-$.

c) 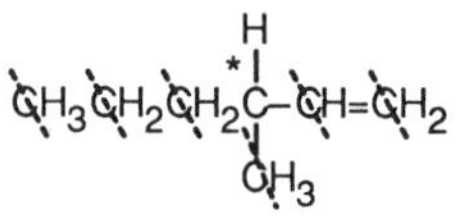
$$\overset{H}{\underset{CH_3}{C\!H_3\,C\!H_2\,C\!H_2\,\overset{*}{C}-C\!H=C\!H_2}}$$
chiral

d)

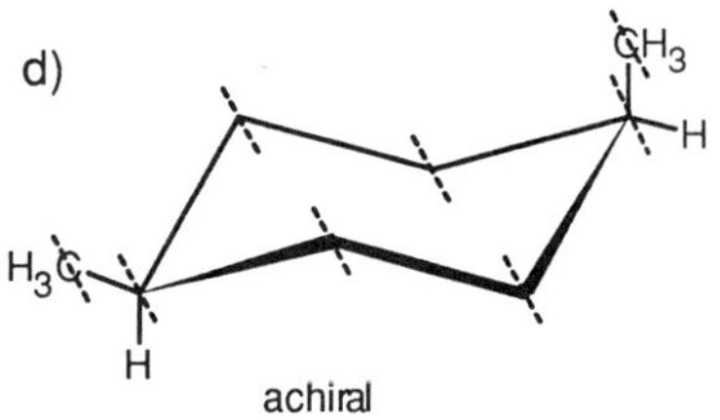

achiral

All carbons are bonded to at least two identical groups.

6.3 Refer to Problem 6.2 for a list of nonstereogenic carbons.

a) 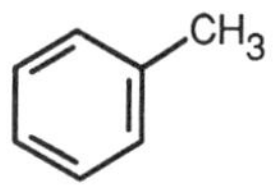 Toluene
achiral

b) 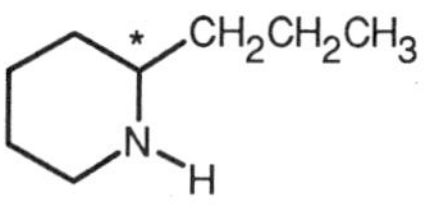 Coniine
chiral

c) 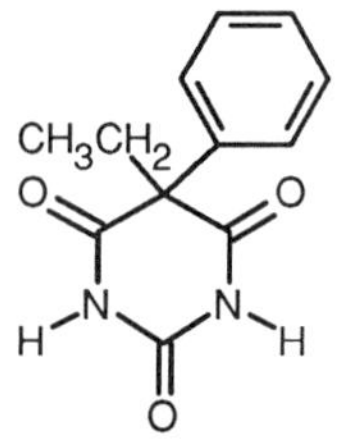 Phenobarbital
achiral

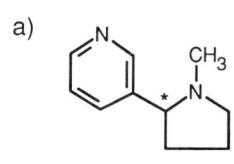

Stereochemistry

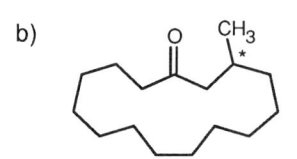

6.4

a) Nicotine

b) Muscone (musk oil)

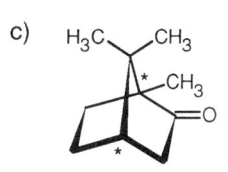

c) Camphor

6.5

Alanine

6.6 Cocaine is levorotatory. (Levorotatory compounds have a minus sign in front of the degree of rotation.)

6.7 Use the formula $[\alpha]_D = \dfrac{\alpha}{l \times C}$, where

$[\alpha]_D$ = specific rotation

α = observed rotation

l = path length of cell (in dm)

C = concentration (in g/mL)

In this problem:

$\alpha = +1.21°$

$l = 5.00$ cm $= 0.500$ dm

$C = 1.50$ g/10.0 mL $= 0.150$ g/mL; thus $[\alpha]_D = \dfrac{+1.21°}{0.500 \text{ dm} \times 0.150 \text{ g/mL}} = 16.1°$

6.8 Use the rules in Section 6.6 to assign priorities.

a) By Rule 1, –H is of lowest priority, and –Br is of highest priority. By Rule 2, –CH₂CH₂OH is of higher priority than –CH₂CH₃.

Highest ⟶ Lowest

–Br, –CH₂CH₂OH, –CH₂CH₃, –H.

b) By Rule 3, –COOH can be considered as having three O atoms singly bonded to the carbon. Since three oxygens are attached to a –COOH carbon and only one oxygen is attached to a –CH₂OH carbon, –COOH is of higher priority than –CH₂OH. –CO₂CH₃ is of higher priority than –COOH by Rule 2, and –OH is of highest priority by Rule 1.

90   Chapter 6

Highest ⟶ Lowest
–OH, –CO₂CH₃, –COOH, –CH₂OH.

c) –Br, –Cl, –CH₂Br, –CH₂Cl.

6.9   The following scheme may be used to assign *R, S* configurations to stereogenic centers:

Step 1. For each stereogenic center, rank substituents by the priority rules. Give the number 4 to the lowest priority substituent. For (a):

Substituent	Priority
–Br	1
–COOH	2
–CH₃	3
–H	4

Step 2. Manipulate the molecule so that the lowest priority group is pointing toward the rear. To avoid errors, use a molecular model of the compound.

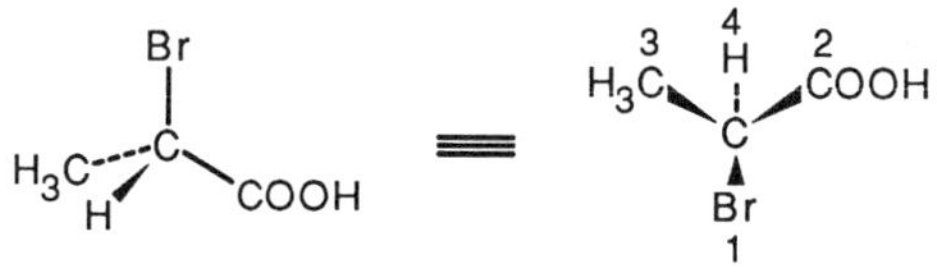

Step 3. Find the direction of rotation of the arrows that go from group 1 to group 2 to group 3. If the arrows have a clockwise rotation, the configuration is *R*; if the arrows have a counterclockwise rotation, the configuration is *S*. Here, the configuration is *S*.

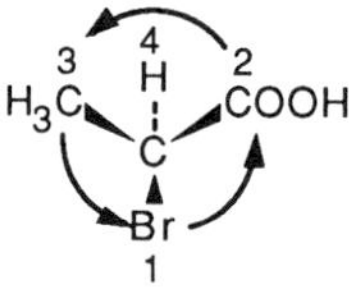

b)

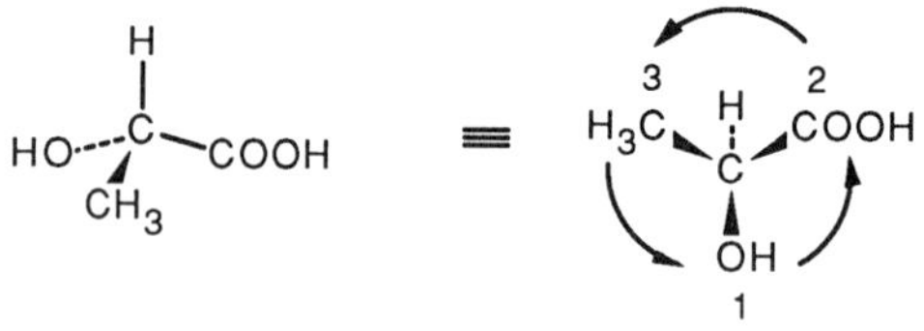

c)

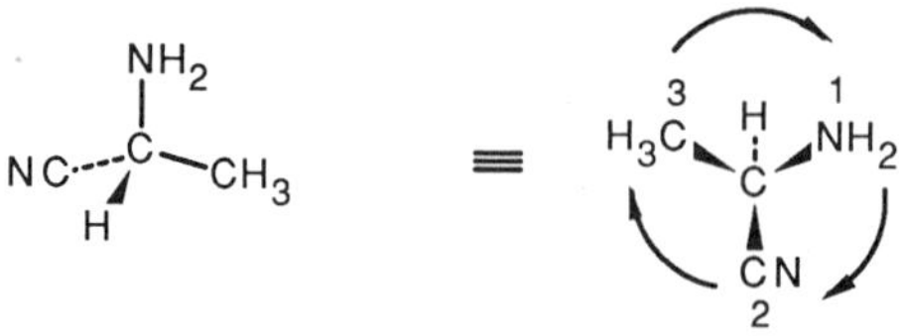

6.10

CH₃CH₂CH₂CHCH₃
         |
         OH

2-Pentanol

```
        3
        CH₃
         |
   4     |  2
   H----C----CH₂CH₂CH₃        S
        |
       HO
        1
```

Substituent	Priority
–OH	1
–CH₂CH₂CH₃	2
–CH₃	3
–H	4

6.11 *R, S* assignments for more complicated molecules can be made by using a slight modification of the rules in Problem 6.9. It is especially important to use molecular models when a compound has more than one stereogenic center.

Step 1. Assign priorities to groups at the *first* stereogenic center.

Substituent	Priority
–Br	1
–CH(OH)CH₃	2
–CH₃	3
–H	4

Step 2. Orient the model so that –H bonded to the top stereogenic center points to the back.

Step 3. Find the direction of the arrows that travel from 1 to 2 to 3 and assign an *R* or *S* configuration to the top stereogenic center.

```
     Br
   H\  ,CH₃                    4
     C                         H   1
     |           ≡        3  \ | /
     C                      H₃C-C-Br       R
   H/  `OH                      |
     CH₃                     CH(OH)CH₃
                                2
```

Step 4. Repeat steps 1-3 for the next stereogenic center.

```
     Br
   H\  ,CH₃                    4
     C                         H   2
     |           ≡        1  \ | /
     C                      HO-C-CH(Br)CH₃    R
   H/  `OH                      |
     CH₃                       CH₃
                                3
```

a) *R, R*   b) *S, R*   c) *R, S*

6.12 Molecules (b) and (c) are enantiomers (mirror images). Molecule (a) is the diastereomer of (b) and (c).

6.13

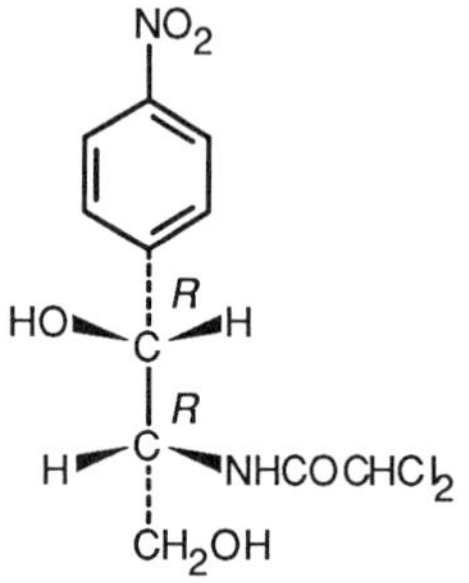

Chloramphenicol

6.14 In order for a molecule to exist as a meso form, it must have a plane of symmetry. 2,3–Dibromobutane can exist as a pair of enantiomers *or* as a meso compound, depending on the configurations at carbons 2 and 3.

a)

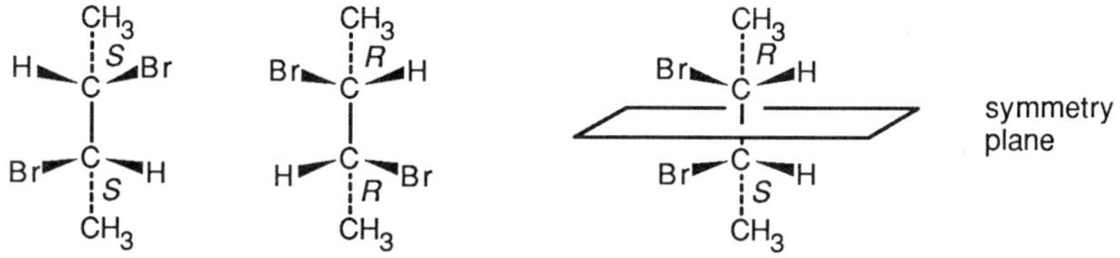

not meso   meso

b) 2,3–Dibromopentane has no symmetry plane and thus can't exist in a meso form.

c) 2,4–Dibromopentane can exist in a meso form.

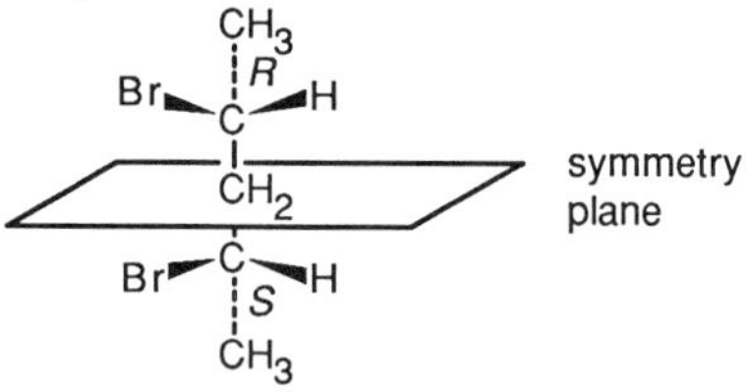

2,4–Dibromopentane can also exist as a pair of enantiomers (2R, 4R and 2S, 4S) that are not meso compounds.

6.15

a) meso

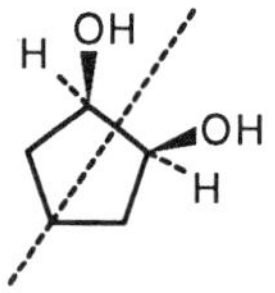

b) not meso

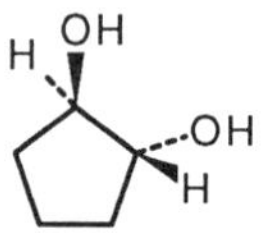

c)

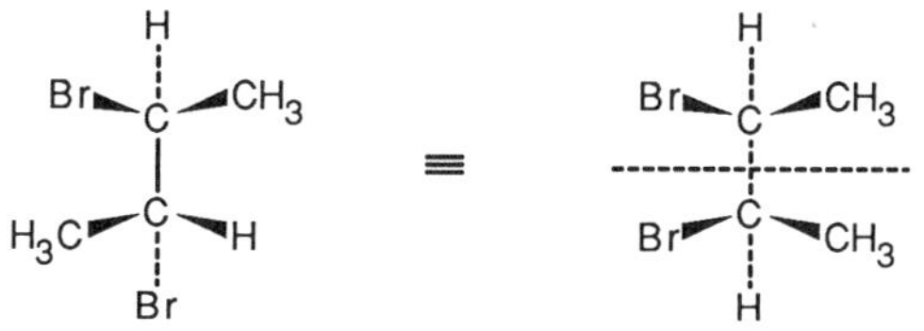

$$\text{[structure]} \equiv \text{[structure with symmetry plane]}$$

symmetry plane

meso

6.16

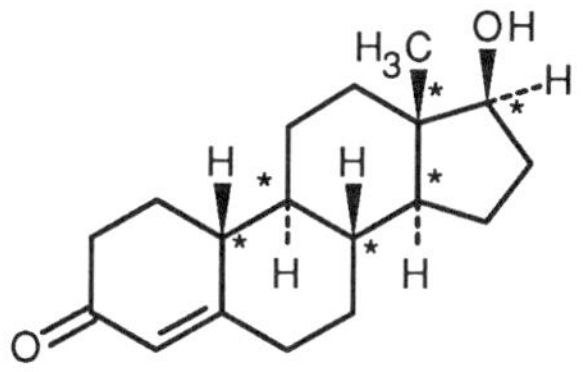

Nandrolone

Nandrolone has six stereogenic centers (starred) and can have, in principle, $2^6 = 64$ stereoisomers.

6.17

a)

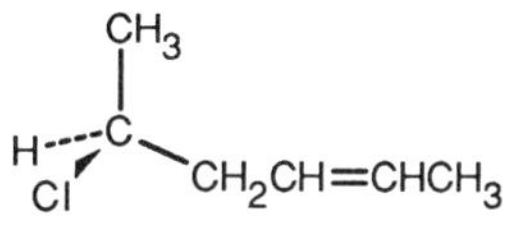

S-5-Chloro-2-hexene    Chlorocyclohexane

The compounds are skeletal isomers because they have different carbon skeletons.

b)

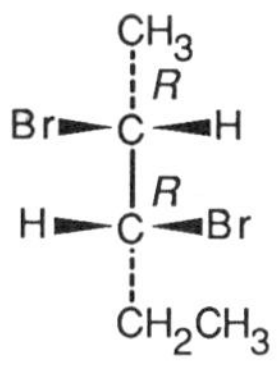

2R,3R-2,3-Dibromopentane   2S,3R-2,3-Dibromopentane

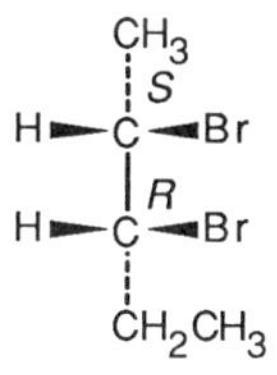

These two compounds are diastereomers — stereoisomers that are not mirror images.

6.18

$$[\alpha]_D = \frac{\alpha}{l \times C}$$

$\alpha = +2.22°$
$l = 1.00 \text{ cm} = 0.100 \text{ dm}$
$C = 3.00 \text{ g} / 5.00 \text{ mL} = 0.600 \text{ g/mL}$

$$[\alpha]_D = \frac{+2.22°}{(0.100 \text{ dm})(0.600 \text{ g/mL})} = +37.0°$$

Stereochemistry 93

6.19

Ecdysone: $[\alpha]_D = \dfrac{+0.087°}{(0.200 \text{ dm})(0.00700 \text{ g/mL})} = +62°$

6.20
a) *Chirality* is the property of "handedness" — the property of an object that causes it to be nonsuperimposable on its mirror image.

b) A *stereogenic center* is an atom that causes chirality in a molecule by being bonded to four different atoms or groups of atoms.

c) A *diastereomer* is a stereoisomer that is not the mirror image of another stereoisomer.

d) A *racemate* is a 50:50 mixture of (+) and (−) enantiomers that behaves as if it were a pure compound and that is optically inactive.

e) A *meso compound* is a compound that contains two or more stereogenic centers yet is optically inactive because it possesses a plane of symmetry.

f) An *enantiomer* is one of a pair of stereoisomers that have a mirror image relationship.

6.21
Chiral: ear, coin, scissors
Achiral: basketball, wine glass, snowflake

6.22

a)

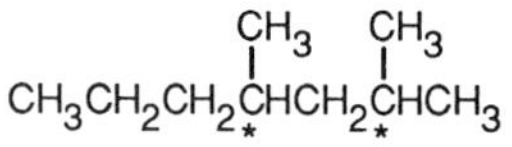

CH₃CH₂CH₂CH(CH₃)CH₂CH(CH₃)CH₃

2,4-Dimethylheptane
chiral

b)

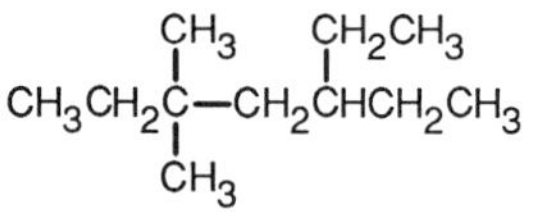

CH₃CH₂C(CH₃)(CH₃)CH₂CH(CH₂CH₃)CH₂CH₃

5-Ethyl-3,3-dimethylheptane
achiral

c)

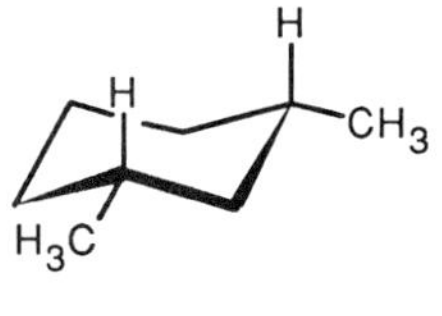

cis-1,3-Dimethylcyclohexane
achiral

6.23

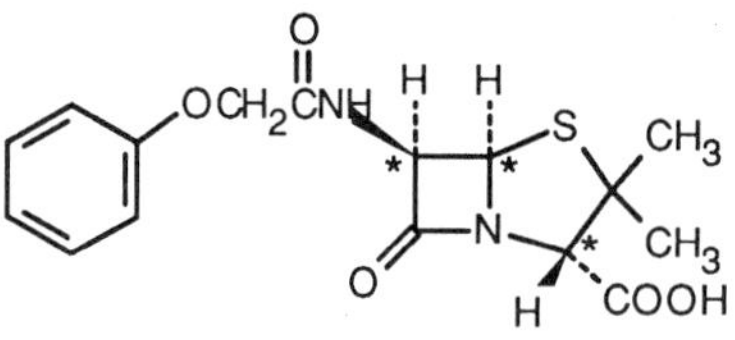

Penicillin V

Stereochemistry 95

6.24 There are several possibilities for most parts of this problem.

a) CH₃CH₂CH₂C*HCH₃ , CH₃CH₂C*HCH₂Cl , CH₃C*HCHCH₃
          | | | |
          Cl   CH₃   CH₃ Cl
                                                   (*on first CH)

b) CH₃CH₂CH₂CH₂C*HCH₃ , CH₃CH₂CH₂C*HCH₂CH₃ , CH₃CH₂CH₂C*HCH₂OH
            | | |
            OH   OH   CH₃

   CH₃CH₂C*H–C*HCH₃ , CH₃C*HCH₂C*HCH₃ , CH₃CH₂C*HCH₂CH₂OH
         | | | | |
         OH CH₃   OH CH₃   CH₃

   CH₃CH₂C*HC*HCH₃ , CH₃C*HC(CH₃)₃ , CH₃C*HC*HCH₂OH
        | | | | |
        CH₃ OH   OH   CH₃ CH₃

c) CH₃CH₂C*HCH=CH₂
          |
          CH₃

d) CH₃CH₂CH₂CH₂C*HCH₂CH₃ , CH₃CH₂CH₂C*HC*HCH₃ , CH₃CH₂C*HCH₂C*HCH₃
                | | | | |
                CH₃   CH₃ CH₃   CH₃   CH₃

   CH₃CH₂C*HC*HCH₂CH₃ , CH₃CH₂C*HC(CH₃)₃
        | | |
        CH₃ CH₃   CH₃

6.25

a) CH₃CH₂C*H–C–CH₂CH₃   b) cyclopentanone with CH₃ at position 3 (chiral)   c) 4-methylcyclohexanone (achiral)
         | |
         H₃C CH₃
             |
             CH₃

   chiral

96   Chapter 6

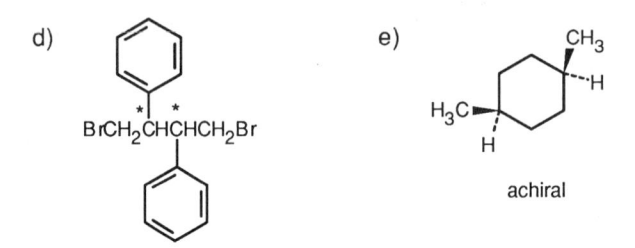

d) BrCH₂CḦCHCH₂Br (with phenyl substituent, stereocenters marked *)

This compound exists as a pair of chiral enantiomers and an achiral *meso* isomer.

e) [cyclohexane with CH₃, H, H₃C, H substituents]

achiral

6.26 If you have trouble with this sort of problem, use the following scheme.

1) Draw all alkanes of the formula C₅H₁₂.

$$CH_3CH_2CH_2CH_2CH_3 \qquad CH_3CH_2\overset{\underset{\mid}{CH_3}}{C}HCH_3 \qquad CH_3\overset{\underset{\mid}{CH_3}}{\underset{\mid}{\underset{CH_3}{C}}}CH_3$$

3 kinds of –H   4 kinds of –H   1 kind of –H

2) Find the number of different kinds of hydrogen for each alkane.
3) Replace one of each different kind of –H with an –OH and draw the resulting structure.
4) Locate all stereogenic centers.

$$CH_3CH_2CH_2CH_2CH_2OH \qquad CH_3CH_2CH_2\overset{\underset{\mid}{OH}}{C}HCH_3 \qquad CH_3CH_2\overset{\underset{\mid}{OH}}{C}HCH_2CH_3 \qquad CH_3CH_2\overset{\underset{\mid}{CH_3}}{C}HCH_2OH$$

achiral   chiral*   achiral   chiral*

$$CH_3CH_2\overset{\underset{\mid}{CH_3}}{\underset{\mid}{\underset{OH}{C}}}CH_3 \qquad CH_3\overset{\underset{\mid}{CH_3}}{C}H\overset{\underset{\mid}{}}{C}HCH_3 \qquad HOCH_2CH_2\overset{\underset{\mid}{CH_3}}{C}HCH_3 \qquad CH_3\overset{\underset{\mid}{CH_3}}{\underset{\mid}{\underset{CH_3}{C}}}CH_2OH$$
                                             OH

achiral   chiral*   achiral   achiral

6.27

a) $CH_3CH_2\overset{\underset{\mid}{OH}}{C}HCH_3$*

b) $CH_3CH_2\overset{\underset{\mid}{CH_3}}{C}HCOOH$*

c) $CH_3\overset{\underset{\mid}{OH}}{C}H\overset{\underset{\mid}{}}{C}HCH_3$*
                  Br

Stereochemistry 97

6.28

Highest priority ⟶ Lowest priority

a) –OCH₃, –OH, –CH₃, –H

b) –Br, –Cl, –CH₂Br, –CH₃

c) –C(CH₃)₃, –CH=CH₂, –CH(CH₃)₂, –CH₂CH₃

d) –OCH₃, –COOCH₃, –COCH₃, –CH₂OCH₃

6.29

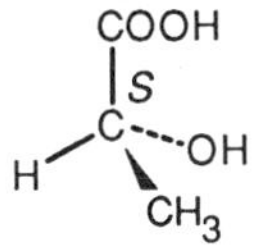

S-Lactic acid 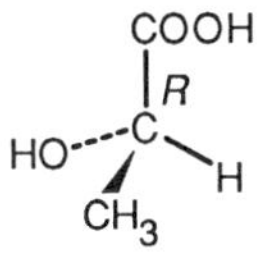   R-Lactic acid

6.30

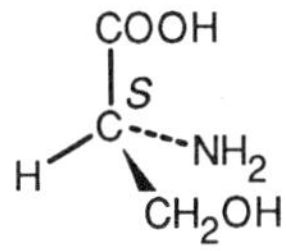

S-Serine 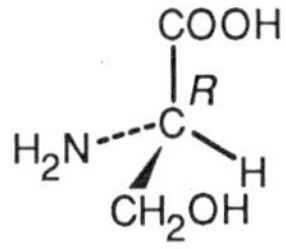   R-Serine

6.31 R–Serine has a specific rotation of +6.83° because enantiomers differ only in the sign of their specific rotations.

6.32

a) 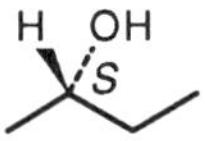   b) 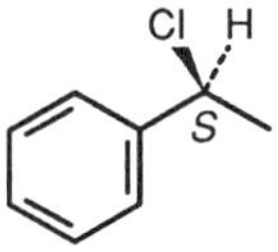   c)

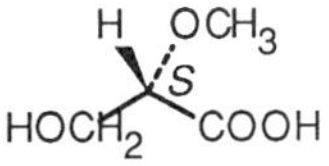

6.33

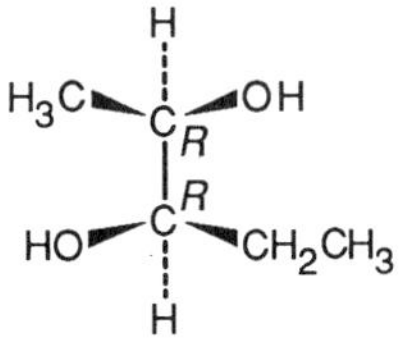

(2R, 3R)   (2S, 3S)   (2R, 3S)

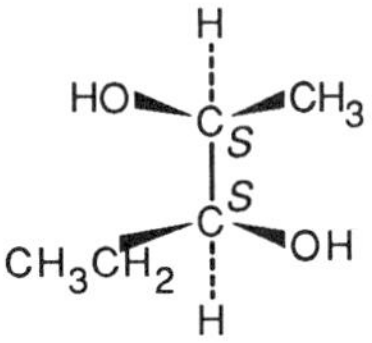

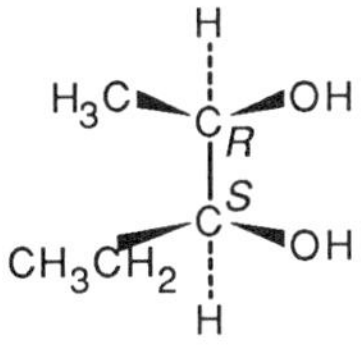

The specific rotations of the (2R, 3R) and (2S, 3S) enantiomers are equal in magnitude and opposite in sign. The specific rotations of the (2R, 3S) and (2R, 3R) diastereomers are not related.

6.34-6.35

[Four Fischer-like projections of 2,4-dibromoheptane stereoisomers]

(2S, 4R) — (2R, 4S) : enantiomers

(2S, 4S) — (2R, 4R) : enantiomers

The (2R, 4S) stereoisomer is the enantiomer of the (2S, 4R) stereoisomer. The (2S, 4S) and (2R, 4R) stereoisomers are diastereomers of the (2S,4R) stereoisomer.

6.36

a) [Structure with H₃C, CH₂CH₃ groups on C(S) and H₃C, CH₂CH₃ on C(R), with H substituents — meso compound indicated by dashed line]

b) [Structure with H₃C, OH on C(R) and CH₃CH₂, CH₃ on C(S), with H substituents]

6.37

[Wedge structure showing OH (S), H, CH₃, CH₂CH₃ groups]   (S)-2-Butanol

6.38

[Newman projection with Cl, H, CH₃ in front; H₃C, H, H in back] ≡ [Wedge structure: Cl, S, CH₃CH₂, H, CH₃]

6.39

[Newman projection with Cl, H₃C, H in front; H, CH₃, H in back] ≡ [Wedge structure: Cl, R, H, CH₂CH₃, CH₃]

6.40

[Newman projection of meso-tartaric acid: front carbon with COOH (top), H (left), OH (right); back carbon with HO (left), H (right), COOH (bottom)] → rotate rear carbon 180° → [eclipsed Newman projection with COOH/COOH eclipsed at top, HO/HO eclipsed on left, H/H eclipsed on right]

meso

The mirror plane of *meso*-tartaric acid is more obvious if the molecule is shown in its eclipsed conformation.

6.41

[Newman projection: front COOH (top), H (left), OH (right); back H (left), OH (right), COOH (bottom)] | [Newman projection: front COOH (top), HO (left), H (right); back HO (left), H (right), COOH (bottom)]

2R, 3R–Tartaric acid   2S, 3S–Tartaric acid

The enantiomeric tartaric acids are mirror images of each other. Unlike the *meso* isomer, neither of the above enantiomers contains a mirror plane.

6.42

$$\text{HOCH}_2\overset{*}{\text{CH}}-\overset{*}{\text{CH}}-\overset{*}{\text{CH}}-\overset{*}{\text{CH}}\text{CHO}$$
$$\phantom{\text{HOCH}_2}\,||||$$
$$\phantom{\text{HOCH}_2}\text{OH OH OH OH}$$

Glucose

The number of stereoisomers of a chiral compound is given by 2^n, where n equals the number of stereogenic centers present. Here, $n = 4$ and $2^n = 16$. Glucose thus has 16 possible stereoisomers.

6.43

[Tetrahedral structure with central C, CH=CH₂ (up), CH₂CH₃ (right), Cl (down-left), H (dashed back); curved arrow R configuration]

(R)-3-Chloro-1-pentene

6.44

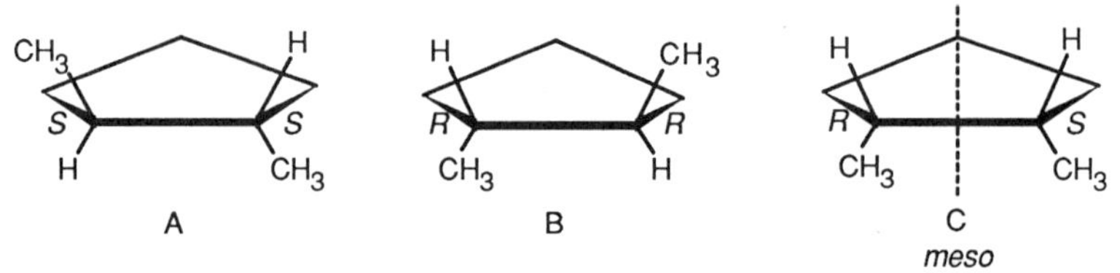

[Structures A, B, C shown as cyclopentane rings with CH₃ and H substituents]

A: (S,S) with CH₃ groups
B: (R,R) with CH₃ groups
C: (R,S) meso

A and B are enantiomers and are chiral. Compound C is their diastereomer and is a *meso* compound.

6.45

a) [Structure showing amino acid with COOH, H, CH₃, NH₂ groups around central C labeled S]

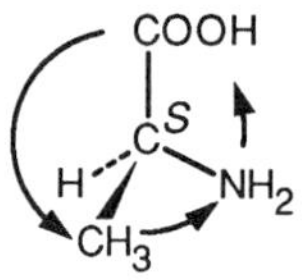

b) [Structure showing phenyl group attached to C(R) with HO and H, connected to C(R) with H₂N, CH₃, and H]

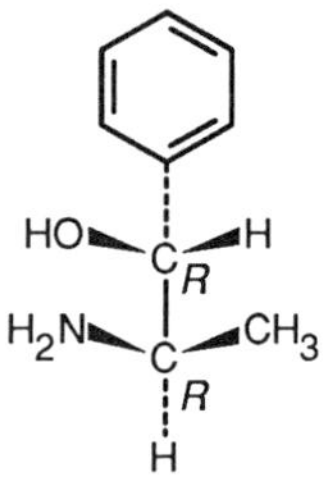

6.46

The intermediate formed in the hydroxylation of a double bond is a cyclic manganate ester. No carbon-oxygen bonds are broken when this manganate ester is cleaved to form the diol; cleavage occurs at manganese-oxygen bonds. Thus, the stereochemistry of the product is the same as that of the initial adduct.

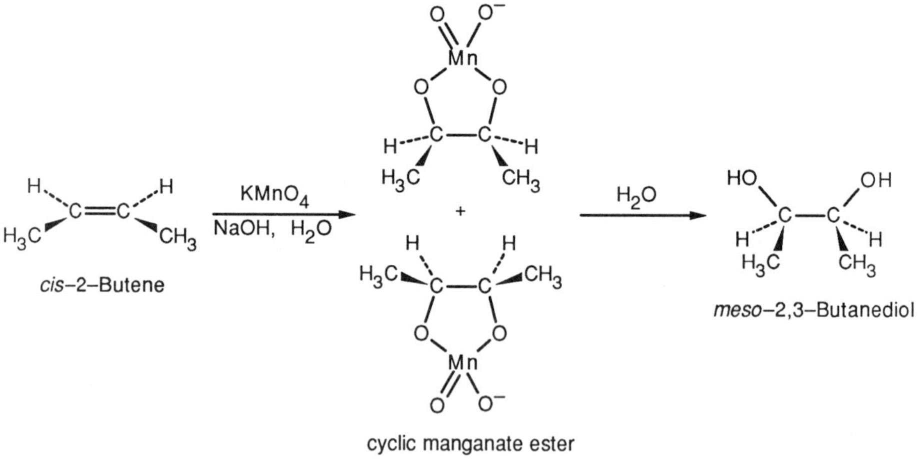

[Reaction scheme: cis-2-Butene + KMnO₄, NaOH, H₂O → cyclic manganate ester (two stereoisomers shown with Mn bridging two oxygens on adjacent carbons bearing H₃C and H groups) → (H₂O) → meso-2,3-Butanediol]

cis-2-Butene

cyclic manganate ester

meso-2,3-Butanediol

Stereochemistry 101

6.47

Hydroxylation of *trans*–2–butene yields a racemic mixture of the enantiomeric 2,3–butanediols.

6.48 A molecule with *n* stereogenic centers can give rise to a maximum of 2^n stereoisomers. Thus, we might predict eight stereoisomers for 2,4–dibromo–3–chloropentane (CH$_3$CHBrCHClCHBrCH$_3$), which has three stereogenic carbons. After drawing the eight possible stereoisomers, it is apparent that only four 2,4–dibromo–3–chloropentanes are unique.

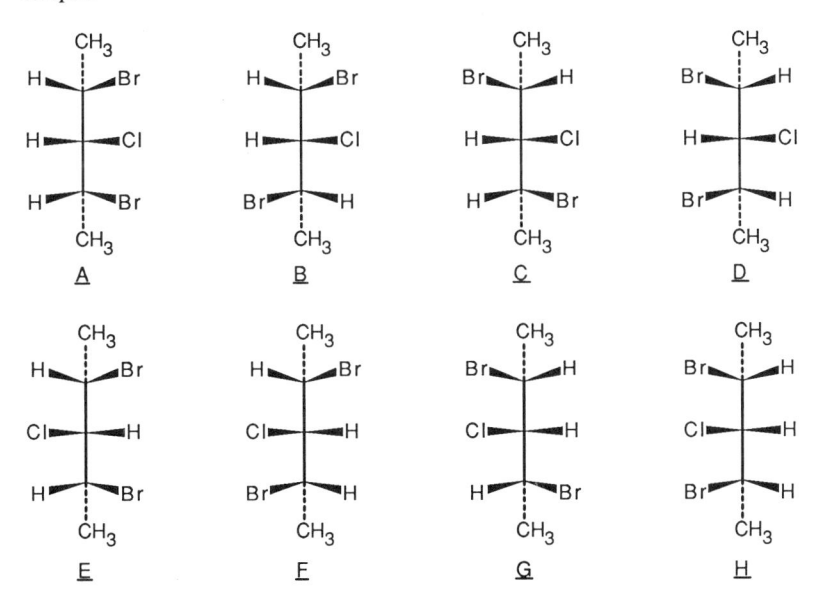

Identical: A and H, B and F, D and E, C and G.

(A,H) and (D,E) are optically inactive *meso* compounds and are diastereomers.

(B,F) and (C,G) are enantiomers and are optically active.

(A,H) and (D,E) are diastereomeric with (B,F) and (C,G).

6.49

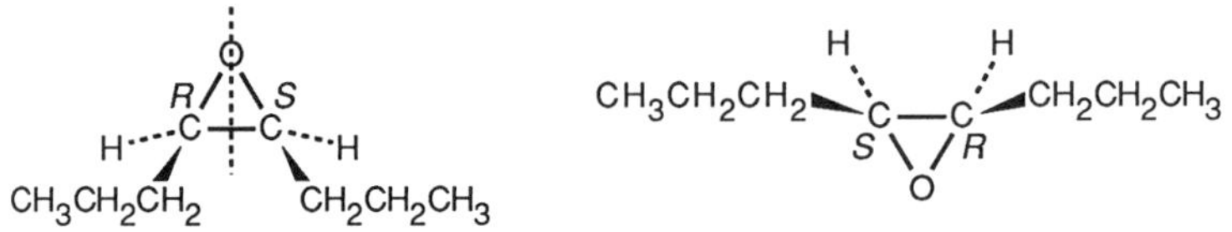

Peroxycarboxylic acids can attack either the "top" side or the "bottom" side of a double bond. The epoxide resulting from "top" side attack, pictured above, has two stereogenic centers, but because it has a plane of symmetry it is a *meso* compound. The epoxide resulting from "bottom" side attack is identical to epoxide resulting from top-side attack.

6.50

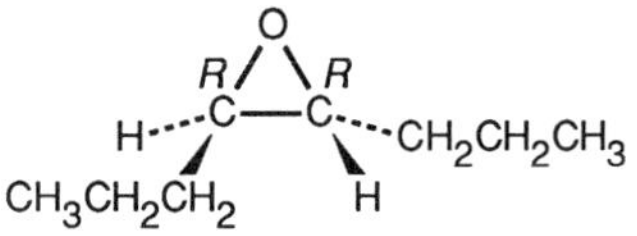

The epoxide formed by "top-side" attack of a peroxyacid on *trans*–4–octene is pictured. This epoxide has two stereogenic centers of R configuration. The epoxide formed by "bottom-side" attach has S,S configuration. The enantiomers are formed in equal amounts and constitute a racemic mixture.

6.51-6.52

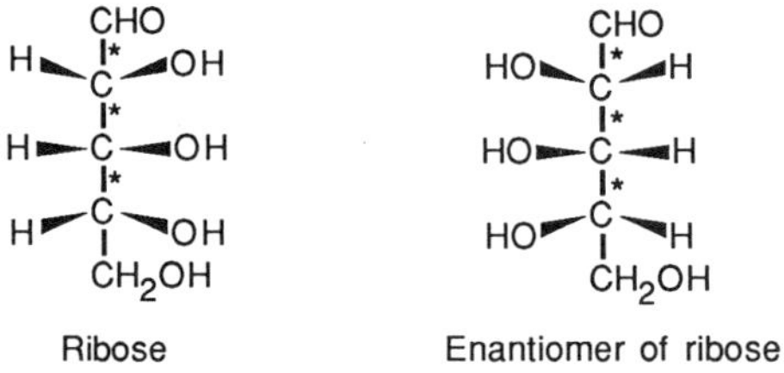

Ribose   Enantiomer of ribose

Ribose has three stereogenic centers, which can give rise to eight stereoisomers.

Stereochemistry   103

6.53   Ribose has six diastereomers.

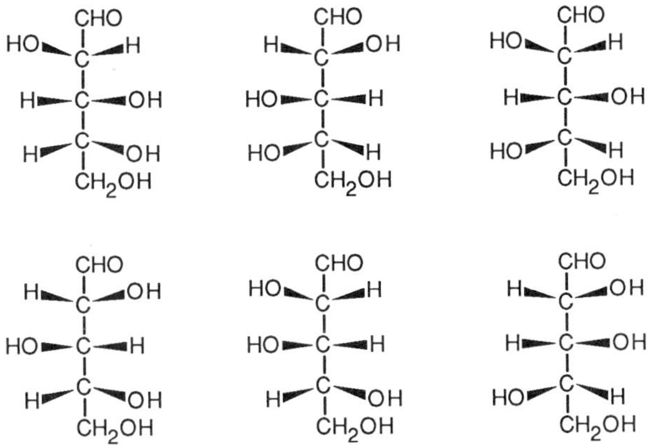

```
   CHO              CHO              CHO
HO-C-H           H-C-OH           HO-C-H
H-C-OH           HO-C-H           H-C-OH
H-C-OH           HO-C-H           HO-C-H
  CH2OH            CH2OH            CH2OH

   CHO              CHO              CHO
H-C-OH           HO-C-H           H-C-OH
HO-C-H           HO-C-H           H-C-OH
H-C-OH           H-C-OH           HO-C-H
  CH2OH            CH2OH            CH2OH
```

6.54   Ribitol is an optically inactive *meso* compound. Catalytic hydrogenation converts the aldehyde functional group into a hydroxyl group and makes the two halves of ribitol mirror images of each other.

6.55

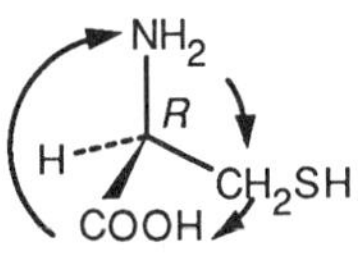

(R)-Cysteine 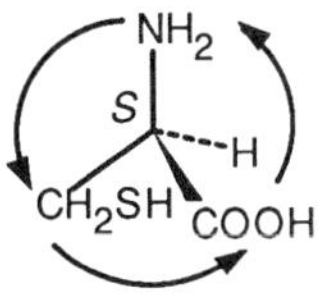   (S)-Cysteine

6.56

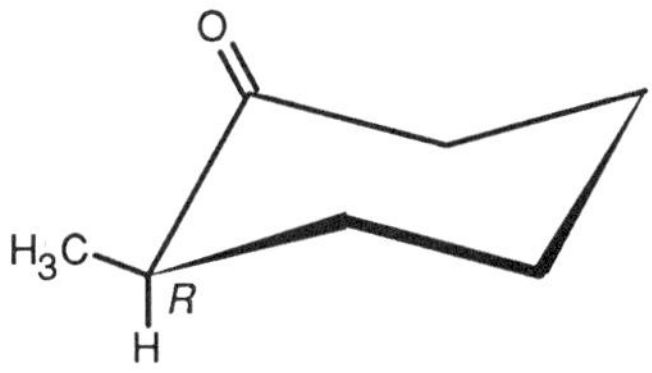

(R)-2-Methylcyclohexanone

6.57

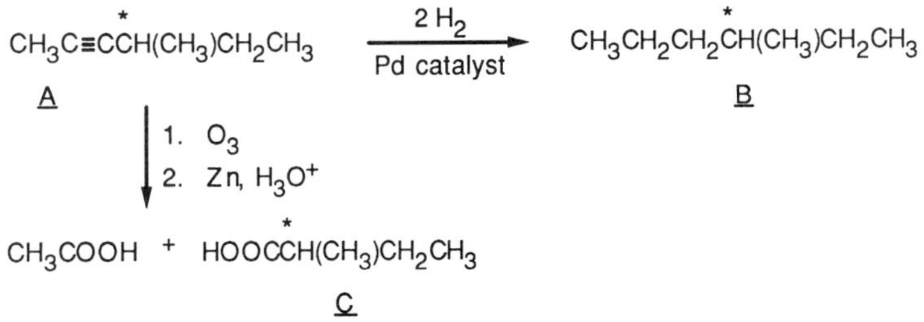

$$CH_3C\equiv CCH(CH_3)CH_2CH_3 \xrightarrow[\text{Pd catalyst}]{2\ H_2} CH_3CH_2CH_2\overset{*}{C}H(CH_3)CH_2CH_3$$
A   B

$$\downarrow \begin{array}{l}1.\ O_3\\ 2.\ Zn,\ H_3O^+\end{array}$$

$$CH_3COOH\ +\ HOOC\overset{*}{C}H(CH_3)CH_2CH_3$$
C

104   Chapter 6

6.58 Make a model of mycomycin. For simplicity, call –CH=CHCH=CHCH$_2$COOH "A" and –C≡CC≡CH "B". The carbon atoms in the allene group are linear and the pi bonds formed are perpendicular to each other. Attach substituents to the sp^2 carbons.

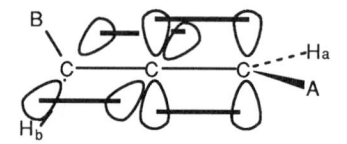

Notice that the substituents A, H$_a$, and all allene carbon atoms lie in a plane that is perpendicular to the plane that contains B, H$_b$, and all allene carbon atoms.

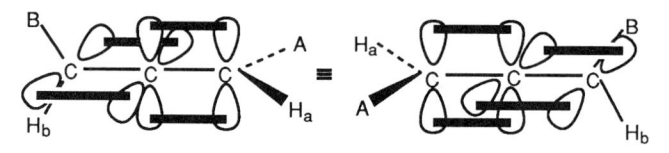

Now, make another model identical to the first, except for an exchange of A and H$_a$. This new allene is not superimposable on the original allene; in fact, the two allenes are mirror images. The two allenes are enantiomers and are chiral for the same reason that tetrasubstituted carbon atoms are chiral — they have no plane of symmetry.

Chapter Outline

I. Chirality (Sections 6.1 - 6.5)
    A. Stereochemistry and the tetrahedral carbon (Section 6.1)
    B. Chirality (Section 6.2)
        1. Stereogenic centers
        2. Symmetry planes
    C. Optical activity (Section 6.3)
        1. Plane-polarized light
        2. Optical rotation
    D. Specific rotation (Section 6.4)
    E. Pasteur's discovery of enantiomers (Section 6.5)
    F. Sequence rules (Section 6.6)
        1. R configuration
        2. S configuration

Stereochemistry   105

II. Compounds with more than one stereogenic center (Sections 6.7 - 6.11)
   A. Diastereomers (Section 6.7)
   B. Meso compounds (Section 6.8)
   C. Molecules with more than two stereogenic centers (Section 6.9)
   D. Racemic mixtures (Section 6.10)
   E. Physical properties of stereoisomers (Section 6.11)
III. A review of isomerism (Section 6.12)
IV. Stereochemistry of reactions -- addition of HBr to alkenes (Section 6.13)

Study Skills for Chapter 6

After studying this chapter, you should be able to:
1. Decide whether objects are chiral (Problems 6.1, 6.21).
2. Locate stereogenic centers in molecules (Problems 6.2, 6.3, 6.4, 6.22, 6.23, 6.25).
3. Calculate the rotation or specific rotation of a solution (Problems 6.7, 6.18, 6.19).
4. Draw chiral molecules in tetrahedral form (Problems 6.5, 6.10, 6.37, 6.43).
5. Draw chiral molecules in Newman projection (Problems 6.38, 6.39, 6.40, 6.41).
6. Draw chiral molecules corresponding to a given structural formula (Problems 6.24, 6.26, 6.27, 6.36).
7. Draw the enantiomer of a given chiral compound (Problems 6.29, 6.30, 6.34, 6.52).
8. Draw the diastereomer of a given chiral compound (Problems 6.35, 6.53).
9. Assign priorities to substituents around a stereogenic carbon (Problems 6.8, 6.28).
10. Assign *R,S* configurations to stereogenic centers (Problems 6.9, 6.11, 6.13, 6.32, 6.44, 6.45, 6.55, 6.56).
11. Decide if a stereoisomer is a meso compound (Problems 6.14, 6.15, 6.44, 6.54).
12. Predict the stereochemistry of reaction products (Problems 6.46, 6.47, 6.49, 6.50).
13. Calculate the number of stereoisomers of a given structure (Problems 6.16, 6.42, 6.48, 6.51).
14. Define the key terms in this chapter (Problem 6.20).
15. Understand the relationship of specific rotations of enantiomers and diastereomers (Problems 6.6, 6.31, 6.33).

Chapter 7 – Alkyl Halides

7.1

a) $CH_3CH_2CHBrCH_3$ 2-Bromobutane

b) $CH_3CH_2CHClCH(CH_3)_2$ 3-Chloro-2-methylpentane

c) $(CH_3)_2CHCH_2CH_2Cl$ 1-Chloro-3-methylbutane

d) $(CH_3)_2CClCH_2CH_2Cl$ 1,3-Dichloro-3-methylbutane

e) $BrCH_2CH_2CH_2CH_2Cl$ 1-Bromo-4-chlorobutane

f) $CH_3CHBrCH_2CH_2CH_2Cl$ 4-Bromo-1-chloropentane

7.2

a) $CH_3CH_2CH_2\underset{\underset{CH_3}{|}}{\overset{\overset{CH_3}{|}}{C}}-\underset{\underset{Cl}{|}}{C}HCH_3$

2-Chloro-3,3-dimethylhexane

b) $CH_3CH_2CH_2\underset{\underset{Cl}{|}}{\overset{\overset{Cl}{|}}{C}}-\underset{\underset{}{|}}{\overset{\overset{CH_3}{|}}{C}}HCH_3$

3,3-Dichloro-2-methylhexane

c) $CH_3CH_2\underset{\underset{Br}{|}}{\overset{\overset{CH_2CH_3}{|}}{C}}CH_2CH_3$

3-Bromo-3-ethylpentane

d) $CH_3\overset{\overset{}{|}}{C}HCH_2\overset{\overset{CH_3}{|}}{C}HCHCH_3$
     Cl   Br

2-Bromo-5-chloro-3-methylhexane

7.3

$CH_3CH_2\overset{\overset{CH_3}{|}}{C}HCH_2CH_3 \xrightarrow{\underset{h\upsilon}{Cl_2}}$

$CH_3CH_2\overset{*}{C}HCH_2CH_2Cl$ + $CH_3CH_2\overset{*}{C}HCHCH_3$ +
          CH₃   CH₃ Cl
          A   **B**

1-Chloro-3-methylpentane   2-Chloro-3-methylpentane

$CH_3CH_2\overset{\overset{CH_3}{|}}{\underset{\underset{Cl}{|}}{C}}CH_2CH_3$ + $CH_3CH_2\overset{\overset{CH_2Cl}{|}}{C}HCH_2CH_3$

      C   **D**

3-Chloro-3-methylpentane   3-(Chloromethyl)pentane

Products <u>A</u> and <u>B</u> contain carbon bonded to four different groups and are chiral.

7.4

$(CH_3)_4C \xrightarrow{\underset{h\upsilon}{Cl_2}} (CH_3)_3CCH_2Cl$

Only one monochloro product is formed on radical chlorination of neopentane because all hydrogens are equivalent.

7.5

a) $(CH_3)_3COH \xrightarrow{HCl} (CH_3)_3CCl$
   2-Chloro-2-methylpropane

b) $(CH_3)_2CHCH_2CH(OH)CH_3 \xrightarrow[\text{Ether}]{PBr_3} (CH_3)_2CHCH_2CH(Br)CH_3$
   2-Bromo-4-methylpentane

c) $HOCH_2CH_2CH_2CH_2CH(CH_3)_2 \xrightarrow[\text{Ether}]{PBr_3} BrCH_2CH_2CH_2CH_2CH(CH_3)_2$
   1-Bromo-5-methylhexane

d) $CH_3CH_2CH(CH_3)CH_2C(OH)(CH_3)_2 \xrightarrow{HCl} CH_3CH_2CH(CH_3)CH_2C(Cl)(CH_3)_2$
   2-Chloro-2,4-dimethylhexane

7.6

a) $CH_3CH_2CH(OH)CH_2CH(CH_3)CH_3 + PBr_3 \longrightarrow CH_3CH_2CH(Br)CH_2CH(CH_3)CH_3$

b) 1-methylcyclohexanol + HCl ⟶ 1-chloro-1-methylcyclohexane

c) 3,3-dimethylcyclopentanol + SOCl₂ ⟶ 1-chloro-3,3-dimethylcyclopentane

7.7

$CH_3CH(Br)CH_2CH_3 \xrightarrow[\text{ether}]{Mg} CH_3CH(MgBr)CH_2CH_3 \xrightarrow{D_2O} CH_3CH(D)CH_2CH_3$

7.8

$CH_3CH(CH_3)CH_2CH_2CH_2OH \xrightarrow{PBr_3} CH_3CH(CH_3)CH_2CH_2CH_2Br \xrightarrow[\text{2. } H_2O]{\text{1. Mg, ether}} CH_3CH(CH_3)CH_2CH_2CH_3$
4-Methylpentanol   2-Methylpentane

7.9

a) $CH_3CH_2CH(Br)CH_3 + Li^+I^- \longrightarrow CH_3CH_2CH(I)CH_3 + Li^+Br^-$

b) $(CH_3)_2CHCH_2Cl + H\ddot{S}:^- \longrightarrow (CH_3)_2CHCH_2SH + :\ddot{C}l:^-$

c) C₆H₅—CH₂Br + Na⁺ ⁻CN ⟶ C₆H₅—CH₂CN + Na⁺Br⁻

7.10
a) CH₃CH₂CH₂CH₂Br + Na⁺ ⁻OH ⟶ CH₃CH₂CH₂CH₂OH + Na⁺Br⁻
b) (CH₃)₂CHCH₂CH₂Br + Na⁺N₃⁻ ⟶ (CH₃)₂CHCH₂CH₂N₃ + Na⁺Br⁻

7.11
a) If [CH₃I] is tripled, the reaction rate is tripled.
b) If both [CH₃I] and [CH₃CO₂Na] are doubled, the reaction rate is quadrupled.

7.12

CH₃CH₂CH₂CH₂\\C(H)(CH₃)—Br + CH₃CO₂⁻ ⟶ CH₃CO—C(R)(H)(CH₃)—CH₂CH₂CH₂CH₃ + :Br:⁻ (S configuration inverts)

7.13

:Br:⁻ + H₃C\\C(H)(R)—Br with CH₂CH₂CH₂CH₃ ⟶ Br—C(S)(H)(CH₃)—CH₂CH₂CH₂CH₃ + :Br:⁻

Reaction of one molecule of (R)–2–bromohexane with one bromide ion produces one molecule of (S)–2–bromohexane. Reaction of 50% of the R starting material gives a mixture of 50% S enantiomer plus 50% unreacted R starting material — a racemic mixture. Thus, after 50% of the R starting material has reacted, the product is 100% racemized.

7.14
a) Reaction of cyanide ion proceeds faster with the primary halide CH₃CH₂CH₂Br than with the secondary halide CH₃CHBrCH₃.
b) Iodide ion reacts faster with (CH₃)₂CHCH₂Cl; H₂C=CHCl is unreactive toward S_N2 displacements.

7.15 Use the previous chart to identify the most reactive leaving groups.

Least reactive ⟶ Most reactive
CH₃–OH < CH₃–OCOCH₃ < CH₃–Br < CH₃–I.

7.16
a) Tripling the HBr concentration has no effect on the rate of reaction. In an S_N1 reaction such as this one, the rate does not depend on the concentration of the nucleophile.
b) Doubling the *tert*–butyl alcohol concentration doubles the rate of reaction. Halving the HBr concentration has no effect on the rate. Thus, the overall rate is doubled.

Alkyl Halides 109

7.17

$$CH_3(CH_2)_3CH_2\underset{\underset{CH_2CH_3}{|}}{\overset{\overset{CH_3}{|}}{C}}\text{—}\overset{..}{\underset{..}{O}}H \;\; \underset{S}{} \;\;\overset{H\text{—}Br}{\curvearrowright}\;\rightleftharpoons\; \left[CH_3(CH_2)_3CH_2\underset{\underset{CH_2CH_3}{|}}{\overset{\overset{CH_3}{|}}{C}}\text{—}\overset{+}{O}H_2 \;\; \rightleftharpoons\;\; \underset{\underset{CH_2CH_3}{|}}{\overset{\overset{CH_3}{|}}{\underset{\,}{C}}^+}\overset{CH_3(CH_2)_3CH_2}{} \;\; :\overset{..}{\underset{..}{Br}}:^- \right]$$

$$+ Br^- \qquad\qquad + H_2O$$

$$\rightleftharpoons \;\; CH_3(CH_2)_3CH_2\underset{\underset{CH_2CH_3}{|}}{\overset{\overset{CH_3}{|}}{\underset{S}{C}}}\text{—}Br \;\; + \;\; Br\text{—}\underset{\underset{CH_2CH_3}{|}}{\overset{\overset{H_3C}{|}}{\underset{R}{C}}}CH_2(CH_2)_3CH_3$$

Attack by Br⁻ can occur on either side of the planar, achiral carbocation intermediate. The resulting product is a racemic mixture.

7.18

a)
$$CH_3CH_2\underset{|}{\overset{Br}{C}}H\underset{|}{\overset{CH_3}{C}}HCH_3 \;\; \xrightarrow{B:} \;\; CH_3CH_2CH=\underset{|}{\overset{CH_3}{C}}CH_3 \;\; + \;\; CH_3CH=\underset{|}{\overset{CH_3}{C}}HCHCH_3$$

major product   minor product

b)
$$CH_3\underset{|}{\overset{CH_3}{C}}HCH_2\text{—}\underset{\underset{CH_3}{|}}{\overset{\overset{Cl}{|}}{C}}\text{—}\underset{|}{\overset{CH_3}{C}}HCH_3 \;\;\xrightarrow{B:}\;\; CH_3\underset{|}{\overset{CH_3}{C}}HCH_2\underset{|}{\overset{CH_3}{C}}=\underset{|}{\overset{CH_3}{C}}CH_3 \;\; + \;\; CH_3\underset{|}{\overset{CH_3}{C}}HCH=\underset{\underset{CH_3}{|}}{\overset{\overset{CH_3}{|}}{C}}\text{—}\underset{|}{\overset{CH_3}{C}}HCH_3$$

major product   minor product

$$+ \;\; CH_3\underset{|}{\overset{CH_3}{C}}HCH_2\text{—}\underset{\underset{CH_2}{||}}{C}\text{—}\underset{|}{\overset{CH_3}{C}}HCH_3$$

minor product

c)

cyclohexyl—CH(Br)CH₃ $\xrightarrow{B:}$ cyclohexylidene=CHCH₃ + cyclohexyl—CH=CH₂

major product   minor product

7.19

$$\underset{\underset{H}{|}}{\overset{\overset{H}{|}}{\underset{Br\,\blacktriangleleft C\blacktriangleright Ph}{Ph\,\blacktriangleleft C\blacktriangleright Br}}}$$

(1R, 2R)–1,2–Dibromo–1,2–diphenylethane

Convert this structure into a Newman projection, and draw the conformation having *anti-periplanar* geometry for –H and –Br.

The alkene resulting from dehydrohalogenation is (Z)–1–bromo–1,2– diphenylethylene.

7.20 The rate of E1 dehydration reaction would triple if the concentration of the alcohol were tripled.

7.21

	Halide +	Nucleophile	Reaction Type	Product
a)	CH₃CH₂CH₂CH₂Br (primary)	Na⁺N₃⁻	S_N2	CH₃CH₂CH₂CH₂N₃ (substitution)
b)	CH₃CH₂CHClCH₂CH₃ (secondary)	KOH (strong base)	E2	CH₃CH₂CH=CHCH₃ (elimination)
c)	(cyclohexyl with Cl and CH₃, tertiary)	CH₃COOH (solvent)	S_N1	(cyclohexyl with OOCCH₃ and CH₃) + HCl (substitution)

7.22
a) CH₃CHBrCHBrCH₂CHCH₃ with CH₃ groups
   3,4–Dibromo–2,6–dimethylheptane

b) CH₃CH=CHCH₂CHICH₃
   5–Iodo–2–hexene

Alkyl Halides   111

c)
$$\text{CH}_3\underset{\underset{\text{Br}}{|}}{\overset{\overset{\text{CH}_3}{|}}{\text{C}}}\text{CH}_2\text{CH}_2\overset{\overset{\text{Cl}}{|}}{\text{CH}}\overset{\overset{\text{CH}_3}{|}}{\text{CH}}\text{CH}_3$$

2-Bromo-5-chloro-2,6-dimethylheptane

d)
$$\text{CH}_3\text{CH}_2\overset{\overset{\text{CH}_2\text{Br}}{|}}{\text{CH}}\text{CH}_2\text{CH}_2\text{CH}_3$$

3-(Bromomethyl)hexane

7.23

a)
$$\text{CH}_3\text{CH}_2\overset{\overset{\text{CH}_3}{|}}{\text{CH}}\underset{\underset{\text{Cl}}{|}}{\overset{\overset{\text{Cl}}{|}}{\text{CH}}}\text{CH}\text{CH}_3$$

2,3-Dichloro-4-methylhexane

b)
$$\text{CH}_3\text{CH}_2\underset{\underset{\text{CH}_2\text{CH}_3}{|}}{\overset{\overset{\text{Br}}{|}}{\text{C}}}\text{CH}_2\overset{\overset{\text{CH}_3}{|}}{\text{CH}}\text{CH}_3$$

4-Bromo-4-ethyl-2-methylhexane

c)
$$\text{CH}_3\underset{\underset{\text{H}_3\text{C}}{|}}{\overset{\overset{\text{H}_3\text{C}}{|}}{\text{C}}}-\text{CH}-\underset{\underset{\text{CH}_3}{|}}{\overset{\overset{\text{CH}_3}{|}}{\text{C}}}\text{CH}_3$$
        I

3-Iodo-2,2,4,4-tetramethylpentane

7.24

$$\text{CH}_3\text{CH}_2\text{CH}_2\overset{\overset{\text{CH}_3}{|}}{\text{CH}}\text{CH}_3 \quad \xrightarrow[hv]{\text{Cl}_2}$$

$$\left[\begin{array}{c}
\text{CH}_3\text{CH}_2\text{CH}_2\overset{\overset{\text{CH}_3}{|}}{\underset{*}{\text{CH}}}\text{CH}_2\text{Cl} \quad + \quad \text{CH}_3\text{CH}_2\text{CH}_2\underset{\underset{\text{Cl}}{|}}{\overset{\overset{\text{CH}_3}{|}}{\text{C}}}\text{CH}_3 \\
\text{1-Chloro-2-methylpentane} \quad\quad \text{2-Chloro-2-methylpentane} \\
+ \\
\text{CH}_3\text{CH}_2\overset{\overset{\text{CH}_3}{|}}{\underset{*}{\text{CH}}}\underset{\underset{\text{Cl}}{|}}{\text{CH}}\text{CH}_3 \quad + \quad \text{CH}_3\overset{\overset{\text{Cl}}{|}}{\text{CH}}\text{CH}_2\overset{\overset{\text{CH}_3}{|}}{\underset{*}{\text{CH}}}\text{CH}_3 \\
\text{3-Chloro-2-methylpentane} \quad\quad \text{2-Chloro-4-methylpentane} \\
+ \\
\text{ClCH}_2\text{CH}_2\text{CH}_2\overset{\overset{\text{CH}_3}{|}}{\text{CH}}\text{CH}_3 \\
\text{1-Chloro-4-methylpentane}
\end{array}\right]$$

Three of the above products are chiral (stereogenic centers are starred). None of the products are optically active; each chiral product is a racemic mixture.

7.25 a) Because the rate-limiting step in an S$_N$2 reaction involves attack of the nucleophile on the substrate, any factor that makes approach of the nucleophile more difficult slows down the rate of reaction. Especially important is the degree of crowding at the reacting carbon atom. Tertiary carbon atoms are too crowded to allow S$_N$2 substitution to occur. Even steric hindrance one carbon atom away from the reacting site causes a drastic slowdown in rate of reaction.

The rate-limiting step in an S$_N$1 reaction involves formation of a carbocation. Any structural factor in the substrate that stabilizes carbocations will increase the rate of reaction. Substrates that are tertiary, allylic, or benzylic react fastest.

112   Chapter 7

b) Good leaving groups (stable anions) increase the rates of both S_N1 and S_N2 reactions.

7.26

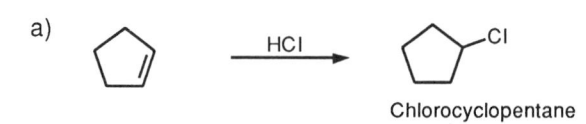

a) cyclopentene $\xrightarrow{HCl}$ chlorocyclopentane (Cl)

Chlorocyclopentane

b) cyclopentene $\xrightarrow{H_3O^+}$ cyclopentanol (OH)

Cyclopentanol

c) cyclopentyl-Cl $\xrightarrow[\text{ether}]{Mg}$ cyclopentyl-MgCl

from (a)   Cyclopentylmagnesium chloride

d) cyclopentyl-MgCl $\xrightarrow{H_2O}$ cyclopentane

from (c)   Cyclopentane

7.27 Use the table in Section 7.7 if you need help.

	Better leaving group	Poorer leaving group
a)	Br^-	F^-
b)	Cl^-	NH_2^-
c)	I^-	OH^-

7.28

a) cyclohexane-CH₃,OH $\xrightarrow{HBr}$ cyclohexane-CH₃,Br

This is a good method for converting a tertiary alcohol into a bromide.

b) $CH_3CH_2CH_2CH_2OH$ $\xrightarrow{SOCl_2}$ $CH_3CH_2CH_2CH_2Cl$

This is a good method for converting a primary or secondary alcohol into a chloride.

c) cyclohexanol (OH) $\xrightarrow{PBr_3}$ bromocyclohexane (Br)

This is a good method for converting a primary or secondary alcohol into a bromide.

Alkyl Halides 113

d) CH₃CH₂CHBrCH₃ $\xrightarrow[\text{ether}]{\text{Mg}}$ CH₃CH₂CH(MgBr)CH₃ $\xrightarrow{\text{H}_2\text{O}}$ CH₃CH₂CH₂CH₃
                                                                     A   B

7.29

Reacts faster	Reacts slower
a) C₆H₅—CH₂Br (primary, benzylic halide)	C₆H₅—Br (aromatic, unreactive)
b) CH₃Cl (primary halide)	(CH₃)₃CCl (tertiary halide)
c) H₂C=CHCH₂Br (primary, allylic halide)	CH₃CH=CHBr (vinylic, unreactive)

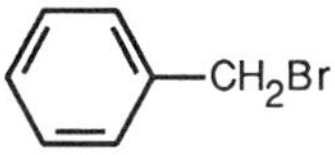

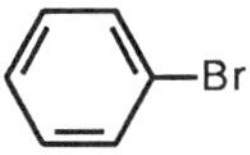

7.30 All these reactions proceed by S_N2 substitution.

a) CH₃CH₂OH + PBr₃ ⟶ CH₃CH₂Br

b) CH₃CH₂CH₂CH₂Br + Na⁺ ⁻CN ⟶ CH₃CH₂CH₂CH₂CN + Na⁺Br⁻

c) CH₃Br + Na⁺ ⁻OC(CH₃)₃ ⟶ CH₃OC(CH₃)₃ + Na⁺Br⁻

    The reaction of CH₃O⁻ with (CH₃)₃CBr causes elimination, not substitution.

d) CH₃CH₂CH₂I + Na⁺ ⁻N=N⁺=N:⁻ ⟶ CH₃CH₂CH₂N=N⁺=N:⁻ + Na⁺I⁻

e) CH₃CH₂I + Na⁺ ⁻SH ⟶ CH₃CH₂SH + Na⁺I⁻

f) CH₃Br + Na⁺ ⁻OC(=O)CH₃ ⟶ CH₃OC(=O)CH₃ + Na⁺Br⁻

7.31

a) CH₃CH₂CH₂Br + NaI ⟶ CH₃CH₂CH₂I + NaBr

b) CH₃CH₂CH₂Br + NaCN ⟶ CH₃CH₂CH₂CN + NaBr

c) CH₃CH₂CH₂Br + NaOH ⟶ CH₃CH₂CH₂OH + NaBr

d) CH₃CH₂CH₂Br + Mg, then H₂O ⟶ CH₃CH₂CH₃

e) CH₃CH₂CH₂Br + NaOCH₃ ⟶ CH₃CH₂CH₂OCH₃ + NaBr

7.32

S_N1 reactivity:

Most reactive ⟶ Least reactive

$(CH_3)_3CCl$ > $C_6H_5CH_2Cl$ (benzyl chloride) > C_6H_5Cl (chlorobenzene)

(most stable carbocation)   (unreactive)

S_N2 reactivity:

Most reactive ⟶ Least reactive

$C_6H_5CH_2Cl$ > $(CH_3)_3CCl$ > C_6H_5Cl

(primary)   (tertiary)   (unreactive)

7.33 S_N2 reactivity:

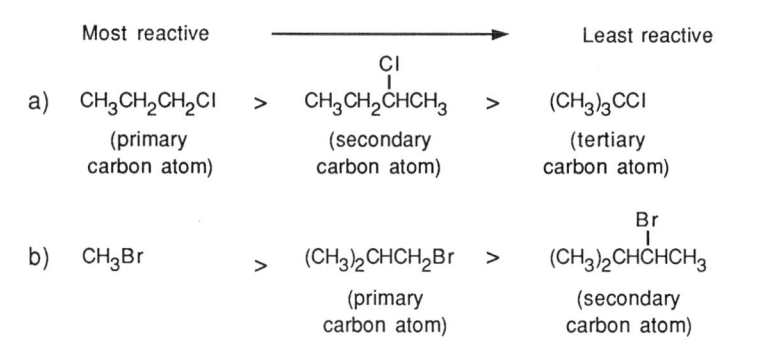

Most reactive ⟶ Least reactive

a) $CH_3CH_2CH_2Cl$ > $CH_3CH_2CHClCH_3$ > $(CH_3)_3CCl$

   (primary carbon atom)   (secondary carbon atom)   (tertiary carbon atom)

b) CH_3Br > $(CH_3)_2CHCH_2Br$ > $(CH_3)_2CHCHBrCH_3$

                                       (primary carbon atom)   (secondary carbon atom)

7.34

a) The reaction of cyanide anion with a tertiary halide is more likely to yield the elimination product 3-methyl-2-pentene than substitution product.

b) Use PBr₃ to convert a primary alcohol into a primary bromide.

c) Reaction of a tertiary alcohol with HBr gives mainly substitution product along with a lesser amount of elimination product.

7.35

H_3C, H, n-C_6H_{13} — C—Br

(R)-2-Bromooctane

Alkyl Halides   115

(R)–2–Bromooctane is a secondary bromoalkane, which can undergo both S_N1 and S_N2 substitution. All of the nucleophiles listed are very reactive, however, and all reactions proceed by an S_N2 mechanism. Since S_N2 reactions proceed with inversion of configuration, the configuration at the stereogenic carbon atom is inverted. (This does not necessarily mean that all R isomers become S isomers after an S_N2 reaction; the R–S designation refers to the priorities of groups, and priorities may change when the nucleophile is varied.)

$$\text{Nu:}^- + \underset{\underset{n-C_6H_{13}}{\text{Nucleophile}}}{\overset{H_3C\quad H}{\underset{}{C-Br}}} \longrightarrow \underset{\underset{n-C_6H_{13}}{\text{Product}}}{\overset{H\quad CH_3}{\underset{}{Nu-C}}} + :Br:^-$$

a)   :CN⁻ $NC-\underset{C_6H_{13}}{\overset{H\ \ CH_3}{C}}$ S

b) $CH_3\overset{O}{\underset{\|}{C}}O:^-$ $CH_3\overset{O}{\underset{\|}{C}}O-\underset{n-C_6H_{13}}{\overset{H\ \ CH_3}{C}}$ S

c)   :Br:⁻ $Br-\underset{n-C_6H_{13}}{\overset{H\ \ CH_3}{C}}$ S   + $\underset{n-C_6H_{13}}{\overset{H_3C\ \ H}{C-Br}}$ R

2–Bromooctane is 100% racemized after 50% of the original (R)–2–bromooctane has reacted with Br⁻.

7.36 S_N2 reactivity:

Most reactive ——————————————————————→ Least reactive

$CH_3CH_2CH_2CH_2Br$ > $CH_3\underset{CH_3}{\overset{|}{C}H}CH_2Br$ > $CH_3CH_2\underset{Br}{\overset{|}{C}H}CH_3$ > $CH_3\underset{\underset{CH_3}{|}}{\overset{\overset{Br}{|}}{C}}CH_3$

1–Bromobutane   1–Bromo–2–methyl–propane   2–Bromobutane   2–Bromo–2–methyl–propane

7.37

$H_2C=CH\dot{C}H_2 \longleftrightarrow H_2\dot{C}CH=CH_2$

Two resonance forms contribute to the relative stability of the allyl radical. Because it is stable, this radical is formed in preference to other radicals.

7.38

Five resonance forms contribute to the stability of the benzyl radical.

7.39

$$\text{C}_6\text{H}_{11}\text{–O}^{:-} + \text{CH}_3\text{Br} \longrightarrow \text{C}_6\text{H}_{11}\text{–OCH}_3$$

This is an excellent method of ether preparation since bromomethane is very reactive in S_N2 displacements.

$$\text{CH}_3\ddot{\text{O}}\text{:}^{-} + \text{C}_6\text{H}_{11}\text{Br} \longrightarrow \text{C}_6\text{H}_{11}\text{OCH}_3 + \text{cyclohexene}$$

Reaction of a secondary haloalkane with a basic nucleophile yields both substitution and elimination products. This is obviously a less satisfactory method of ether preparation.

7.40

(1) $\text{CH}_3\text{CH}_2\text{OH} + \text{PBr}_3 \longrightarrow \text{CH}_3\text{CH}_2\text{Br}$

(2) $\text{CH}_3\text{CH}_2\ddot{\text{O}}\text{:}^{-} + \text{CH}_3\text{CH}_2\text{Br} \longrightarrow \text{CH}_3\text{CH}_2\text{OCH}_2\text{CH}_3 + \text{:Br:}^{-}$

7.41

3-bromocyclohexene $\xrightarrow{\text{H}_2,\ \text{Pd catalyst}}$ bromocyclohexane $\xrightarrow{\text{Mg, ether}}$ cyclohexyl-MgBr $\xrightarrow{\text{H}_2\text{O}}$ cyclohexane

7.42

3-bromo-1,3-propanediol derivative $\xrightarrow{\text{:Base}}$ [alkoxide intermediate] $\xrightarrow{S_N2}$ tetrahydrofuran + :Br:⁻

7.43

$$\text{BrCH}_2\text{CH}_2\text{Br} + 2\ \text{NaOH} \longrightarrow \text{HOCH}_2\text{CH}_2\text{OH}$$

diol + dibromide $\xrightarrow{\text{:Base}}$ 1,4-Dioxane

Alkyl Halides 117

7.44

$$\underset{\underset{\text{3-Bromo-3-methylhexane}}{\text{CH}_3\text{CH}_2\text{CH}_2\overset{\overset{\text{CH}_3}{|}}{\underset{\underset{\text{Br}}{|}}{\text{C}}}\text{CH}_2\text{CH}_3}}{} \xrightarrow{\text{:Base}} \underset{\text{3-Methyl-3-hexene}}{\text{CH}_3\text{CH}_2\text{CH}=\overset{\overset{\text{CH}_3}{|}}{\text{C}}\text{CH}_2\text{CH}_3} + \underset{\text{3-Methyl-2-hexene}}{\text{CH}_3\text{CH}_2\text{CH}_2\overset{\overset{\text{CH}_3}{|}}{\text{C}}=\text{CHCH}_3}$$

$$+ \underset{\text{2-Ethyl-1-pentene}}{\text{CH}_3\text{CH}_2\text{CH}_2\overset{\overset{\text{CH}_2\text{CH}_3}{|}}{\text{C}}=\text{CH}_2}$$

7.45 Both *tert*–butyl chloride and *tert*–butyl bromide dissociate to form the same carbocation intermediate. In ethanol, this carbocation yields the same mixture of products in the same ratio, regardless of the starting material.

7.46

$$\underset{\text{(primary halide)}}{\text{CH}_3\text{CH}_2\overset{\overset{\text{CH}_3}{|}}{\text{CH}}\text{CH}_2\text{I}} + {}^-\text{:CN} \longrightarrow \text{CH}_3\text{CH}_2\overset{\overset{\text{CH}_3}{|}}{\text{CH}}\text{CH}_2\text{CN} + \text{:I:}^-$$

This is a S_N2 reaction, in which reaction rate depends on the concentration of both alkyl halide and of nucleophile.

a) Halving the concentration of cyanide and doubling the concentration of alkyl halide will not change the reaction rate.

b) Tripling the concentrations of both cyanide and alkyl halide will cause a ninefold increase in reaction rate.

7.47

$$\underset{\text{(tertiary halide)}}{\text{CH}_3\text{CH}_2\overset{\overset{\text{CH}_3}{|}}{\underset{\underset{\text{I}}{|}}{\text{C}}}\text{CH}_3} + \text{CH}_3\text{CH}_2\text{OH} \longrightarrow \text{CH}_3\text{CH}_2\overset{\overset{\text{CH}_3}{|}}{\underset{\underset{\text{OCH}_2\text{CH}_3}{|}}{\text{C}}}\text{CH}_3 + \text{HI}$$

This is an S_N1 reaction, whose reaction rate depends only on the concentration of 2–iodo–2–methylbutane. Tripling the concentration of alkyl halide will triple the rate of reaction.

7.48

a)

(secondary benzylic halide) → (elimination product), reagent KOH, E2

The reaction is an E2 reaction.

b) 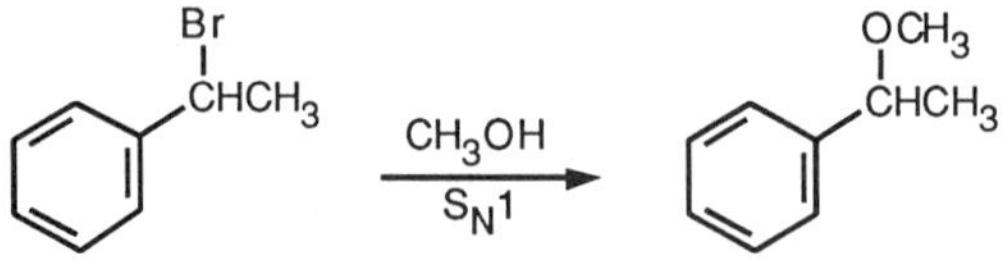
Br
CHCH3
CH3OH
SN1
OCH3
CHCH3

(secondary, benzylic halide) (substitution product)

This is an S_N1 reaction.

7.49

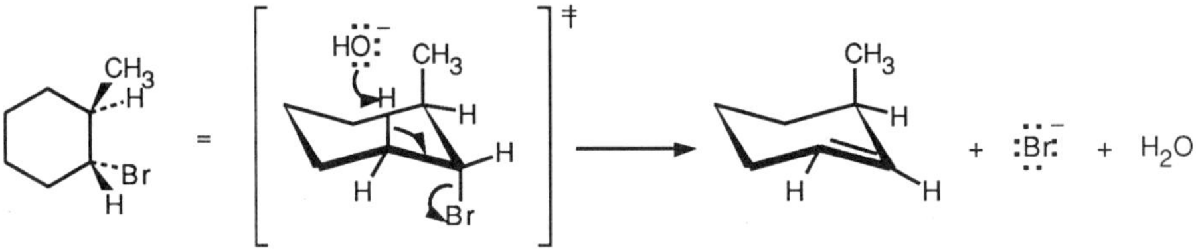
CH3
H
Br
H
=
HO:
CH3
H
H
H
H
Br
‡
CH3
H
H
H
+ :Br:
+ H2O

Only this alkene can result from *anti*-periplanar E_2 elimination.

7.50

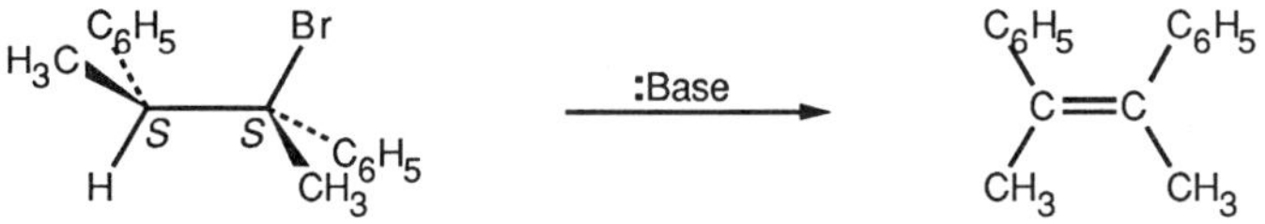
C6H5
H3C
Br
S
S
C6H5
H
CH3
:Base
C6H5
C6H5
C=C
CH3
CH3

(2*S*, 3*S*)–2–Bromo–2,3–diphenylbutane (or 2*R*, 3*R* enantiomer) (*Z*)-2,3-Diphenyl-2-butene

7.51

a) 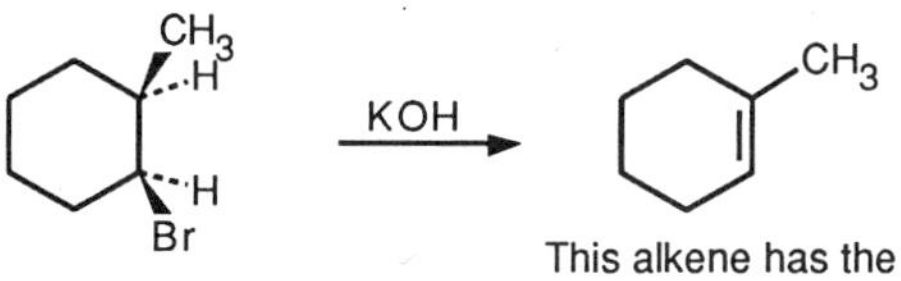
CH3
H
H
Br
KOH
CH3

This alkene has the more substituted double bond.

b) 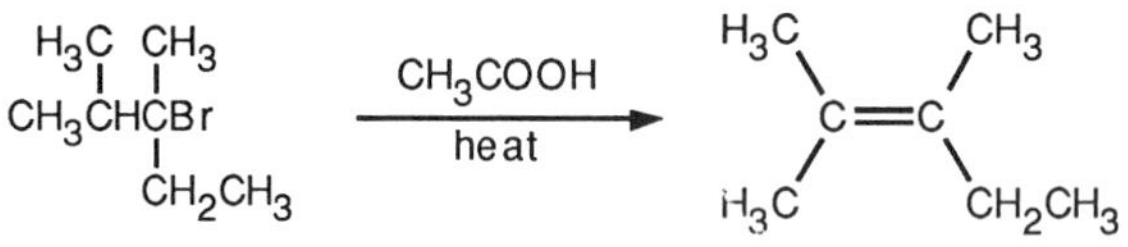
H3C CH3
CH3CHCBr
CH2CH3
CH3COOH
heat
H3C
CH3
C=C
H3C
CH2CH3

This alkene has the most substituted double bond.

7.52

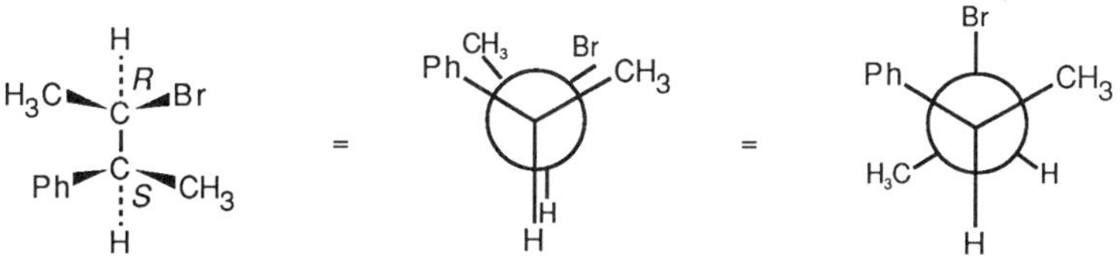

Draw (2*R*,3*S*)–2–bromo–3–phenylbutane; then draw its Newman projection. The Newman projection can be rotated until the –Br and the –H on the adjacent carbon atom are *anti*-periplanar. Even though this is not the most stable conformation, it is the only conformation in which –Br and –H are 180° apart.

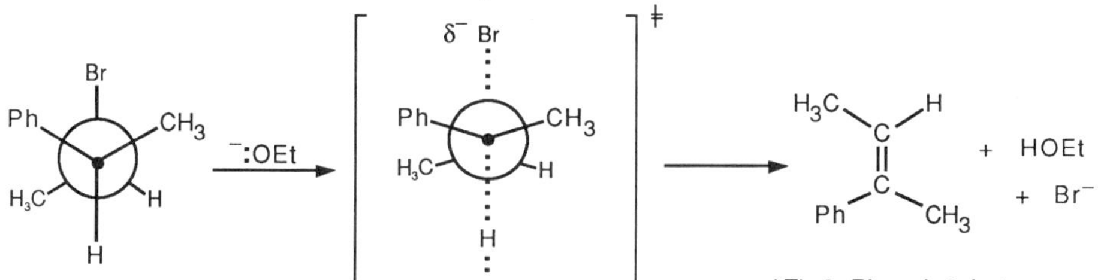

(*Z*)–2–Phenyl–2–butene

Elimination yields the *Z* isomer of 2–phenyl–2–butene. Refer to Section 3.4 for the method of assigning *E*,*Z* designation.

7.53 By the same arguments used in Problem 7.52, you can show that elimination from (2*R*,3*R*)–2–bromo–3–phenylbutane gives the *E*–alkene.

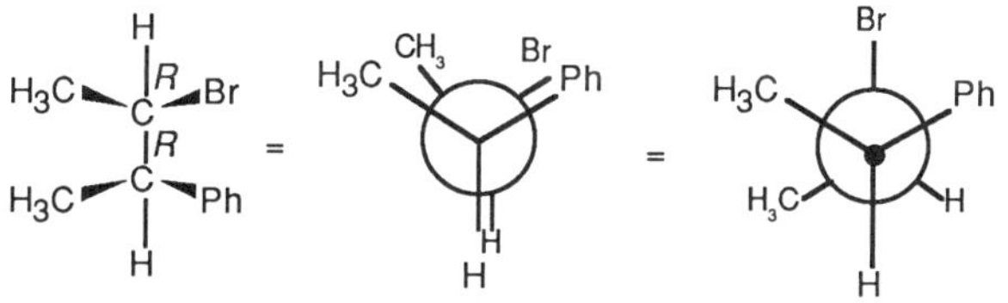

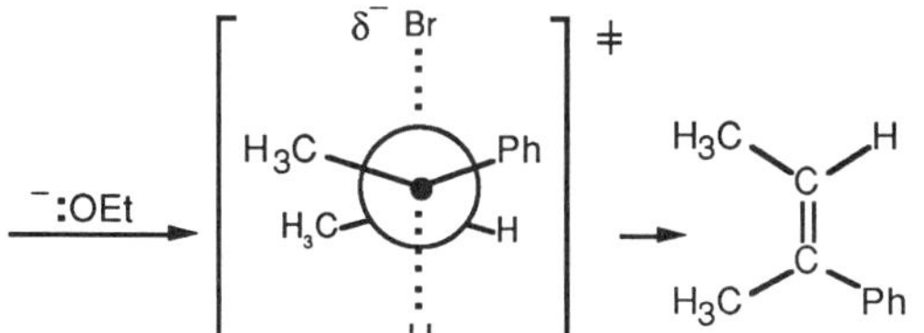

(*E*)–2-Phenyl–
2–butene

The 2*S*,3*S* isomer also forms the *E*–alkene; the 2*S*,3*R* isomer forms *Z*–alkene.

120   Chapter 7

7.54 The chiral tertiary alcohol (*R*)-3-methyl-3-hexanol reacts with HBr by an S$_N$1 pathway. HBr protonates the hydroxyl group, which dissociates to yield a planar, achiral carbocation. Attack by the nucleophilic bromide anion can occur from either side of the carbocation to produce (±)3-bromo-3-methylhexane.

7.55

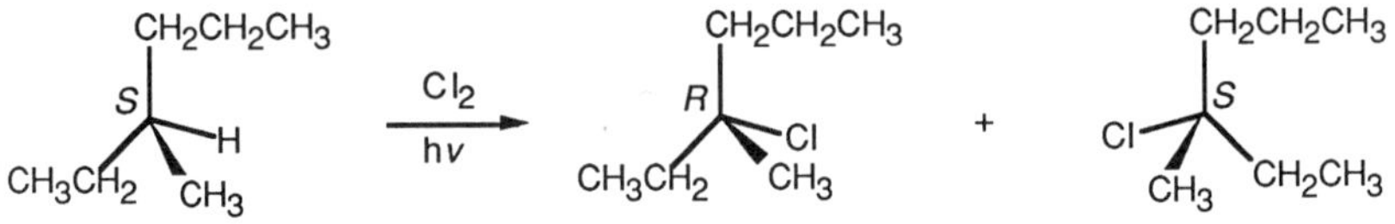

Abstraction of a hydrogen at the stereogenic center of *S*-3-methylhexane produces an achiral radical intermediate, which reacts with chlorine to form a 1:1 mixture of *R* and *S* enantiomeric, chiral chloroalkanes. The product mixture is optically inactive.

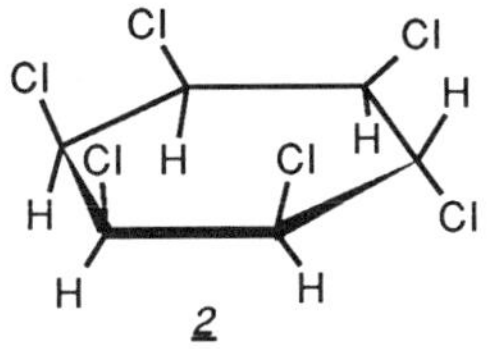

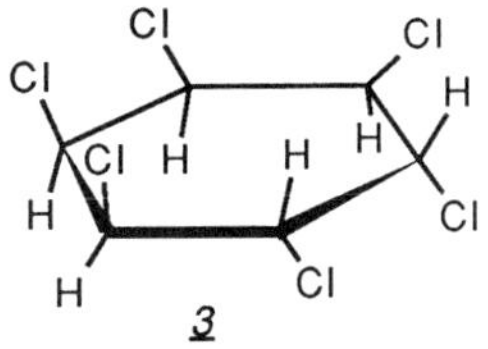

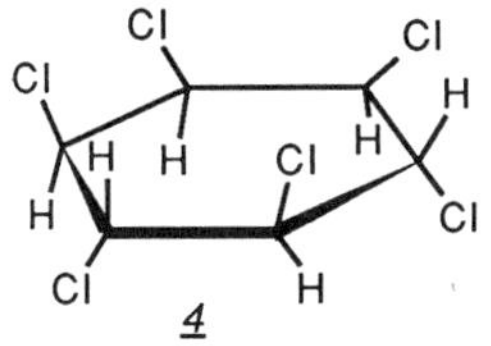

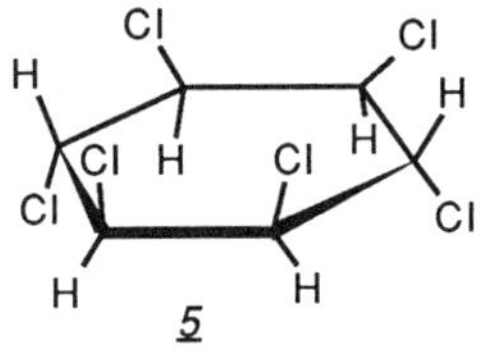

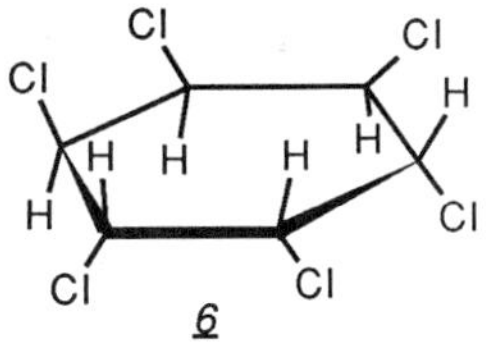

7.56

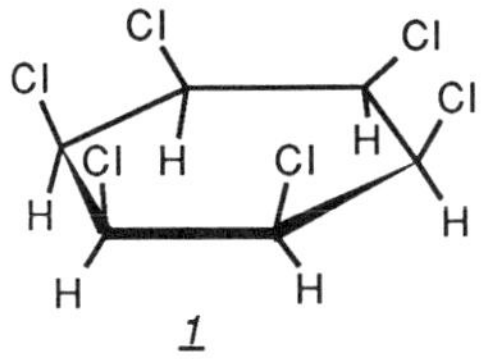

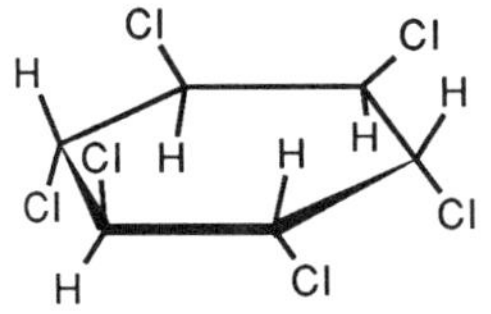

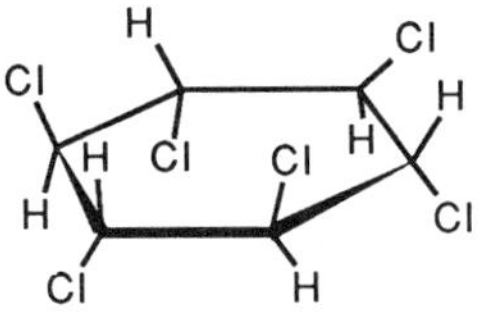

Diastereomer *8* reacts very slowly in an E2 reaction. No pair of hydrogen and chlorine atoms can assume the *anti*-periplanar orientation preferred for E2 elimination.

Alkyl Halides 121

7.57 Two optically inactive structures are possible for compound <u>A</u>. Any other structure of the formula C$_{16}$H$_{16}$Br$_2$ that undergoes the series of reactions is optically active.

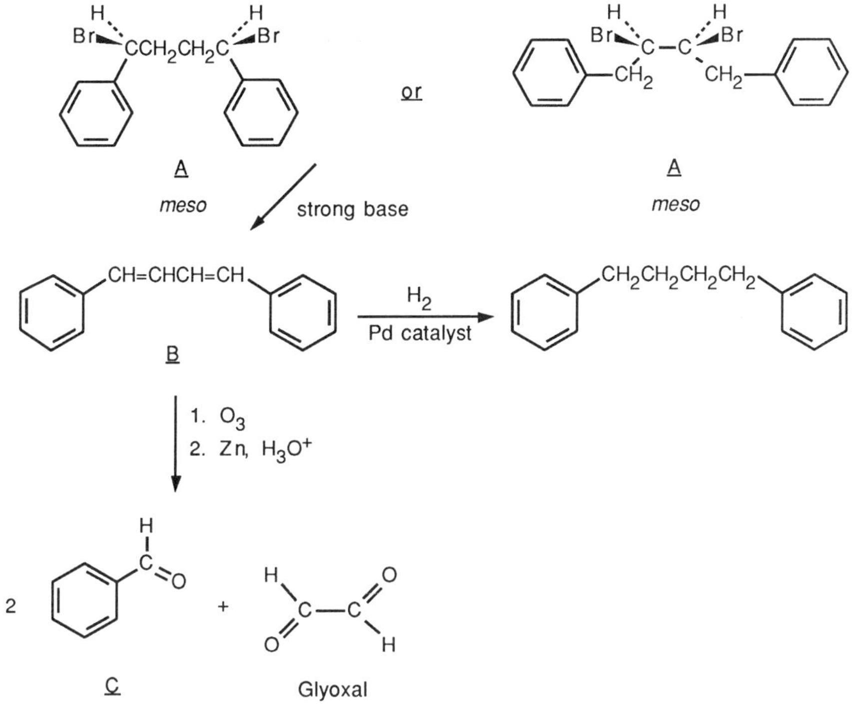

[Structure: Ph-CHBr-CH$_2$-CH$_2$-CHBr-Ph (meso), labeled <u>A</u>]   or   [Structure: Ph-CHBr-CH$_2$-CH$_2$-CHBr-Ph drawn differently (meso), labeled <u>A</u>]

↓ strong base

Ph-CH=CH-CH=CH-Ph, labeled <u>B</u>

—H$_2$, Pd catalyst→ Ph-CH$_2$CH$_2$CH$_2$CH$_2$-Ph

<u>B</u>:
1. O$_3$
2. Zn, H$_3$O$^+$

↓

2 Ph-CHO (labeled <u>C</u>)   +   OHC-CHO (Glyoxal)

Chapter Outline

I. Alkyl halides (Sections 7.1-7.4)
   A. Naming alkyl halides (Section 7.1)
   B. Preparation of alkyl halides (Sections 7.2-7.3)
      1. Radical chlorination of alkanes (Section 7.2)
      2. From alcohols (Section 7.3)
   C. Grignard reagents from alkyl halides (Section 7.4)
II. Nucleophilic substitution reactions (Sections 7.5 - 7.8)
   A. General features (Sections 7.5 - 7.6)
      1. Discovery of Walden inversion (Section 7.5)
      2. Kinds of nucleophilic substitutions (Section 7.6)

    B. S$_N$2 reactions (Section 7.7)
        1. Rates of S$_N$2 reactions
        2. Stereochemistry of S$_N$2 reactions
        3. Steric effects in S$_N$2 reactions
        4. Leaving groups in S$_N$2 reactions
    C. S$_N$1 reactions (Section 7.8)
        1. Rate-limiting step in S$_N$1 reactions
        2 Stereochemistry of S$_N$1 reactions
        3. Leaving groups in S$_N$1 reactions
III. Elimination reactions (Sections 7.9 - 7.10)
    A. E2 reactions (Section 7.9)
    B. E1 reactions (Section 7.10)
IV. A summary of reactivity : S$_N$1, S$_N$2, E1, E2 (Section 7.11)
V. Biological substitution reactions (Section 7.12)

Study Skills for Chapter 7

After studying this chapter, you should be able to:

1. Name alkyl halides (Problems 7.1, 7.22).
2. Draw structures of alkyl halides corresponding to given names (Problems 7.2, 7.23).
3. Synthesize alkyl halides (Problems 7.3, 7.4, 7.5, 7.8, 7.24, 7.26, 7.28).
4. Predict the products of reactions of alkyl halides (Problems 7.6, 7.7, 7.9, 7.12, 7.18, 7.19, 7.31, 7.39, 7.40, 7.41, 7.42, 7.43, 7.45, 7.51, 7.56).
5. Predict the effects of changes in reaction conditions on substitution and elimination reactions (Problems 7.11, 7.14, 7.15, 7.16, 7.20, 7.25, 7.27, 7.29, 7.32, 7.33, 7.35, 7.36, 7.46, 7.47).
6. Identify reactions as to type — S$_N$1, S$_N$2, E1, E2 (Problems 7.21, 7.48).
7. Choose the correct alkyl halide to yield a specific product (Problems 7.10, 7.30, 7.44, 7.50).
8. Formulate mechanisms of reactions of alkyl halides (Problems 7.13, 7.17, 7.49, 7.52, 7.53).

Chapter 8 – Alcohols, Ethers, and Phenols

8.1 - 8.2

a) CH₃CH(OH)CH₂CH(OH)CH(CH₃)₂
5-Methyl-2,4-hexanediol
secondary alcohol

b) 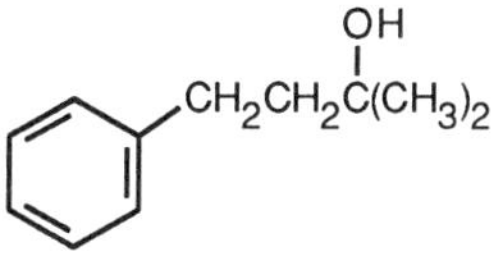 C₆H₅CH₂CH₂C(CH₃)₂OH
2-Methyl-4-phenyl-2-butanol
tertiary alcohol

c) 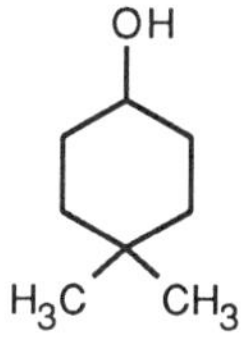 4,4-Dimethylcyclohexanol (with OH on ring, H₃C and CH₃ on opposite carbon)
secondary alcohol

d) 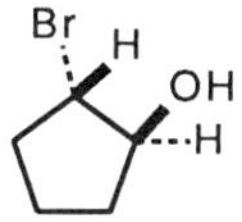 trans-2-Bromocyclopentanol
secondary alcohol

8.3

a) CH₃CH₂CH₂CH₂C(OH)(CH₃)₂
2-Methyl-2-hexanol

b) CH₃CH(OH)CH₂CH₂CH₂CH₂OH
1,5-Hexanediol

c) CH₃CH=C(CH₂CH₃)CH₂OH
2-Ethyl-2-buten-1-ol

d) 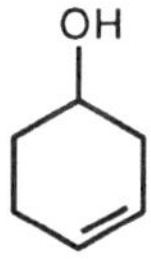 3-Cyclohexen-1-ol

e) 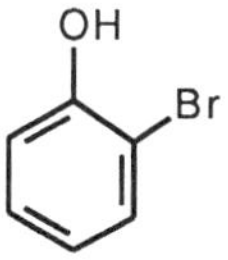 o-Bromophenol

f) 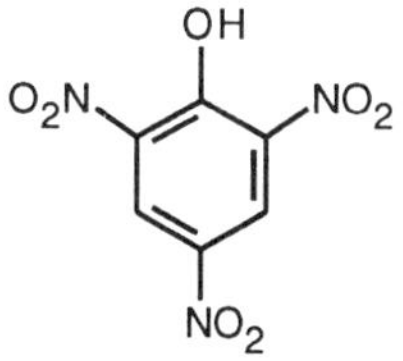 2,4,6-Trinitrophenol

8.4

a) (CH₃)₂CH–O–CH(CH₃)₂
Diisopropyl ether

b) 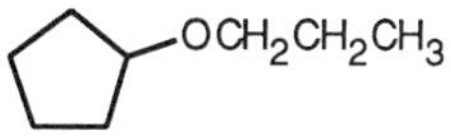 Cyclopentyl–OCH₂CH₂CH₃
Cyclopentyl propyl ether

124   Chapter 8

c) Br—⟨benzene ring⟩—OCH₃
1-Bromo-4-methoxybenzene
(p-Bromoanisole)

d) (CH₃)₂CHCH₂OCH₂CH₃
Ethyl isobutyl ether
(Ethyl 2-methylpropyl ether)

8.5 Remember:
1) Phenols are generally more acidic than alcohols.
2) Electron-withdrawing substituents increase phenol acidity; electron-donating substituents decrease phenol acidity.

Least acidic ⟶ Most acidic

a) Methanol < p-methylphenol < phenol < p-nitrophenol
b) Benzyl alcohol < p-methoxyphenol < p-bromophenol < 2,4-dibromophenol

8.6

[Six resonance structures showing delocalization of negative charge in p-cyanophenolate, with the charge on oxygen, ortho, para carbons, and finally on the nitrogen of the cyano group.]

8.7

a) $CH_3\overset{O}{\underset{\|}{C}}CH_2CH_2\overset{O}{\underset{\|}{C}}OCH_3$ $\xrightarrow[2.\ H_3O^+]{1.\ NaBH_4}$ $CH_3\overset{OH}{\underset{|}{C}}HCH_2CH_2\overset{O}{\underset{\|}{C}}OCH_3$

NaBH₄ reduces aldehydes and ketones without disturbing other functional groups.

b) $CH_3\overset{O}{\underset{\|}{C}}CH_2CH_2\overset{O}{\underset{\|}{C}}OCH_3$ $\xrightarrow[2.\ H_3O^+]{1.\ LiAlH_4}$ $CH_3\overset{OH}{\underset{|}{C}}HCH_2CH_2CH_2OH$

LiAlH₄ reduces both ketones and esters.

8.8

a) PhCHO   or   PhCOOH   or   PhCOOR $\xrightarrow[2.\ H_3O^+]{1.\ LiAlH_4}$ PhCH₂OH

b) PhCOCH₃ $\xrightarrow[2.\ H_3O^+]{1.\ LiAlH_4}$ PhCH(OH)CH₃

Alcohols, Ethers, and Phenols 125

c) Cyclohexanone $\xrightarrow{\text{1. LiAlH}_4}{\text{2. H}_3\text{O}^+}$ Cyclohexanol (OH, H)

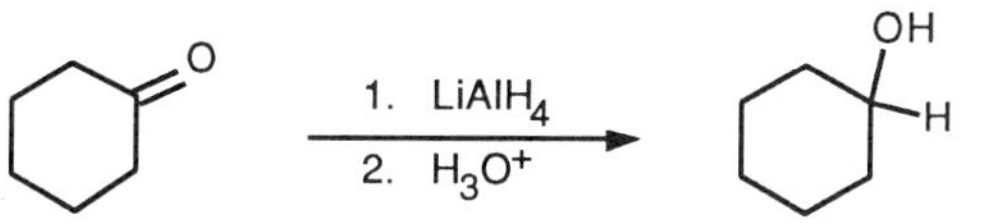

8.9

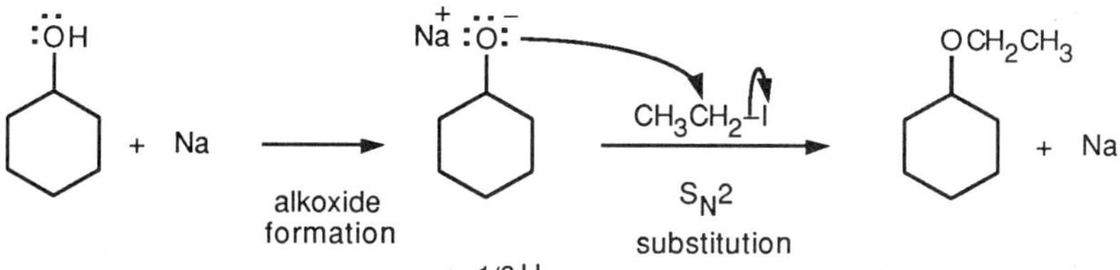

Cyclohexanol + Na → Sodium cyclohexoxide (alkoxide formation) + ½ H$_2$

Sodium cyclohexoxide + CH$_3$CH$_2$–I $\xrightarrow{S_N2 \text{ substitution}}$ Cyclohexyl ethyl ether (OCH$_2$CH$_3$) + NaI

8.10

a) CH$_3$Br + CH$_3$CH$_2$CH$_2$O$^-$ Na$^+$

or

CH$_3$O$^-$ Na$^+$ + CH$_3$CH$_2$CH$_2$Br

→ CH$_3$OCH$_2$CH$_2$CH$_3$ + NaBr
Methyl propyl ether

b) Sodium phenoxide (O$^-$ Na$^+$) + CH$_3$Br → Anisole (OCH$_3$) + NaBr

c) Benzyl bromide (CH$_2$Br) + (CH$_3$)$_2$CHO$^-$ Na$^+$ → Benzyl isopropyl ether (CH$_2$OCH(CH$_3$)$_2$) + NaBr

8.11

Least reactive ⟶ Most reactive

$\underset{\text{CH}_3}{\text{CH}_3\text{CCH}_3}$ (Cl) < CH$_3$CHCH$_3$ (Br) < CH$_3$CH$_2$Cl < CH$_3$CH$_2$Br

2-Chloro-2-methylpropane < 2-bromopropane < chloroethane < bromoethane

The reactivity of alkyl halides in the Williamson ether synthesis is the same as their reactivity in any S$_N$2 reaction.

8.12

a) Ph-CH(OH)CH₃ → (PCC, CH₂Cl₂) → Ph-C(=O)CH₃

b) CH₃CH(CH₃)CH₂OH → (PCC, CH₂Cl₂) → CH₃CH(CH₃)CHO

c) cyclopentanol → (PCC, CH₂Cl₂) → cyclopentanone

8.13

a) cyclohexanol → (CrO₃, H₃O⁺) → cyclohexanone

b) CH₃CH₂CH₂CH₂CH₂CH₂OH → (CrO₃, H₃O⁺) → CH₃CH₂CH₂CH₂CH₂COOH

c) CH₃CH₂CH₂CH₂CH(OH)CH₃ → (CrO₃, H₃O⁺) → CH₃CH₂CH₂CH₂C(=O)CH₃

8.14

a) cyclohexanone   b) CH₃CH₂CH₂CH₂CH₂CHO   c) CH₃CH₂CH₂CH₂C(=O)CH₃

8.15

benzene → (CH₃Cl, AlCl₃) → toluene → (SO₃, H₂SO₄) → p-toluenesulfonic acid → (NaOH, 200°C) → p-cresol

8.16

a) CH₃CH₂OCH₂CH₃ → (HI) → CH₃CH₂OH + ICH₂CH₃

b) cyclohexyl-O-CH₂CH₃ → (HI) → cyclohexanol + ICH₂CH₃

Remember that oxygen stays with the more hindered alkyl group in an S$_N$2 cleavage.

c) 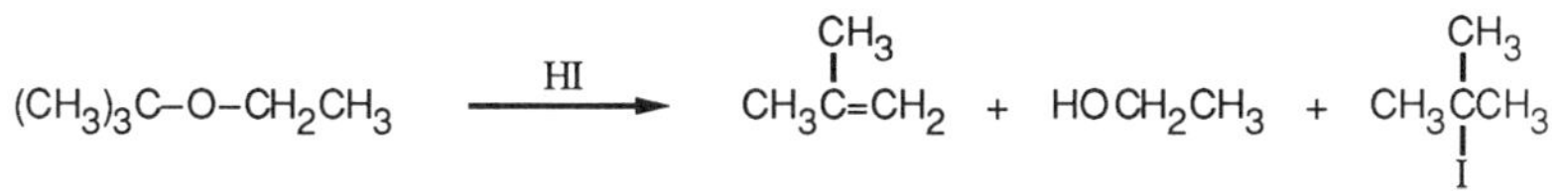 $(CH_3)_3C-O-CH_2CH_3 \xrightarrow{HI} CH_3\underset{CH_3}{\overset{CH_3}{C}}=CH_2 + HOCH_2CH_3 + CH_3\underset{I}{\overset{CH_3}{C}}CH_3$

Oxygen stays with the less hindered alkyl group in an S_N1 cleavage.

8.17

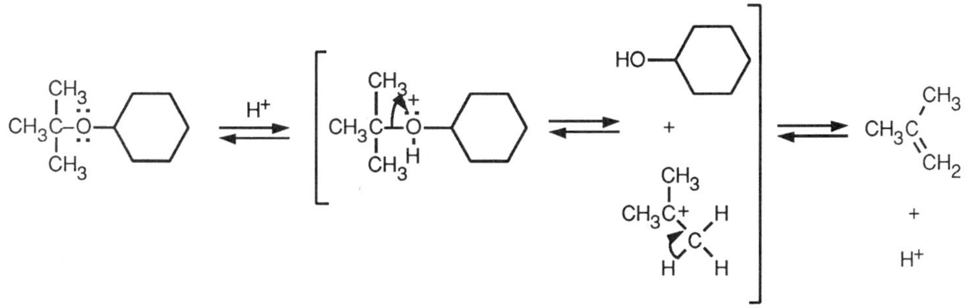

[Reaction mechanism scheme showing protonation of tert-butyl cyclohexyl ether, formation of cyclohexanol and tert-butyl cation, followed by elimination to 2-methylpropene]

The first step of acid-catalyzed cleavage of ethers is protonation of the ether oxygen. The protonated intermediate collapses to form cyclohexanol and a tertiary carbocation. The carbocation loses a proton to form 2–methylpropene. This is an example of an E1 elimination.

8.18

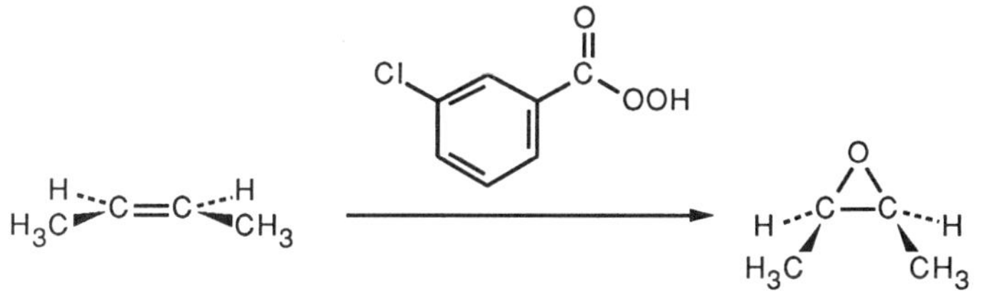

[Reaction of cis-2-Butene with m-chloroperoxybenzoic acid giving cis-2,3-Epoxybutane]

cis-2–Butene   →   cis-2,3–Epoxybutane

The product of the reaction of cis–2–butene with m-chloroperoxybenzoic acid is cis–2,3–epoxybutane, a meso compound.

8.19

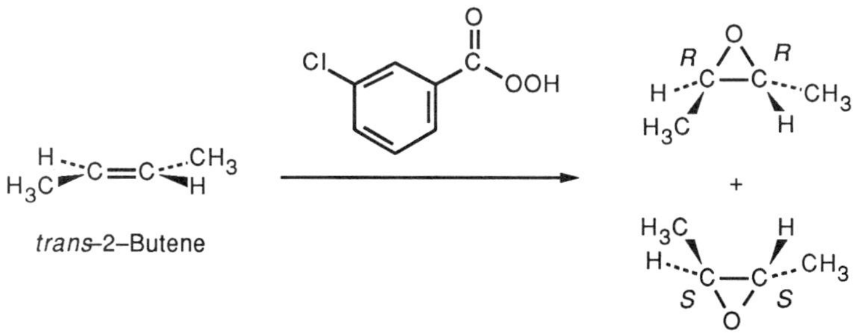

[Reaction of trans-2-Butene with m-chloroperoxybenzoic acid giving trans-2,3-Epoxybutane as R,R and S,S enantiomers]

trans-2–Butene

trans-2,3–Epoxybutane

128   Chapter 8

The product of the reaction of *trans*–2–butene with *m*-chloroperoxybenzoic acid is a racemic mixture of the enantiomeric *trans*–2,3–epoxybutanes. The epoxidation reaction occurs with syn stereochemistry and retains the configuration of the double bond.

8.20

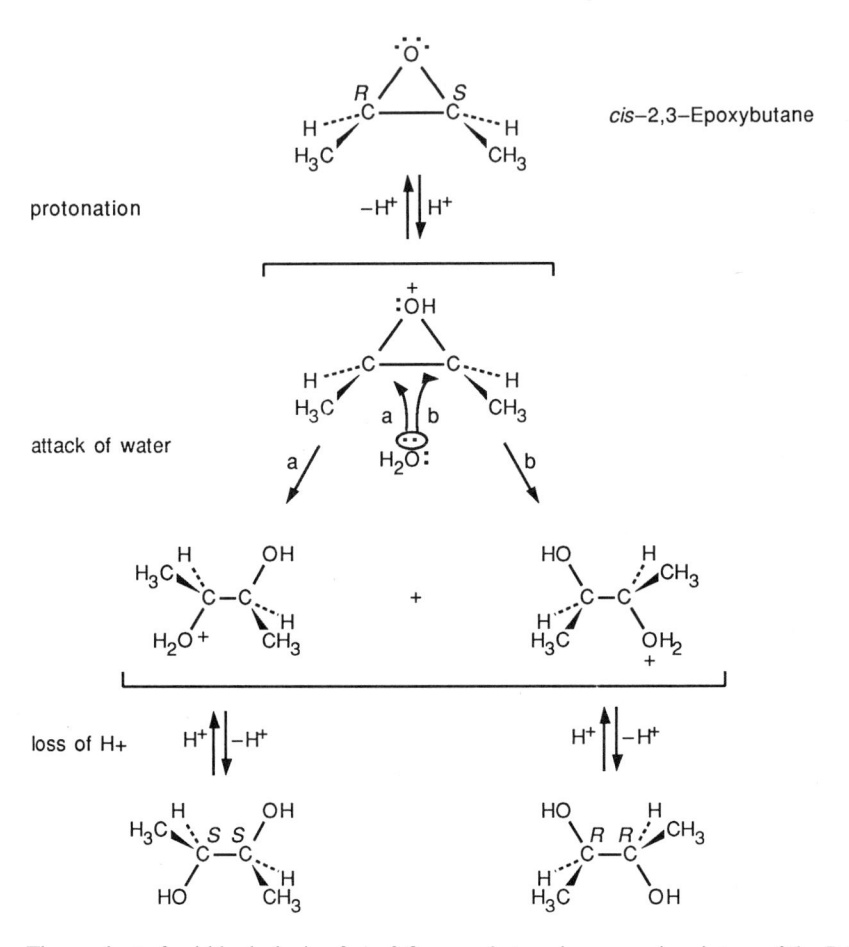

cis–2,3–Epoxybutane

protonation

attack of water

loss of H+

The product of acid hydrolysis of *cis*–2,3–epoxybutane is a racemic mixture of the *R,R* and *S,S* diol enantiomers.

8.21

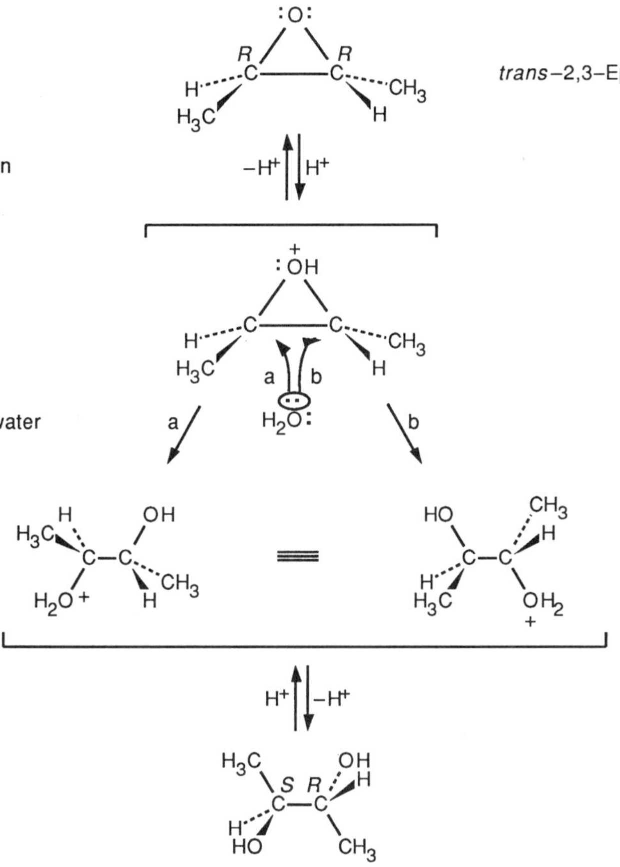

$$\text{trans-2,3-Epoxybutane}$$

[Structure: trans-2,3-epoxybutane with R,R configuration, oxygen on top, H and CH₃ on each carbon]

protonation $-H^+ \updownarrow H^+$

[Protonated epoxide intermediate with :ÖH⁺, showing attack paths a and b by H₂O]

attack of water

[Two product structures shown, equivalent by rotation, with H₂O⁺ and OH groups on adjacent carbons]

loss of H+ $H^+ \updownarrow -H^+$

[Final product structure showing S,R meso configuration with CH₃, OH, H groups]

The product of acid hydrolysis of the *R,R* enantiomer of *trans*–2,3–epoxybutane is a meso compound. Acid hydrolysis of the *S,S* enantiomer yields the same meso compound.

8.22

a) $\underset{\text{2-Butanethiol}}{\text{CH}_3\text{CH}_2\overset{\underset{|}{\text{SH}}}{\text{C}}\text{HCH}_3}$

b) $\underset{\text{2,2,6-Trimethyl-4-heptanethiol}}{(\text{CH}_3)_3\text{CCH}_2\overset{\underset{|}{\text{SH}}}{\text{C}}\text{HCH}_2\text{CH}(\text{CH}_3)_2}$

c) [cyclopentene ring with −SH substituent]

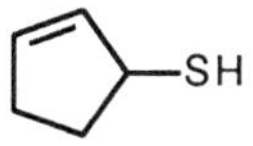

  2-Cyclopentene-1-thiol

8.23

a) CH₃CH₂SCH₃
   Ethyl methyl sulfide

b) (CH₃)₃CSCH₂CH₃
   tert-Butyl ethyl sulfide

c) 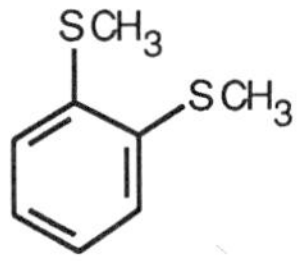 [benzene ring with SCH₃ and SCH₃ groups ortho]
   o-Di(methylthio)benzene

8.24

CH₃CH=CHCH₂OH →[PBr₃] CH₃CH=CHCH₂Br →[Na⁺ ⁻:SH] CH₃CH=CHCH₂SH + NaBr
2-Buten-1-ol   2-Butene-1-thiol

↑ 1. LiAlH₄
  2. H₃O⁺

CH₃CH=CHCOOCH₃
Methyl 2-butenoate

8.25

a) (CH₃)₂CHOCH₂CH₃
   Ethyl isopropyl ether

b) 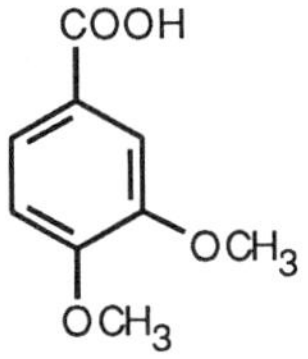 [benzene ring with COOH, and two OCH₃ groups at 3,4 positions]
   3,4-Dimethoxybenzoic acid

c) CH₃CH₂CH(OH)CH₂CH₂C(OH)(CH₃)₂
   2-Methyl-2,5-heptanediol

d) 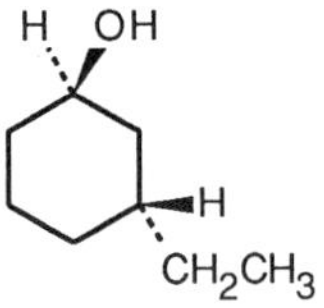 [cyclohexane ring with H, OH and H, CH₂CH₃ shown with stereochemistry]
   trans-3-Ethylcyclohexanol

e) 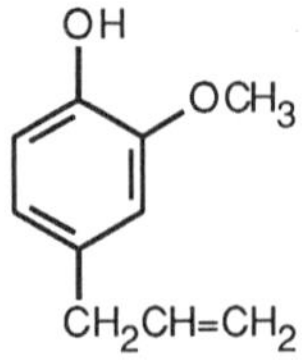 [benzene ring with OH, OCH₃, and CH₂CH=CH₂ substituents]
   Eugenol

8.26

a) HOCH₂CH₂CH(CH₃)CH₂OH
2-Methyl-1,4-butanediol

b) CH₃CH(OH)CH(CH₂CH₃)CH₂CH₂CH₃
3-Ethyl-2-hexanol

c) cis-3-Phenylcyclopentanol (Ph and OH on same face of cyclopentane ring, H's on opposite face)

d) (CH₃)₂CHC(SH)(CH₃)CH₂CH₂CH₃
2,3-Dimethyl-3-hexanethiol

8.27

CH₃CH₂CH₂CH₂CH₂OH
1-Pentanol

CH₃CH₂CH₂CH(OH)CH₃
2-Pentanol

CH₃CH₂CH(OH)CH₂CH₃
3-Pentanol

CH₃CH₂CH(CH₃)CH₂OH
2-Methyl-1-butanol

CH₃CH(CH₃)CH₂CH₂OH
3-Methyl-1-butanol

CH₃CH₂C(OH)(CH₃)CH₃
2-Methyl-2-butanol

CH₃CH(OH)CH(CH₃)CH₃
3-Methyl-2-butanol

CH₃C(CH₃)₂CH₂OH
2,2-Dimethyl-1-propanol

8.28 Primary alcohols react with aqueous acidic CrO₃ to form carboxylic acids, secondary alcohols yield ketones, and tertiary alcohols are unreactive to oxidation. Of the eight alcohols in the previous problem, only 2-methyl-2-butanol is unreactive to aqueous acidic CrO₃.

CH₃CH₂CH₂CH₂CH₂OH $\xrightarrow{\text{CrO}_3,\ \text{H}_2\text{O}}_{\text{H}_2\text{SO}_4}$ CH₃CH₂CH₂CH₂COOH

CH₃CH₂CH₂CH(OH)CH₃ $\xrightarrow{\text{CrO}_3,\ \text{H}_2\text{O}}_{\text{H}_2\text{SO}_4}$ CH₃CH₂CH₂C(=O)CH₃

CH₃CH₂CH(OH)CH₂CH₃ $\xrightarrow{\text{CrO}_3,\ \text{H}_2\text{O}}_{\text{H}_2\text{SO}_4}$ CH₃CH₂C(=O)CH₂CH₃

CH₃CH₂CH(CH₃)CH₂OH $\xrightarrow{\text{CrO}_3,\ \text{H}_2\text{O}}_{\text{H}_2\text{SO}_4}$ CH₃CH₂CH(CH₃)COOH

132 Chapter 8

$$CH_3CHCH_2CH_2OH \xrightarrow{CrO_3,\ H_2O}_{H_2SO_4} CH_3CHCH_2COOH$$
(with CH₃ substituent on both)

$$CH_3CHCHCH_3 \text{ (with OH and CH}_3\text{)} \xrightarrow{CrO_3,\ H_2O}_{H_2SO_4} CH_3CHCCH_3 \text{ (with =O and CH}_3\text{)}$$

$$CH_3CCH_2OH \text{ (with two CH}_3\text{)} \xrightarrow{CrO_3,\ H_2O}_{H_2SO_4} CH_3CCOOH \text{ (with two CH}_3\text{)}$$

8.29 Of the alcohols in Problem 8.27, only 2-pentanol, 2-methyl-1-butanol, and 3-methyl-2-butanol are chiral.

8.30

$CH_3CH_2OCH_2CH_2CH_3$ — Ethyl propyl ether

$CH_3CH_2OCHCH_3$ (with CH₃) — Ethyl isopropyl ether

$CH_3CH_2CH_2CH_2OCH_3$ — Butyl methyl ether

$CH_3CHCH_2OCH_3$ (with CH₃) — Isobutyl methyl ether

$CH_3CH_2CHOCH_3$ (with CH₃) — sec-Butyl methyl ether

CH_3COCH_3 (with two CH₃) — tert-Butyl methyl ether

Only *sec*-butyl methyl ether is chiral.

8.31

$$CH_3CH_2CH_2OCH_2CH_3 \xrightarrow{HI} CH_3CH_2CH_2OH + ICH_2CH_3$$
$$+$$
$$CH_3CH_2CH_2I + HOCH_2CH_3$$

$$(CH_3)_2CHOCH_2CH_3 \xrightarrow{HI} (CH_3)_2CHOH + ICH_2CH_3$$

$$CH_3CH_2CH_2CH_2OCH_3 \xrightarrow{HI} CH_3CH_2CH_2CH_2OH + ICH_3$$

$$(CH_3)_2CHCH_2OCH_3 \xrightarrow{HI} (CH_3)_2CHCH_2OH + ICH_3$$

$$CH_3CH_2CHOCH_3 \text{ (with CH}_3\text{)} \xrightarrow{HI} CH_3CH_2CHOH \text{ (with CH}_3\text{)} + ICH_3$$

$$(CH_3)_3COCH_3 \xrightarrow{HI} (CH_3)_2C=CH_2 + ICH_3 + H_2O$$

8.32

a) $CH_3CH_2O\overset{\overset{\displaystyle CH_3}{|}}{C}HCH_3$ $\xrightarrow{HI, H_2O}$ CH_3CH_2I + $HO\overset{\overset{\displaystyle CH_3}{|}}{C}HCH_3$

b) $(CH_3)_3CCH_2OCH_3$ $\xrightarrow{HI, H_2O}$ $(CH_3)_3CCH_2OH$ + CH_3I

8.33

a) cyclohexanol $\xrightarrow{PCC}$ cyclohexanone

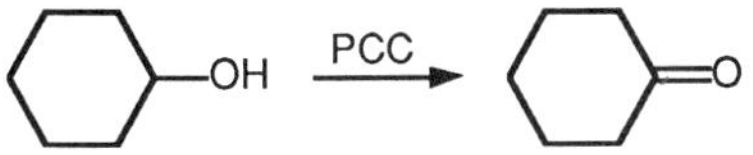

b) cyclohexanol $\xrightarrow{PBr_3}$ cyclohexyl bromide

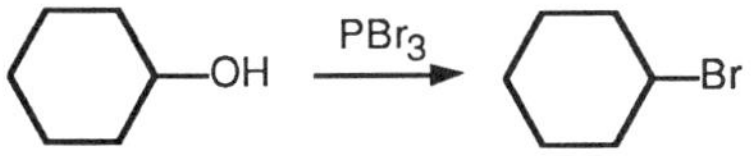

c) $CH_3CH_2CH_2OH$ $\xrightarrow{PCC}$ CH_3CH_2CHO

d) $CH_3CH_2CH_2OH$ $\xrightarrow[H_3O^+]{CrO_3}$ CH_3CH_2COOH

e) $CH_3CH_2CH_2OH$ $\xrightarrow{Na}$ $CH_3CH_2CH_2O^-Na^+$ + $1/2\, H_2$

f) $CH_3CH_2CH_2OH$ $\xrightarrow{SOCl_2}$ $CH_3CH_2CH_2Cl$

8.34

a) Ph-CH₂CH₂OH $\xrightarrow[H_2O]{KMnO_4}$ Ph-COOH

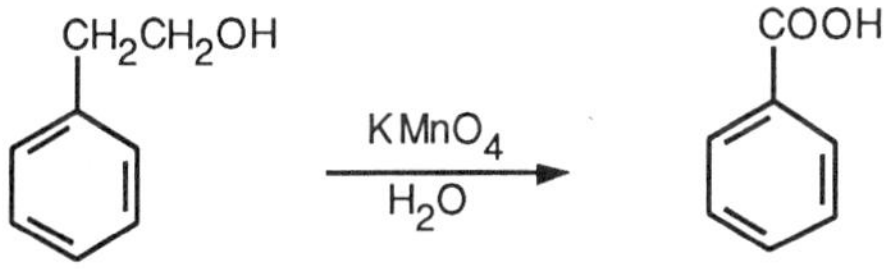

b) Ph-CH₂CH₂OH $\xrightarrow{PBr_3}$ Ph-CH₂CH₂Br $\xrightarrow[\text{ether}]{Mg}$ Ph-CH₂CH₂MgBr $\xrightarrow{H_2O}$ Ph-CH₂CH₃

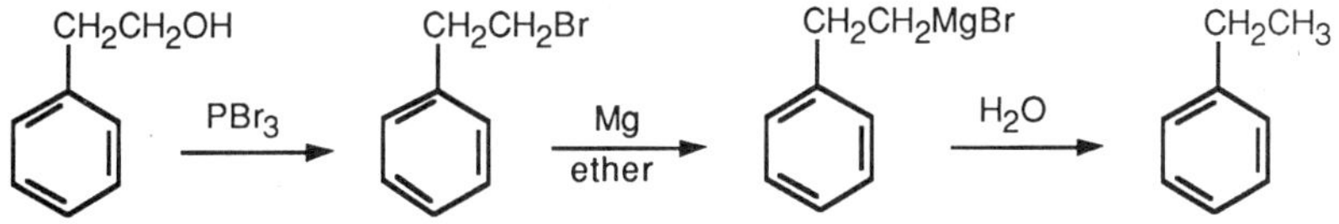

c) Ph-CH₂CH₂OH $\xrightarrow{PBr_3}$ Ph-CH₂CH₂Br

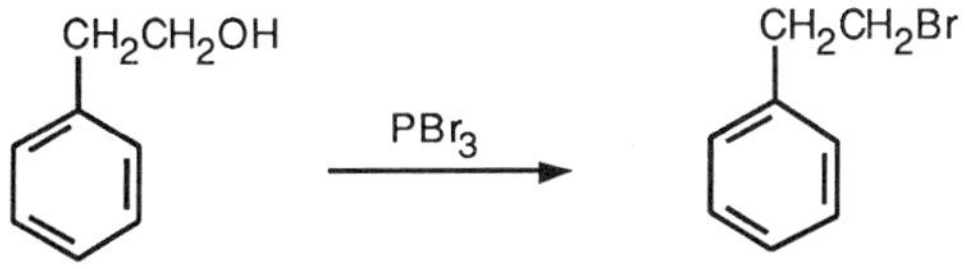

d) PhCH₂CH₂OH → (CrO₃, H₂O, H₂SO₄) → PhCH₂COOH

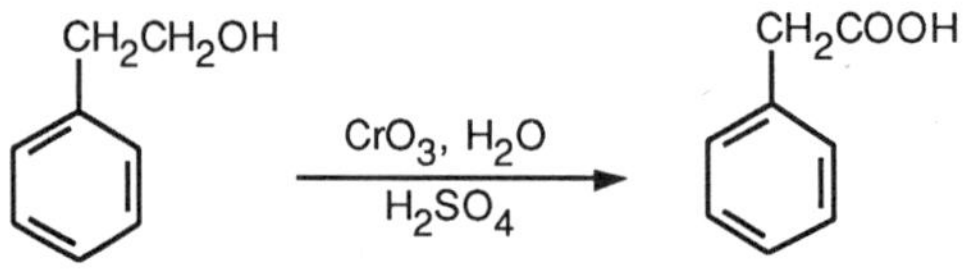

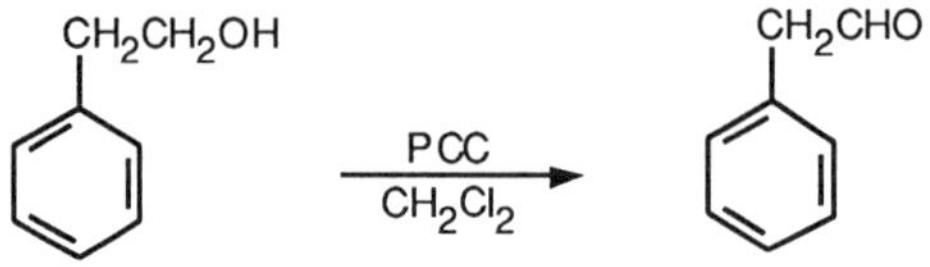

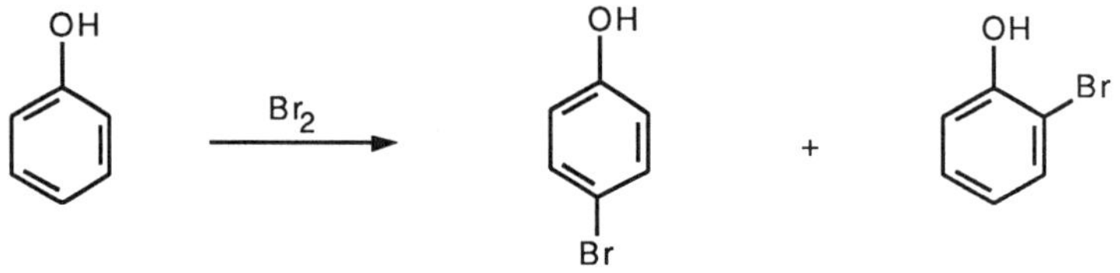

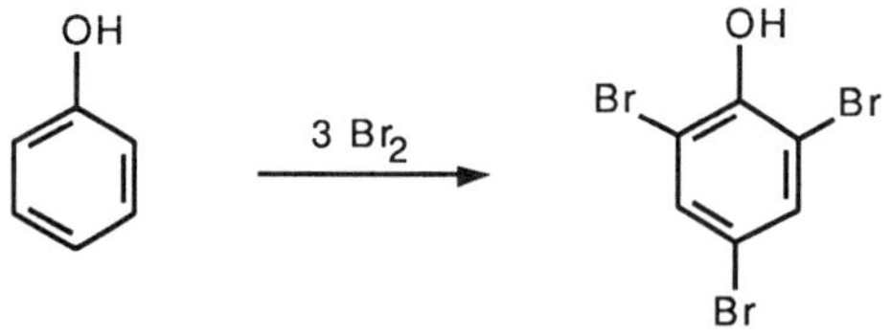

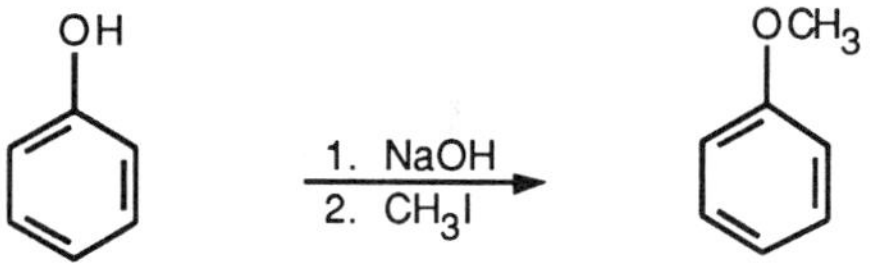

e) PhCH₂CH₂OH → (PCC, CH₂Cl₂) → PhCH₂CHO

8.35

a) PhOH → (Br₂) → 4-bromophenol + 2-bromophenol

b) PhOH → (3 Br₂) → 2,4,6-tribromophenol

c) PhOH → (1. NaOH, 2. CH₃I) → PhOCH₃

d) PhOH → (Na₂Cr₂O₇, H₃O⁺) → 1,4-benzoquinone

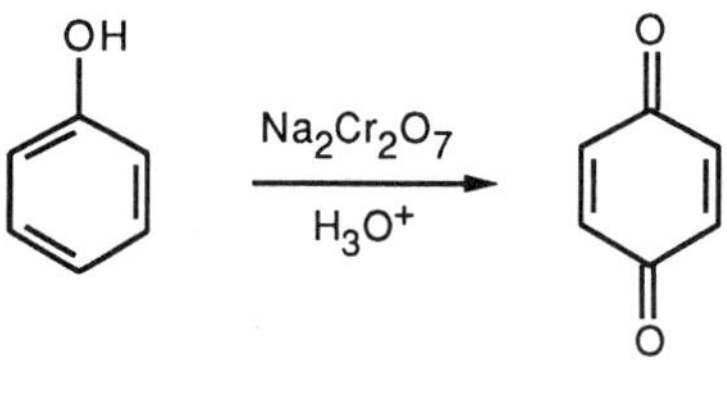

8.36

a) CH₃CH₂CH₂CH₂OH → (PBr₃) → CH₃CH₂CH₂CH₂Br

b) CH₃CH₂CH₂CH₂OH → (CrO₃, H₂O, H₂SO₄) → CH₃CH₂CH₂COOH

c) CH₃CH₂CH₂CH₂OH → (Na) → CH₃CH₂CH₂CH₂O⁻Na⁺ + 1/2 H₂

Alcohols, Ethers, and Phenols   135

d) $CH_3CH_2CH_2CH_2OH \xrightarrow{PCC} CH_3CH_2CH_2CHO$

8.37

a) 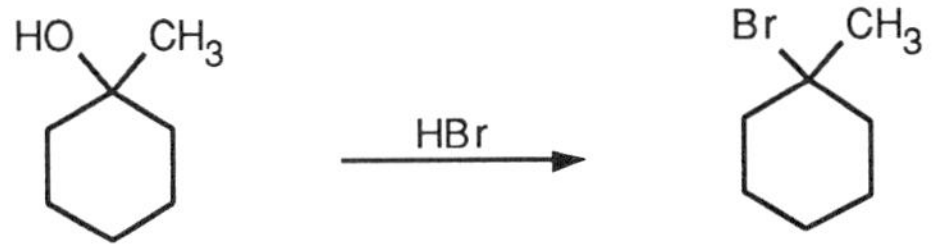 1-hydroxy-1-methylcyclohexane $\xrightarrow{HBr}$ 1-bromo-1-methylcyclohexane

b) 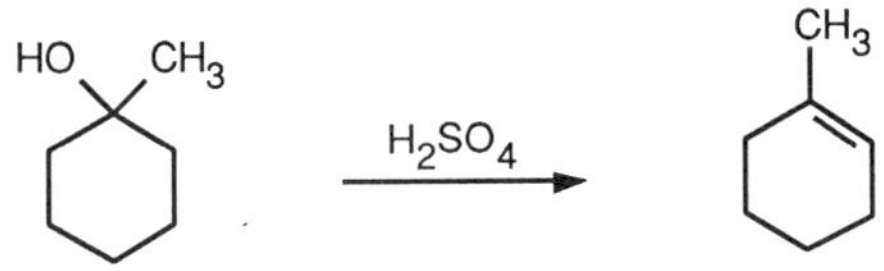 1-hydroxy-1-methylcyclohexane $\xrightarrow{H_2SO_4}$ 1-methylcyclohexene

c) 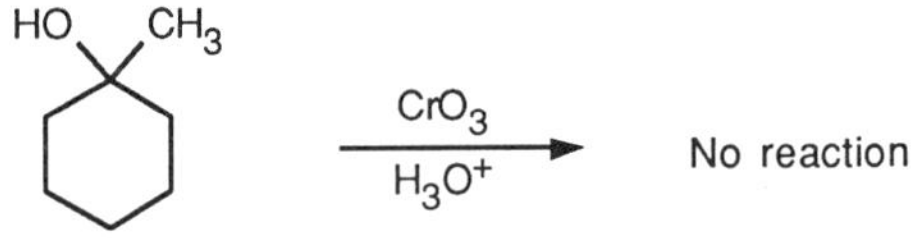 1-hydroxy-1-methylcyclohexane $\xrightarrow[H_3O^+]{CrO_3}$ No reaction

d) 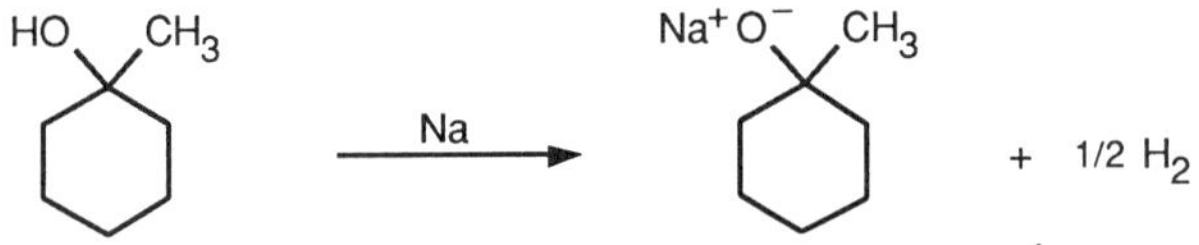 1-hydroxy-1-methylcyclohexane $\xrightarrow{Na}$ 1-(sodium oxide)-1-methylcyclohexane + 1/2 H$_2$

e) 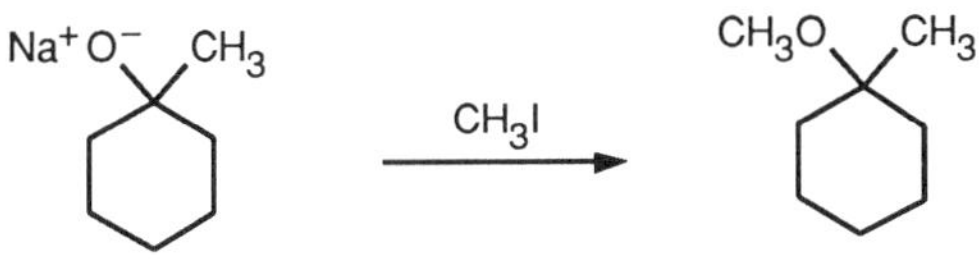 1-(Na$^+$O$^-$)-1-methylcyclohexane $\xrightarrow{CH_3I}$ 1-methoxy-1-methylcyclohexane

8.38

a) 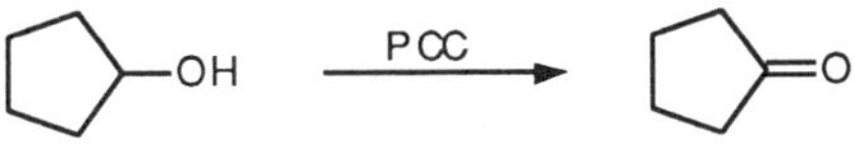 cyclopentanol $\xrightarrow{PCC}$ cyclopentanone

b) 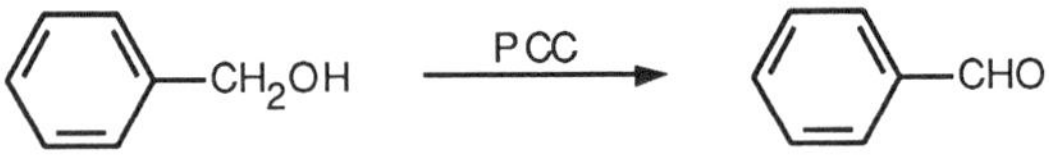 PhCH$_2$OH $\xrightarrow{PCC}$ PhCHO

c) 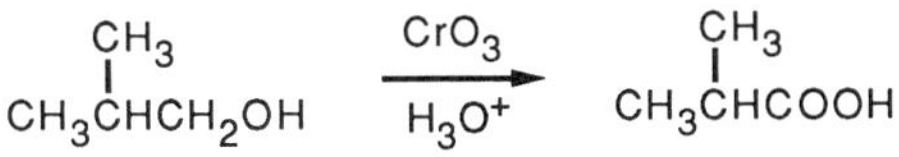 $(CH_3)_2CHCH_2OH \xrightarrow[H_3O^+]{CrO_3} (CH_3)_2CHCOOH$

8.39

a) 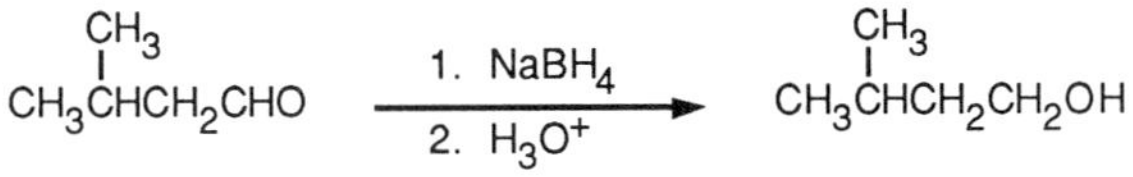 $(CH_3)CH(CH_3)CH_2CHO \xrightarrow[\text{2. } H_3O^+]{\text{1. NaBH}_4} (CH_3)CH(CH_3)CH_2CH_2OH$

136   Chapter 8

b) 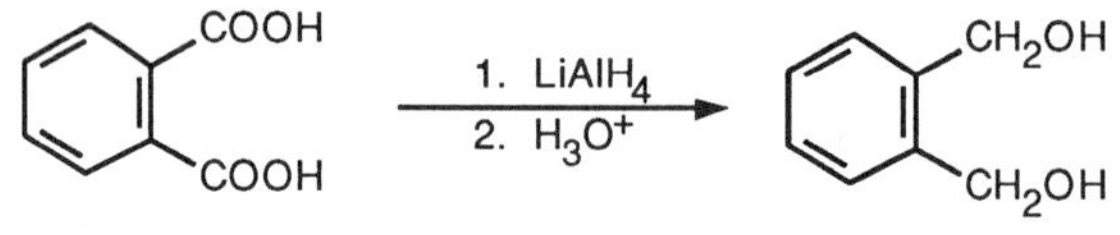 Phthalic acid → 1. LiAlH$_4$; 2. H$_3$O$^+$ → 1,2-bis(hydroxymethyl)benzene

c) 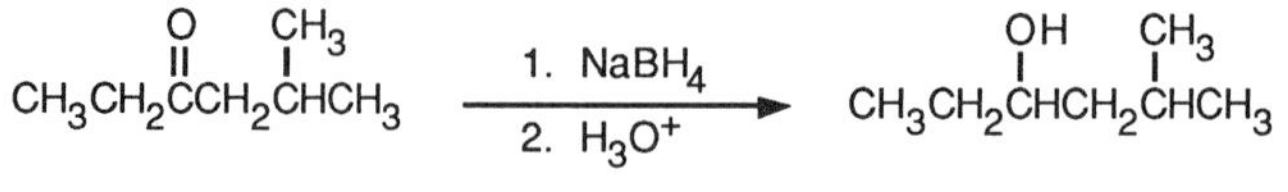 CH$_3$CH$_2$COCH$_2$CH(CH$_3$)CH$_3$ → 1. NaBH$_4$; 2. H$_3$O$^+$ → CH$_3$CH$_2$CH(OH)CH$_2$CH(CH$_3$)CH$_3$

8.40

Top pathway: CH$_3$(CH$_3$CH$_2$)C=O with H$^-$ attack → [CH$_3$(CH$_2$CH$_3$)CH–O$^-$] → H$_3$O$^+$ → (S)-CH$_3$(CH$_2$CH$_3$)CH–OH + H$_2$O

Bottom pathway: CH$_3$(CH$_3$CH$_2$)C=O with H$^-$ attack from opposite face → [CH$_3$CH$_2$(CH$_3$)CH–O$^-$] → H$_3$O$^+$ → (R)-CH$_3$CH$_2$(CH$_3$)CH–OH + H$_2$O

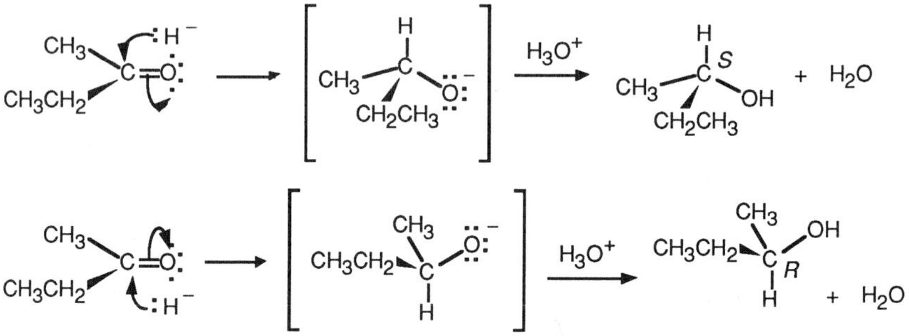

Attack occurs from both sides of the planar carbonyl group to yield a racemic product mixture.

8.41

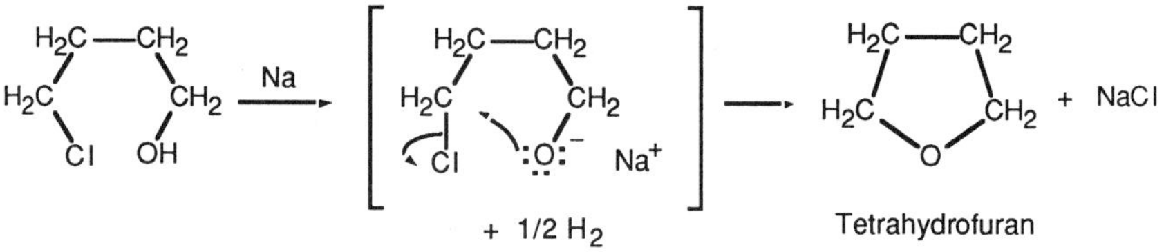

5-chloropentan-1-ol + Na → [cyclic alkoxide intermediate with Cl and O$^-$ Na$^+$] + ½ H$_2$ → Tetrahydrofuran + NaCl

8.42

p-bromophenol ; p-methylphenol

Since electron-withdrawing groups increase phenol acidity, *p*-bromophenol is more acidic than *p*-methylphenol.

8.43

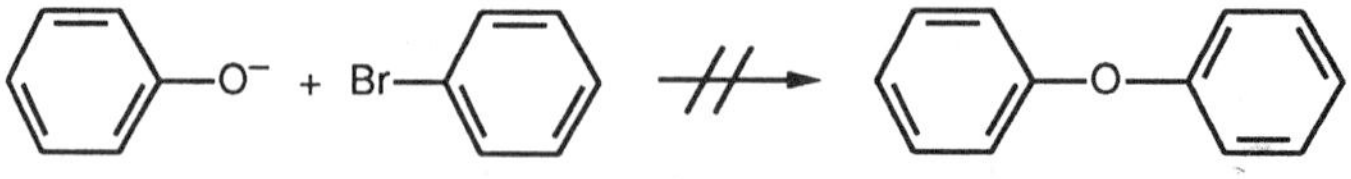

C$_6$H$_5$–O$^-$ + Br–C$_6$H$_5$ ↛ C$_6$H$_5$–O–C$_6$H$_5$

The Williamson ether synthesis is an S$_N$2 reaction between an alkoxide or phenoxide anion and an alkyl halide. It can't be used to synthesize diphenyl ether because an aryl halide (C$_6$H$_5$Br) doesn't undergo S$_N$2 reactions.

Alcohols, Ethers, and Phenols   137

8.44

Least acidic ⟶ Most acidic

$$CH_3\overset{O}{\overset{\|}{C}}CH_3 \;<\; C_6H_5\text{—OH} \;<\; CH_3\overset{O}{\overset{\|}{C}}CH_2\overset{O}{\overset{\|}{C}}CH_3 \;<\; CH_3\overset{O}{\overset{\|}{C}}OH$$

$pK_a \approx 20$ $pK_a \approx 10$ $pK_a \approx 9$ $pK_a \approx 4.7$

8.45 To react completely with NaOH, an acid must have a pK_a lower than the pK_a of H_2O. Thus, all substances in the previous problem except acetone will react completely with NaOH.

8.46

$$\underset{\substack{|\\CH_3}}{\overset{\substack{CH_3\\|}}{CH_3\text{–C–O}^-}} + H_2O \longrightarrow \underset{\substack{|\\CH_3}}{\overset{\substack{CH_3\\|}}{CH_3\text{–C–OH}}} + HO^-$$

$pK_a \approx 15.7$ $pK_a \approx 18$
stronger acid   weaker acid

The reaction will take place as written because water is a stronger acid than *tert*-butyl alcohol.

8.47 Only acetic acid will react with sodium bicarbonate.

8.48 Sodium bicarbonate reacts with acetic acid to produce carbonic acid, which breaks down to form CO_2. The resulting CO_2 bubbles indicate the presence of acetic acid. Phenol does not react with sodium bicarbonate.

8.49

a) benzene $\xrightarrow{\text{CH}_3\text{Cl}, \text{AlCl}_3}$ toluene $\xrightarrow{\text{KMnO}_4, \text{H}_2\text{O}}$ benzoic acid (PhCOOH) $\xrightarrow{\text{1. LiAlH}_4,\ \text{2. H}_3\text{O}^+}$ PhCH$_2$OH $\xrightarrow{\text{PBr}_3}$ PhCH$_2$Br

b) benzene $\xrightarrow{\text{SO}_3, \text{H}_2\text{SO}_4}$ PhSO$_3$H $\xrightarrow{\text{NaOH}, 200°}$ PhOH $\xrightarrow{\text{NaOH}}$ PhO$^-$ Na$^+$

138 Chapter 8

c)

C₆H₅−Ö:⁻ Na⁺ + BrCH₂−C₆H₅ ⟶ C₆H₅−OCH₂−C₆H₅

Benzyl phenyl ether

8.50

2 CH₃Br + 2 NaSH ⟶ 2 CH₃SH + 2 NaBr

2 CH₃SH + Br₂ ⟶ CH₃SSCH₃ + 2 HBr

Dimethyl disulfide

8.51

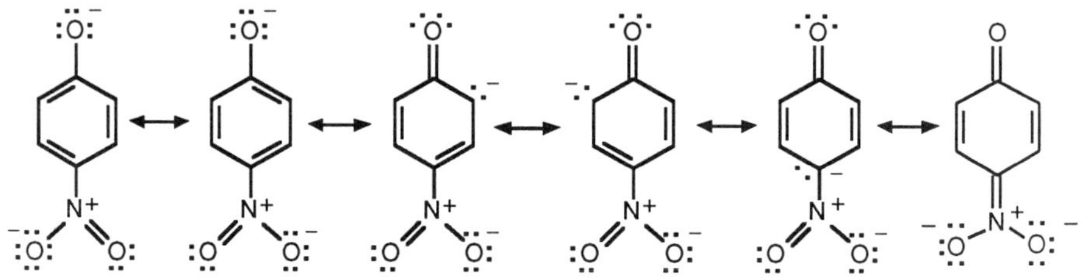

[Resonance structures of p-nitrophenoxide anion showing delocalization of negative charge through the ring and onto the nitro group oxygens]

p-Nitrophenol is more acidic than phenol because the *p*-nitrophenoxide anion is stabilized by extensive delocalization of the negative charge by the *p*-nitro group.

8.52

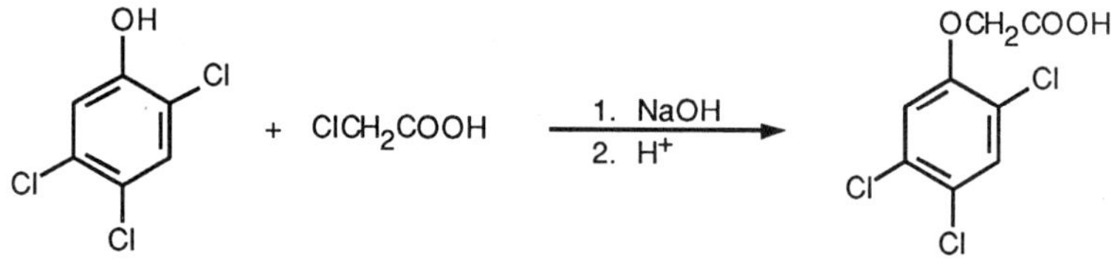

2,4,5-trichlorophenol + ClCH₂COOH $\xrightarrow{\text{1. NaOH} \\ \text{2. H}^+}$ 2,4,5-T

Treatment of 2,4,5–trichlorophenol with NaOH produces 2,4,5–trichlorophenoxide anion, which displaces chlorine from ClCH₂COO⁻ ⁺Na in an S_N2 reaction to yield 2,4,5–T.

8.53

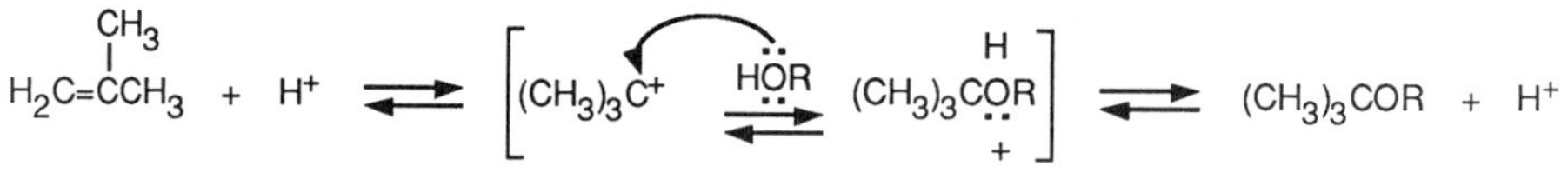

H₂C=C(CH₃)CH₃ + H⁺ ⇌ [(CH₃)₃C⁺ $\xrightarrow{\text{HÖR}}$ (CH₃)₃CÖR(H)⁺] ⇌ (CH₃)₃COR + H⁺

This mechanism is the reverse of the mechanism illustrated in Problem 8.17.

8.54

a) C₆H₅OH $\xrightarrow{\text{Na}}$ C₆H₅Ö:⁻ Na⁺ (+ ½ H₂) $\xrightarrow{\text{CH}_3\text{CH}_2\text{Br}}$ C₆H₅OCH₂CH₃ + NaBr

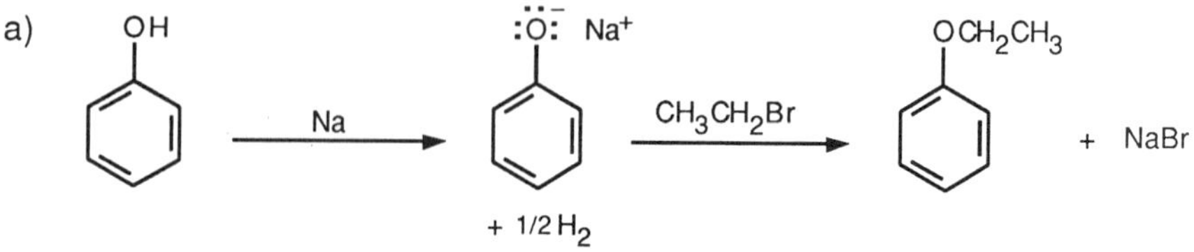

Alcohols, Ethers, and Phenols   139

b)

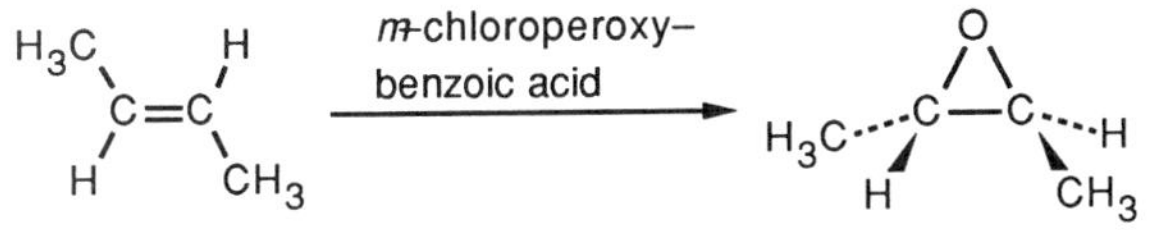

H₃C, H, C=C, H, CH₃ →(m-chloroperoxy-benzoic acid)→ epoxide with H₃C, H, CH₃

8.55

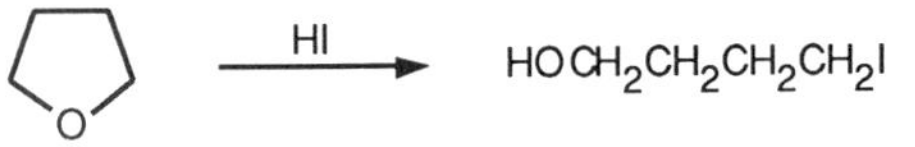

tetrahydrofuran →(HI)→ HOCH₂CH₂CH₂CH₂I

8.56

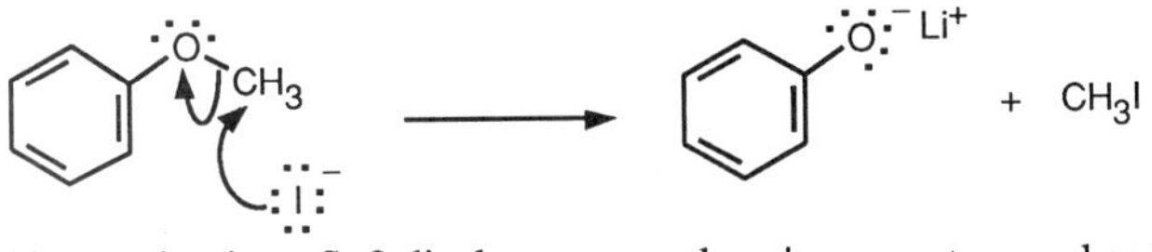

PhO-CH₃ with I⁻ attacking → PhO⁻ Li⁺ + CH₃I

This reaction is an S_N2 displacement and can't occur at an aryl carbon.

8.57

$$\frac{1.06 \text{ g vanillin}}{152 \text{ g/mol}} = 6.97 \times 10^{-3} \text{ mol vanillin}$$

$$\frac{1.60 \text{ g AgI}}{235 \text{ g/mol}} = 6.81 \times 10^{-3} \text{ mol AgI}$$

6.81×10^{-3} mol AgI → 6.81×10^{-3} mol :I⁻ → 6.81×10^{-3} mol CH₃I → 6.81×10^{-3} mol ⁻OCH₃

Thus, 6.97×10^{-3} moles of vanillin contains 6.81×10^{-3} moles of methoxyl groups. Since the ratio of moles vanillin to moles methoxyl is approximately 1:1, each vanillin contains one methoxyl group.

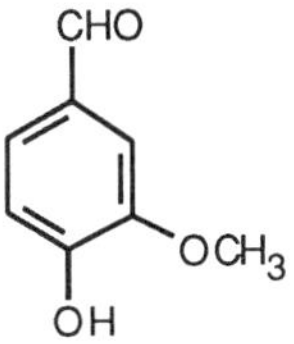

Vanillin (CHO, OCH₃, OH on benzene ring)

Chapter Outline

I. Naming alcohols, phenols, and ethers (Section 8.1)
II. Properties (Sections 8.2 - 8.3)
    A. Hydrogen bonding (Section 8.2)
    B. Acidity (Section 8.3)
III. Alcohols and phenols (Sections 8.4 - 8.8)
    A. Preparation (Sections 8.4 - 8.5)
        1. Alcohols (Sections 8.4 - 8.5)
            a. From alkenes (Section 8.4)
            b. By reduction of carbonyl compounds (Section 8.5)
        2. Phenols (Section 8.8)
    B. Reactions (Sections 8.6 - 8.7)
        1. Alcohols (Sections 8.6 - 8.7)
            a. Williamson ether synthesis (Section 8.6)
            b. Dehydration to yield alkenes (Section 8.7)
            c. Conversion into alkyl halides (Section 8.7)
            d. Oxidation to yield carbonyl compounds (Section 8.7)
        2. Phenols (Section 8.8)
            a. Williamson ether synthesis
            b. Electrophilic aromatic substitution
            c. Oxidation to yield quinones
IV. Ethers (Sections 8.9 - 8.11)
    A. Preparation by Williamson ether synthesis (Section 8.6)
    B. Reactions -- Acidic cleavage (Section 8.9)
    C. Epoxides (Sections 8.10 - 8.11)
        1. Preparation (Section 8.10)
        2. Ring-opening reactions of epoxides (Section 8.11)
V. Thiols and sulfides (Section 8.12)
    A. Nomenclature
    B. Preparation from alkyl halides
    C. Reactions

Study Skills for Chapter 8

After studying this chapter, you should be able to:

1. Name alcohols, ethers, phenols, thiols, and sulfides (Problems 8.1, 8.4, 8.22, 8.23, 8.26, 8.27).
2. Draw structures corresponding to names given for alcohols, ethers, phenols, thiols, and sulfides (Problems 8.3, 8.25, 8.27, 8.30).
3. Predict relative acidities of molecules (Problems 8.6, 8.42, 8.44).
4. Use pK_a values to predict the likelihood of a reaction taking place (Problems 8.45, 8.46, 8.47, 8.48).
5. Synthesize alcohols phenols, and ethers (Problems 8.7, 8.8, 8.9, 8.10, 8.15, 8.18, 8.19, 8.39, 8.49, 8.54).
6. Predict the products of reactions involving alcohols and phenols (Problems 8.12, 8.13, 8.14, 8.28, 8.33, 8.34, 8.35, 8.36, 8.37, 8.38, 8.52).
7. Predict the products of ether cleavage (Problems 8.16, 8.20, 8.21, 8.31, 8.32, 8.55).
8. Formulate the mechanisms of simple reactions involving alcohols, ethers, and phenols (Problems 8.17, 8.41, 8.53, 8.56).
9. Prepare simple thiols, sulfides, and disulfides (Problems 8.24, 8.50).

Chapter 9 – Aldehydes and Ketones

9.1

a) 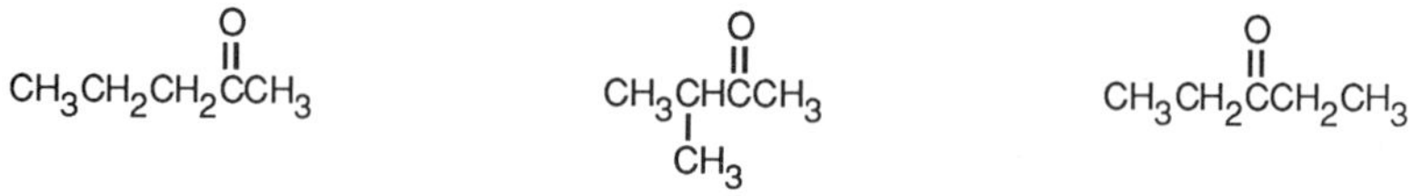 CH$_3$CH$_2$CH$_2$COCH$_3$ CH$_3$CH(CH$_3$)COCH$_3$ CH$_3$CH$_2$COCH$_2$CH$_3$

b) 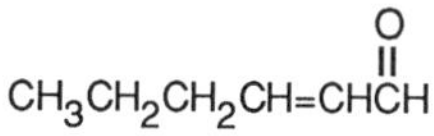 CH$_3$CH$_2$CH$_2$CH=CHCHO   and many other possible answers.

c) 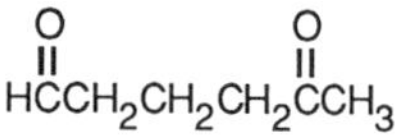 HCOCH$_2$CH$_2$CH$_2$COCH$_3$ and many other possibilities.

d) 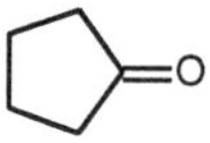 cyclopentanone   and many other possibilities.

9.2

a) 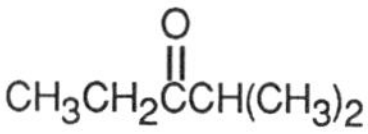 CH$_3$CH$_2$COCH(CH$_3$)$_2$
2-Methyl-3-pentanone

b) 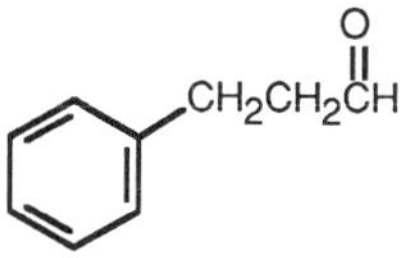 PhCH$_2$CH$_2$CHO
3-Phenylpropanal

c) 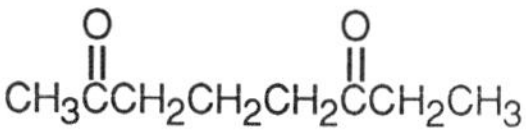 CH$_3$COCH$_2$CH$_2$CH$_2$COCH$_2$CH$_3$
2,6-Octanedione

d) 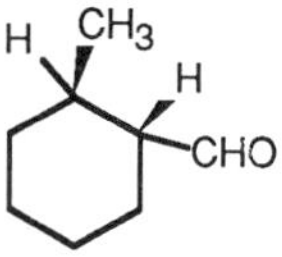 trans-2-Methylcyclohexane-carbaldehyde

e) 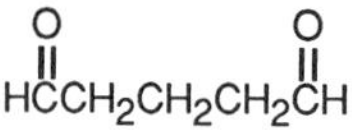 HCOCH$_2$CH$_2$CH$_2$CHO
Pentanedial

f) 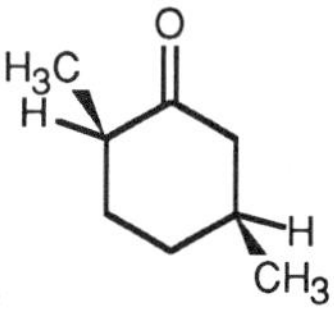 cis-2,5-Dimethylcyclohexanone

9.3

a) 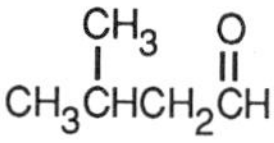 (CH$_3$)$_2$CHCH$_2$CHO
3-Methylbutanal

b) 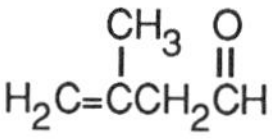 H$_2$C=C(CH$_3$)CH$_2$CHO
3-Methyl-3-butenal

c) 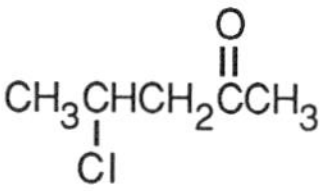 CH$_3$CHClCH$_2$COCH$_3$
4-Chloro-2-pentanone

d) 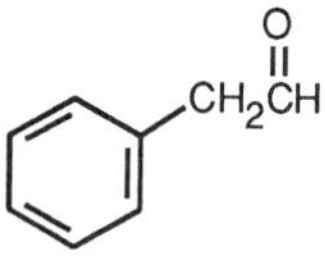 PhCH$_2$CHO
Phenylacetaldehyde

e) 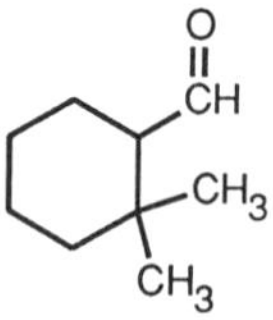 2,2-Dimethylcyclo-hexanecarbaldehyde

f) 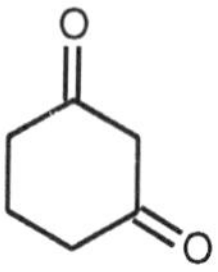 1,3-Cyclohexanedione

Aldehydes and Ketones   143

9.4

a) $CH_3CH_2CH_2CH_2CH_2OH$ $\xrightarrow{\text{PCC}}{CH_2Cl_2}$ $CH_3CH_2CH_2CH_2\overset{\overset{\displaystyle O}{\|}}{C}H$
   1-Pentanol   Pentanal

b) $CH_3CH_2CH_2CH_2CH=CH_2$ $\xrightarrow[\text{2. Zn, }H_3O^+]{\text{1. }O_3}$ $CH_3CH_2CH_2CH_2\overset{\overset{\displaystyle O}{\|}}{C}H$ + H_2CO
   1-Hexene

c) $CH_3CH_2CH_2CH_2CH=CHCH_2CH_2CH_3$ $\xrightarrow[\text{2. Zn, }H_3O^+]{\text{1. }O_3}$ 2 $CH_3CH_2CH_2CH_2\overset{\overset{\displaystyle O}{\|}}{C}H$
   5-Decene

d) $CH_3CH_2CH_2CH_2\overset{\overset{\displaystyle O}{\|}}{C}OH$ $\xrightarrow[\text{2. PCC}]{\text{1. LiAH}_4}$ $CH_3CH_2CH_2CH_2\overset{\overset{\displaystyle O}{\|}}{C}H$

9.5

a) $CH_3CH_2CH_2CH_2\overset{\overset{\displaystyle OH}{|}}{C}HCH_3$ $\xrightarrow{\text{PCC}}{CH_2Cl_2}$ $CH_3CH_2CH_2CH_2\overset{\overset{\displaystyle O}{\|}}{C}CH_3$
   2-Hexanol   2-Hexanone

b) $CH_3CH_2CH_2CH_2C\equiv CH$ $\xrightarrow[\text{HgSO}_4]{H_3O^+}$ $CH_3CH_2CH_2CH_2\overset{\overset{\displaystyle O}{\|}}{C}CH_3$

c) $CH_3CH_2CH_2CH_2\overset{\overset{\displaystyle CH_3}{|}}{C}=CH_2$ $\xrightarrow[\text{2. Zn, }H_3O^+]{\text{1. }O_3}$ $CH_3CH_2CH_2CH_2\overset{\overset{\displaystyle O}{\|}}{C}CH_3$

9.6

a) $CH_3CH_2CH=CHCH_2CH_3$ $\xrightarrow{H_3O^+}$ $CH_3CH_2CH_2\overset{\overset{\displaystyle OH}{|}}{C}HCH_2CH_3$ $\xrightarrow{\text{PCC}}{CH_2Cl_2}$ $CH_3CH_2CH_2\overset{\overset{\displaystyle O}{\|}}{C}CH_2CH_3$
   3-Hexene   3-Hexanone

b) C_6H_6 + $CH_3\overset{\overset{\displaystyle O}{\|}}{C}Cl$ $\xrightarrow{AlCl_3}$ $C_6H_5\overset{\overset{\displaystyle O}{\|}}{C}CH_3$ $\xrightarrow[\text{2. }H_3O^+]{\text{1. NaBH}_4}$ $C_6H_5\overset{\overset{\displaystyle OH}{|}}{C}HCH_3$
                                                                                                       1-Phenylethanol

9.7

a) $CH_3CH_2CH_2CH_2\overset{\overset{\displaystyle O}{\|}}{C}H$ $\xrightarrow{\text{Tollens' reagent}}$ $CH_3CH_2CH_2CH_2\overset{\overset{\displaystyle O}{\|}}{C}OH$
   Pentanal   Pentanoic acid

144   Chapter 9

b) $CH_3CH_2CH_2CH_2C(CH_3)_2CH{=}O$ $\xrightarrow{\text{Tollens' reagent}}$ $CH_3CH_2CH_2CH_2C(CH_3)_2COOH$

2,2–Dimethylhexanal   2,2–Dimethylhexanoic acid

c) cyclohexanone $\xrightarrow{\text{Tollens' reagent}}$ No reaction

9.8

$CH_3-C(=O)-CH_3$ + :CN$^-$ $\rightleftharpoons$ [$H_3C-C(O^-)(CH_3)(CN)$] $\xrightarrow{H_3O^+}$ $H_3C-C(OH)(CH_3)(CN)$

9.9

$R-C(=O)-R'$ + :H$^-$ $\rightleftharpoons$ [$R-C(O^-)(R')(H)$] $\xrightarrow{H_3O^+}$ $R-C(OH)(R')(H)$ + H_2O

9.10 Propanal should be more reactive than 2,2–dimethylpropanal for two reasons. (1) The two methyl groups in 2,2–dimethylpropanal are electron-donating and decrease the reactivity of the carbonyl carbon toward nucleophiles. (2) These methyl groups also crowd the reaction site and make approach of the nucleophile more difficult.

9.11

$Cl_3CC(=O)-H$ + :OH$_2$ $\rightleftharpoons$ [$Cl_3C-C(O^-)(H)(OH_2^+)$] $\rightleftharpoons$ $Cl_3C-C(OH)(H)(OH)$

Chloral hydrate

9.12

$R-C(=O)-R$ + :*OH$_2$ $\rightleftharpoons$ [$R-C(O^-)(R)(*OH_2^+)$] $\rightleftharpoons$ $R-C(OH)(R)(*OH)$

The above mechanism is similar to other nucleophilic addition mechanisms we have written. Since all of the steps are reversible, we can write the reaction in reverse to show how labeled oxygen is incorporated into an aldehyde or ketone.

Aldehydes and Ketones 145

$$R\cdots\overset{OH}{\underset{\underset{*}{R}}{\overset{|}{C}}}-OH \rightleftharpoons \left[R\cdots\overset{\overset{+}{O}H_2}{\underset{R}{\overset{|}{C}}}\cdots\overset{\cdot\cdot}{\underset{*}{O}}{:}^{-} \right] \rightleftharpoons R\overset{\overset{*}{\overset{\cdot\cdot}{O}}}{\underset{}{\overset{\|}{C}}}R + H_2O$$

This exchange is very slow in water but proceeds more rapidly when either acid or base is present.

9.13

cyclohexanone =O $\xrightarrow{\text{CH}_3\text{CH}_2\text{OH}}_{\text{H}^+ \text{ catalyst}}$ cyclohexane with two OCH$_2$CH$_3$ groups

9.14

$$\underset{H_3C}{\overset{O}{\overset{\|}{C}}}\underset{CH_3}{} + \underset{HO}{\overset{H_2C-CH_2}{}}\underset{OH}{} \underset{}{\overset{H^+}{\rightleftharpoons}} \underset{H_3C}{\overset{OH}{\underset{CH_3}{\overset{|}{C}}}}-OCH_2CH_2OH$$

hemiacetal

$$\overset{H^+}{\rightleftharpoons} \quad \text{H}_3\text{C}\underset{CH_3}{\overset{}{\overset{}{\diagdown}}}\overset{O-CH_2}{\underset{O}{\overset{}{\diagup}}}\overset{}{\underset{}{\diagdown}}CH_2 \; + \; H_2O$$

acetal

9.15

$$\overset{O}{\overset{\|}{H}C}CH_2CH_2\overset{O}{\overset{\|}{C}}CH_3 \xrightarrow[\text{H}^+ \text{ catalyst}]{\text{HOCH}_2\text{CH}_2\text{OH}} \overset{\overset{O\diagup O}{\diagdown \diagup}}{HCCH_2CH_2}\overset{O}{\overset{\|}{C}}CH_3$$

$$\Big\downarrow \begin{array}{l} 1. \; \text{LiAlH}_4 \\ 2. \; \text{H}_3\text{O}^+ \end{array}$$

$$\overset{O}{\overset{\|}{H}C}CH_2CH_2CH_2OH \xleftarrow{\text{H}_3\text{O}^+} \overset{\overset{O\diagup O}{\diagdown \diagup}}{HCCH_2CH_2CH_2OH}$$

9.16 In parts (a) - (c), draw the reagent next to the ketone carbon of cyclohexanone. Remove oxygen from cyclohexanone and two hydrogens from the reagent, and draw the new bonds.

a) cyclohexanone=O + H$_2$NOH $\longrightarrow$ cyclohexane=NOH + H$_2$O

b)

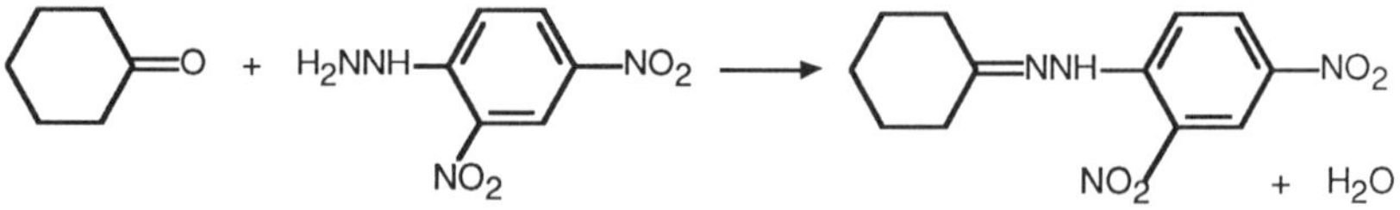

cyclohexanone + H₂NNH—C₆H₃(NO₂)₂ → cyclohexylidene=NNH—C₆H₃(NO₂)₂ + H₂O

c)

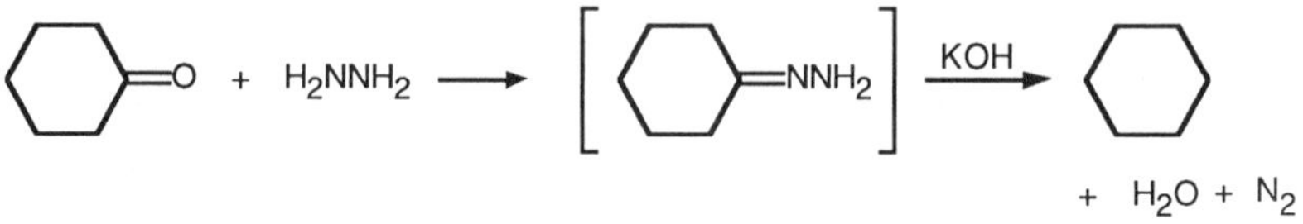

cyclohexanone + H₂NNH₂ → [cyclohexylidene=NNH₂] —KOH→ cyclohexane + H₂O + N₂

d)

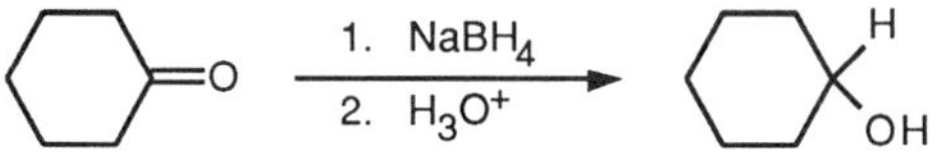

cyclohexanone —1. NaBH₄; 2. H₃O⁺→ cyclohexanol

9.17

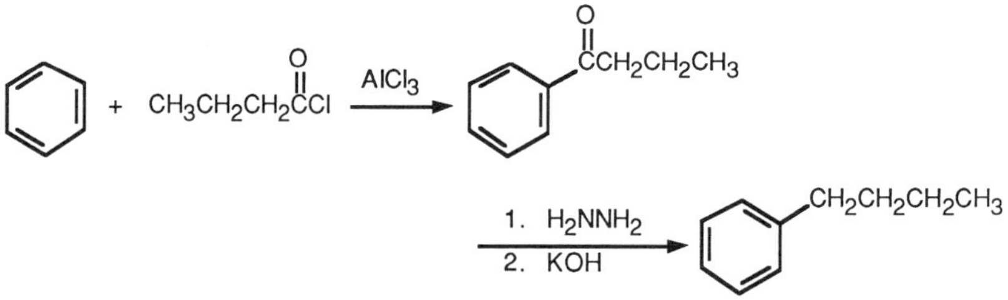

benzene + CH₃CH₂CH₂COCl —AlCl₃→ PhCOCH₂CH₂CH₃ —1. H₂NNH₂; 2. KOH→ PhCH₂CH₂CH₂CH₃

9.18

a) cyclopentanone —1. CH₃MgBr; 2. H₃O⁺→ 1-methylcyclopentanol

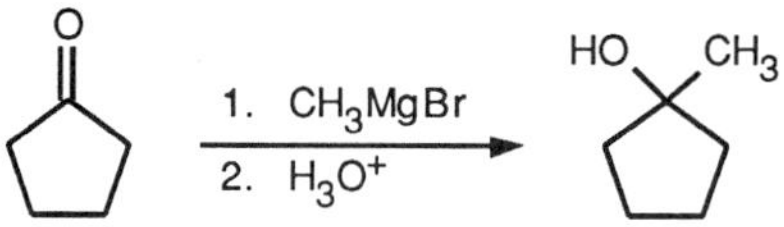

b) benzophenone —1. CH₃MgBr; 2. H₃O⁺→ 1,1-diphenyl-1-methylmethanol (Ph₂C(OH)CH₃)

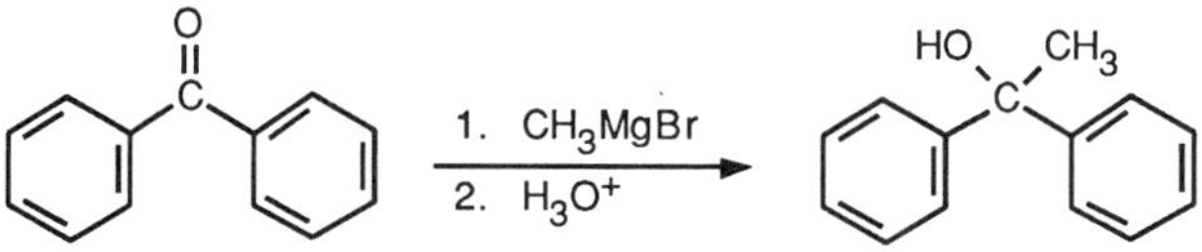

c) CH₃CH₂CH₂COCH₂CH₃ —1. CH₃MgBr; 2. H₃O⁺→ CH₃CH₂CH₂C(CH₃)(OH)CH₂CH₃

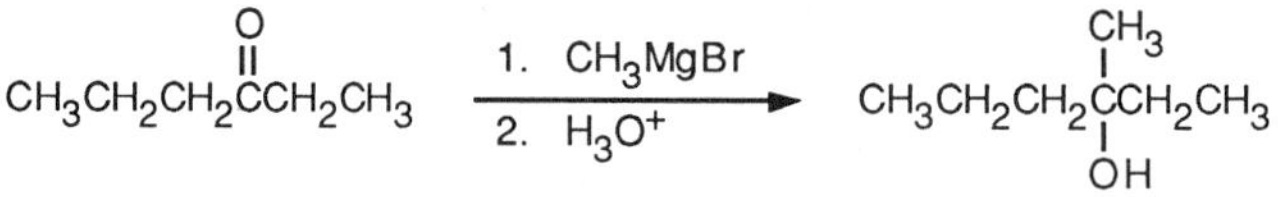

9.19 When choosing the starting materials for a Grignard reaction, two reminders may be helpful.

    1. Ketone + Grignard reagent ⟶ tertiary alcohol

       Aldehyde + Grignard reagent ⟶ secondary alcohol

       Formaldehyde + Grignard reagent ⟶ primary alcohol

    2. More than one combination of carbonyl compound plus Grignard reagent may yield the same product.

Aldehydes and Ketones 147

	Carbonyl Compound	Grignard Reagent	Product
a.	CH$_3$CCH$_3$ (with C=O)	CH$_3$MgBr	CH$_3$C(OH)(CH$_3$)CH$_3$
b.	cyclohexanone	CH$_3$MgBr	1-methylcyclohexanol (HO, CH$_3$)
c.	CH$_3$CH$_2$CCH$_2$CH$_3$ (with C=O)	CH$_3$MgBr	CH$_3$CH$_2$C(OH)(CH$_3$)CH$_2$CH$_3$
	CH$_3$CH$_2$CCH$_3$ (with C=O)	CH$_3$CH$_2$MgBr	

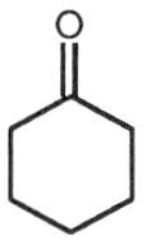

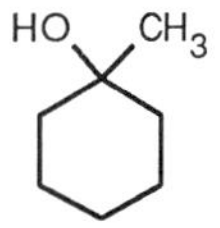

9.20

a) 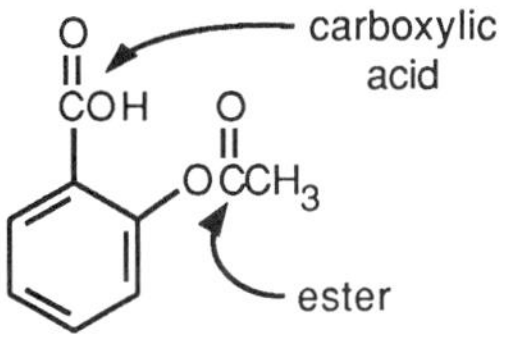 Aspirin — COH (carboxylic acid); OCCH$_3$ (ester)

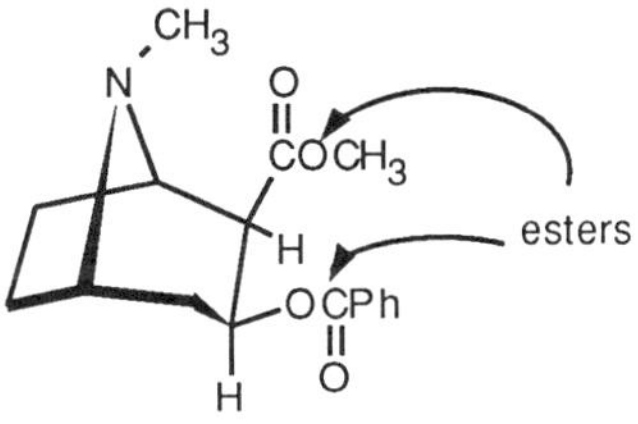

b) Cocaine — COCH$_3$ (esters); OCPh

c) 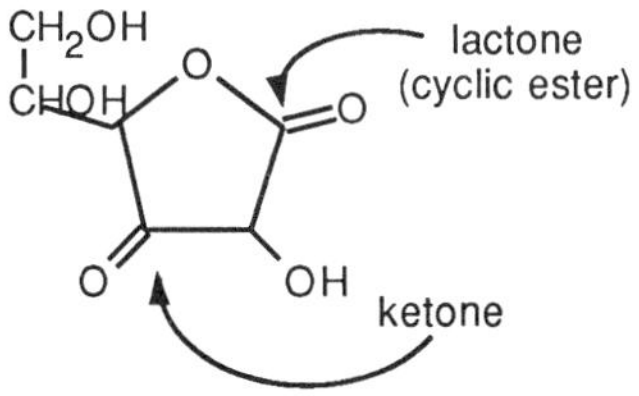 Ascorbic acid — lactone (cyclic ester); ketone

9.21 The carbonyl carbon of an aldehyde is bonded to carbon and hydrogen; the carbonyl carbon of a ketone is bonded to two carbons.

9.22

a) $CH_3\overset{O}{\underset{\|}{C}}CH_2Br$

b) $CH_3\overset{O}{\underset{\|}{C}}H\overset{O}{\underset{\|}{C}}CH_3$
    $|$
    CH_3

c) 1-CHO, 3,5-dinitro substituted benzene (CHO with O_2N and NO_2 groups on ring)

d) cyclohexanone with two methyl groups (H_3C and CH_3) on the ring

e) $(CH_3)_3C\overset{O}{\underset{\|}{C}}C(CH_3)_3$

f) $H\overset{O}{\underset{\|}{C}}CH_2CH_2\overset{O}{\underset{\|}{C}}H$

g) structure with C=O, H, S substituents, and H—C(CH_3)—OH

h) Ph—CH=CH—CHO

9.23

$CH_3CH_2CH_2CH_2\overset{O}{\underset{\|}{C}}H$
Pentanal

$CH_3CH_2\overset{O}{\underset{\|}{C}}H$
$|$
CH_3
2-Methylbutanal

$CH_3\overset{CH_3}{\underset{|}{C}}HCH_2\overset{O}{\underset{\|}{C}}H$
3-Methylbutanal

$(CH_3)_3\overset{O}{\underset{\|}{C}}CH$
2,2-Dimethylpropanal

$CH_3CH_2CH_2\overset{O}{\underset{\|}{C}}CH_3$
2-Pentanone

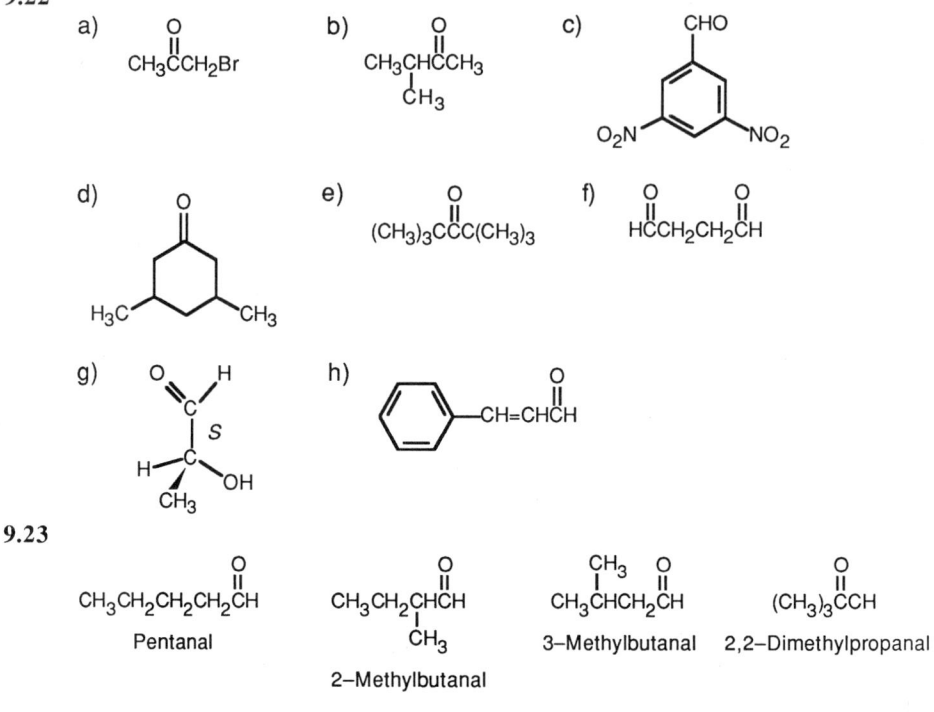

$CH_3CH_2\overset{O}{\underset{\|}{C}}CH_2CH_3$
3-Pentanone

$CH_3\overset{O}{\underset{\|}{C}}H\overset{O}{\underset{\|}{C}}CH_3$
$|$
CH_3
3-Methyl-2-butanone

9.24
Of the compounds shown in Problem 9.23, only 2-methylbutanal is chiral.

9.25

a) cyclohexenone

b) $CH_3CH_2\overset{O}{\underset{\|}{C}}CH_2\overset{O}{\underset{\|}{C}}CH_3$

c) 4-methylacetophenone (H_3C—C$_6H_4$—$\overset{O}{\underset{\|}{C}}CH_3$)

d) $CH_3CH_2CH_2\overset{O}{\underset{\|}{C}}H$
   $|$
   Br

9.26

a) 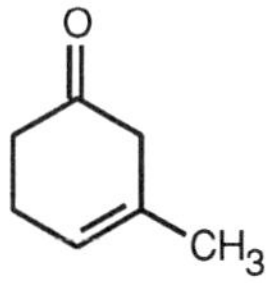 3-Methyl-3-cyclohexenone

b) 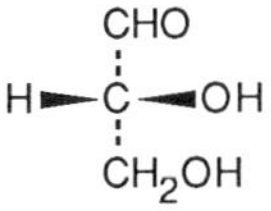 (2R)-2,3-Dihydroxypropanal

$$\begin{array}{c} CHO \\ H-C-OH \\ CH_2OH \end{array}$$

c) 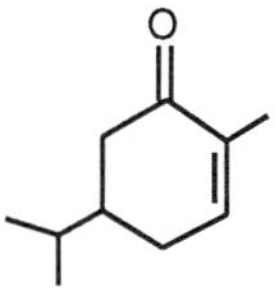 5-Isopropyl-2-methyl-2-cyclohexenone

d) $CH_3CH(CH_3)COCH_2CH_3$
2-Methyl-3-pentanone

e) $CH_3CH(OH)CH_2CHO$
3-Hydroxybutanal

f) 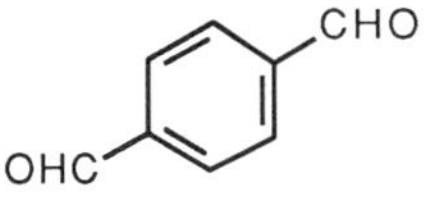 p-Benzenedicarbaldehyde (OHC–C$_6$H$_4$–CHO)

9.27

a) 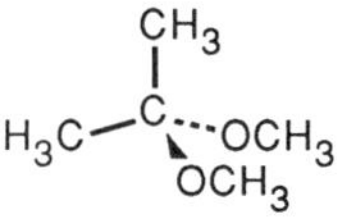 Acetal
$$H_3C-\underset{OCH_3}{\overset{CH_3}{C}}-OCH_3$$

b) 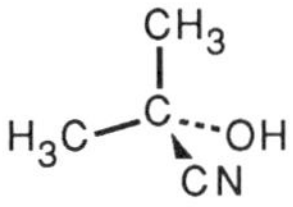 Cyanohydrin
$$H_3C-\underset{CN}{\overset{CH_3}{C}}-OH$$

c) 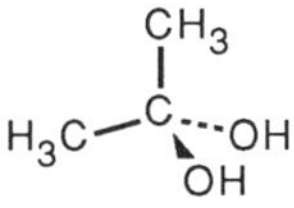 Gem diol
$$H_3C-\underset{OH}{\overset{CH_3}{C}}-OH$$

d) 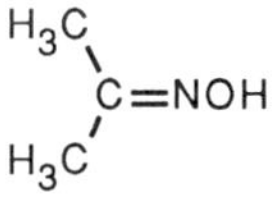 Oxime
$$\underset{H_3C}{\overset{H_3C}{>}}C=NOH$$

e) 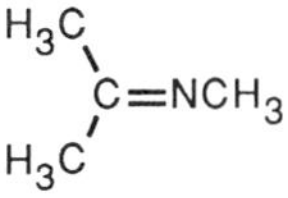 Imine
$$\underset{H_3C}{\overset{H_3C}{>}}C=NCH_3$$

f) 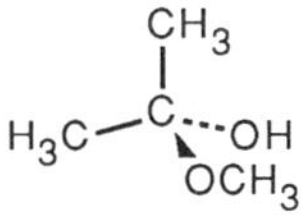 Hemiacetal
$$H_3C-\underset{OCH_3}{\overset{CH_3}{C}}-OH$$

9.28

a) 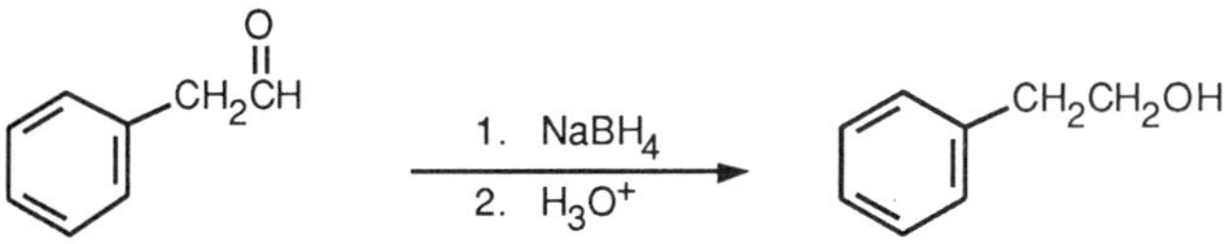 Phenylacetaldehyde (PhCH$_2$CHO) $\xrightarrow{\text{1. NaBH}_4 \;\; 2. \text{H}_3\text{O}^+}$ PhCH$_2$CH$_2$OH

b)  PhCH$_2$CHO $\xrightarrow{\text{Tollens' reagent}}$ PhCH$_2$COOH

150   Chapter 9

c) PhCH$_2$CHO + :NH$_2$OH → PhCH$_2$CH=N-OH

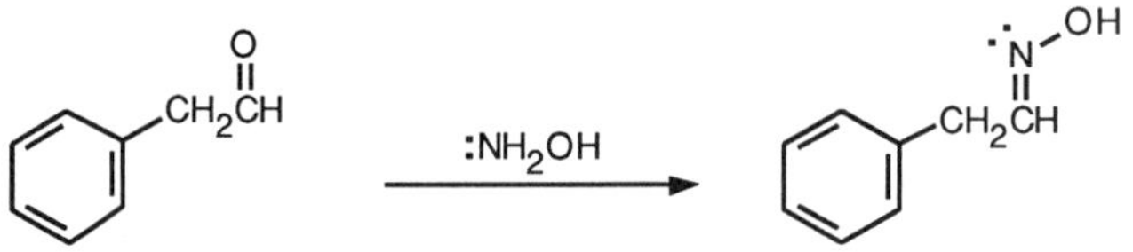

d) PhCH$_2$CHO + 1. CH$_3$MgBr, 2. H$_3$O$^+$ → PhCH$_2$CH(OH)CH$_3$

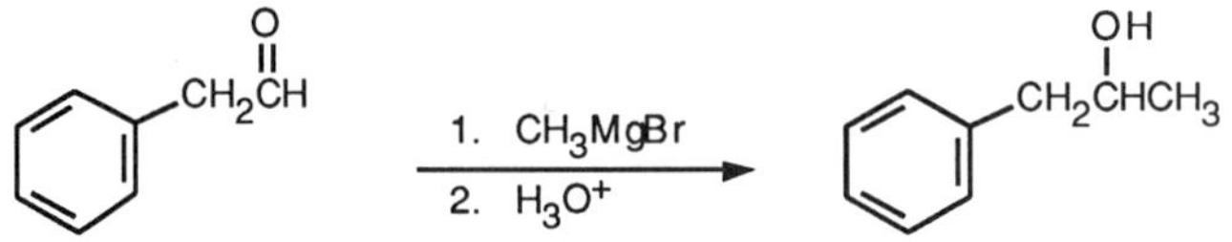

e) PhCH$_2$CHO + CH$_3$OH, H$^+$ catalyst → PhCH$_2$CH(OCH$_3$)$_2$

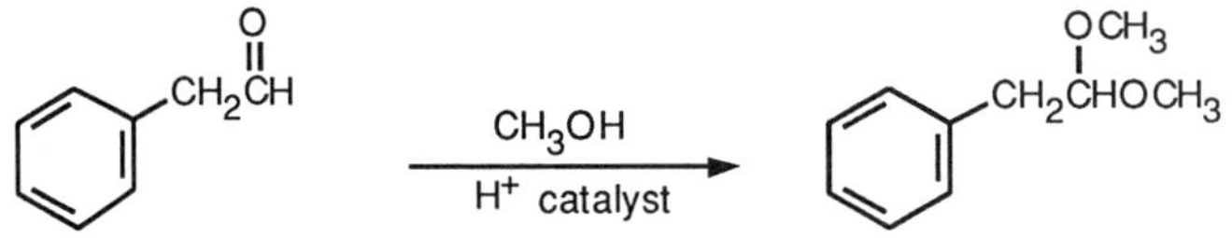

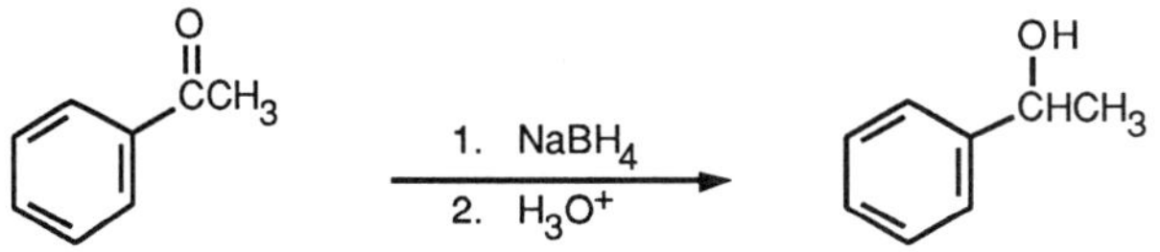

f) PhCH$_2$CHO + H$_2$NNH$_2$, KOH → PhCH$_2$CH$_3$

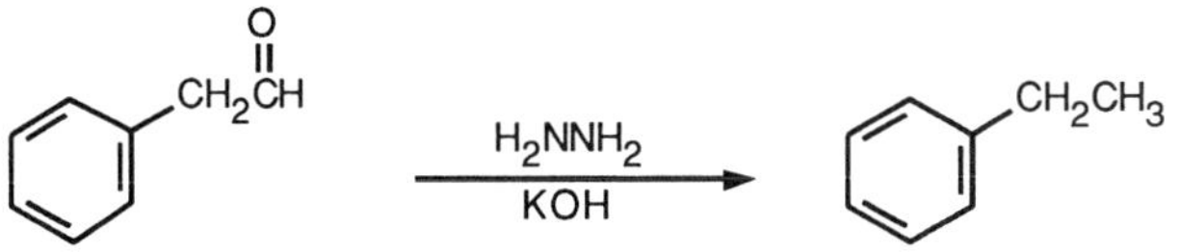

9.29

a) PhCOCH$_3$ (Acetophenone) + 1. NaBH$_4$, 2. H$_3$O$^+$ → PhCH(OH)CH$_3$

b) PhCOCH$_3$ + Tollens' reagent → no reaction

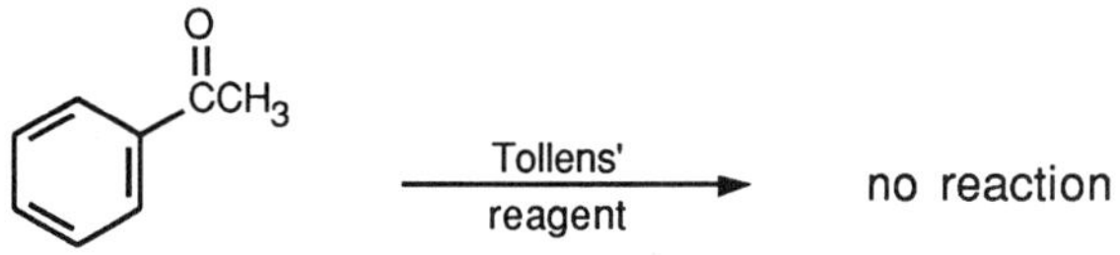

c) PhCOCH$_3$ + :NH$_2$OH → PhC(CH$_3$)=N-OH

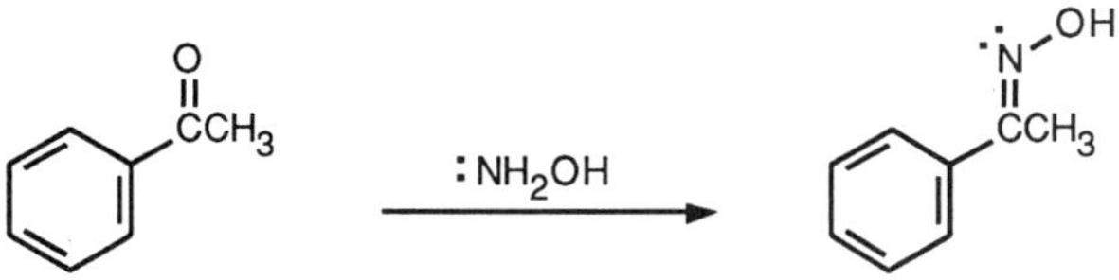

d) PhCOCH$_3$ + 1. CH$_3$MgBr, 2. H$_3$O$^+$ → PhC(OH)(CH$_3$)C(CH$_3$)$_2$... [PhC(OH)(CH$_3$)$_2$]

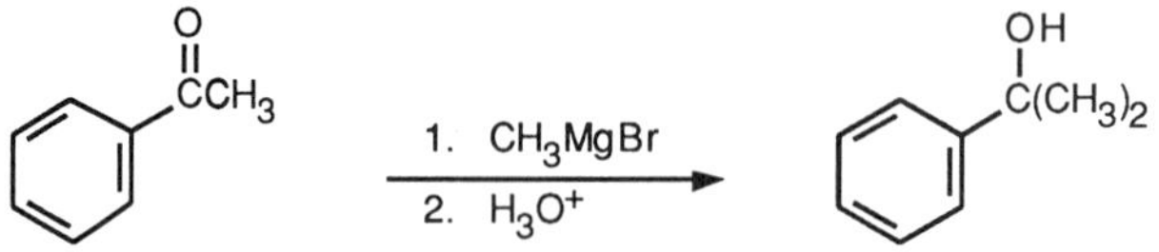

Aldehydes and Ketones 151

e) [PhC(O)CH₃] + CH₃OH, H⁺ catalyst → PhC(OCH₃)₂CH₃

f) [PhC(O)CH₃] + H₂NNH₂, KOH → PhCH₂CH₃

9.30

$$\text{CH}_3\text{-C(=O)-CH}_2\text{CH}_3 + :CN^- \rightleftharpoons \left[\begin{array}{c}\text{CN}\\ \text{CH}_3\text{-C-O}^-\\ \text{CH}_2\text{CH}_3\end{array}\right] \xrightarrow{H_3O^+} \begin{array}{c}\text{CN}\\ \text{CH}_3\text{-}\overset{S}{C}\text{-OH}\\ \text{CH}_2\text{CH}_3\end{array} + H_2O$$

$$\text{CH}_3\text{-C(=O)-CH}_2\text{CH}_3 + :CN^- \rightleftharpoons \left[\begin{array}{c}\text{CH}_3\\ \text{CH}_3\text{CH}_2\text{-C-O}^-\\ \text{CN}\end{array}\right] \xrightarrow{H_3O^+} \begin{array}{c}\text{CH}_3\\ \text{CH}_3\text{CH}_2\text{-}\overset{R}{C}\text{-OH}\\ \text{CN}\end{array} + H_2O$$

Attack occurs from both sides of the planar carbonyl group to yield a racemic product mixture.

9.31

$$\text{CH}_3\text{-C(=O)-CH}_2\text{CH}_3 + \text{BrMg}^+ {}^-C_6H_5 \rightarrow \left[\begin{array}{c}C_6H_5\\ \text{CH}_3\text{-C-O}^-\\ \text{CH}_2\text{CH}_3\end{array}\right] \xrightarrow{H_3O^+} \begin{array}{c}C_6H_5\\ \text{CH}_3\text{-}\overset{S}{C}\text{-OH}\\ \text{CH}_2\text{CH}_3\end{array}$$

$$\text{CH}_3\text{-C(=O)-CH}_2\text{CH}_3 + \text{BrMg}^+ {}^-C_6H_5 \rightarrow \left[\begin{array}{c}\text{CH}_3\\ \text{CH}_3\text{CH}_2\text{-C-O}^-\\ C_6H_5\end{array}\right] \xrightarrow{H_3O^+} \begin{array}{c}\text{CH}_3\\ \text{CH}_3\text{CH}_2\text{-}\overset{R}{C}\text{-OH}\\ C_6H_5\end{array}$$

The product is a racemic mixture of R and S enantiomers.

9.32

a) cyclohex-2-enone → (H$_2$NNH$_2$ / KOH) → cyclohexene

b) cyclohex-2-enone → (1. CH$_3$MgBr; 2. H$_3$O$^+$) → 1-methylcyclohex-2-en-1-ol → (H$_2$, Pd catalyst) → 1-methylcyclohexan-1-ol

c) cyclohex-2-enone → (1. NaBH$_4$; 2. H$_3$O$^+$) → cyclohex-2-en-1-ol → (H$_2$, Pd catalyst) → cyclohexanol

d) cyclohex-2-enone → (1. C$_6$H$_5$MgBr; 2. H$_3$O$^+$) → 1-phenylcyclohex-2-en-1-ol

9.33

HO:$^-$ attacks C$_6$H$_5$CHBr$_2$ (SN2) → HO–CH(C$_6$H$_5$)–Br + :Br:$^-$

HO:$^-$ attacks HO–CH(C$_6$H$_5$)–Br (SN2) → [HO–CH(OH)–C$_6$H$_5$ + :Br:$^-$] (gem diol) ⇌ O=CH–C$_6$H$_5$ + H$_2$O (Benzaldehyde)

The initial product of two successive S$_N$2 reactions of C$_6$H$_5$CHBr$_2$ with hydroxide ion is a gem diol. Since the equilibrium between the gem diol and aldehyde favors the aldehyde, benzaldehyde is the observed product.

9.34

	Carbonyl Compound	Grignard Reagent	Product
a.	CH₃CHO	CH₃CH₂CH₂MgBr	CH₃CH₂CH₂CH(OH)CH₃
	CH₃CH₂CH₂CHO	CH₃MgBr	
b.	CH₃CH₂COCH₃	C₆H₅MgBr	CH₃CH₂C(OH)(CH₃)(C₆H₅)
	CH₃CH₂CO-C₆H₅	CH₃MgBr	
	CH₃CO-C₆H₅	CH₃CH₂MgBr	
c.	cyclohexanone	CH₃CH₂MgBr	1-ethylcyclohexan-1-ol (HO, CH₂CH₃)
d.	C₆H₅CHO (benzaldehyde)	C₆H₅MgBr	diphenylmethanol (OH, CH with two phenyls)

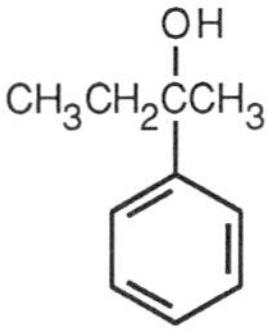

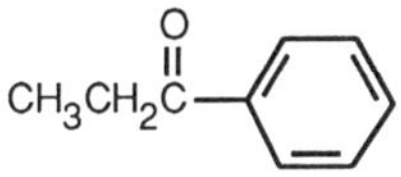

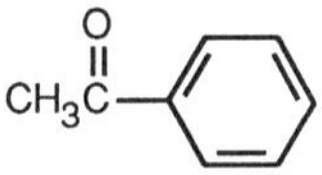

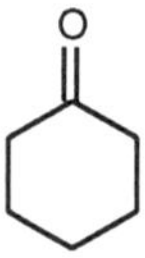

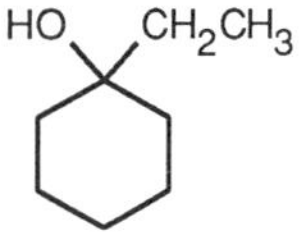

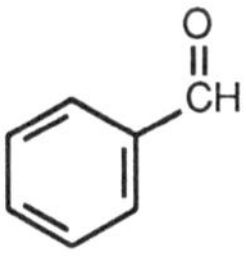

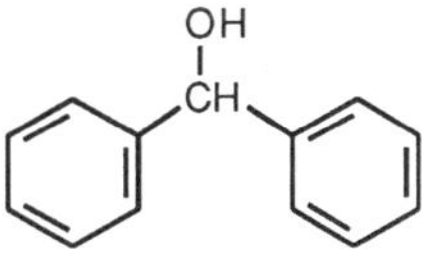

9.35

a) CH₂O $\xrightarrow{\text{1. C}_6\text{H}_5\text{MgBr} \quad \text{2. H}_3\text{O}^+}$ C₆H₅CH₂OH

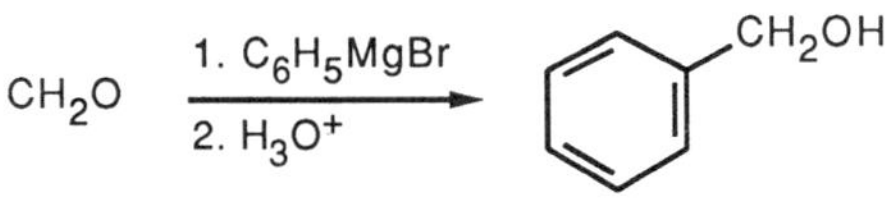

Benzyl alcohol

b)

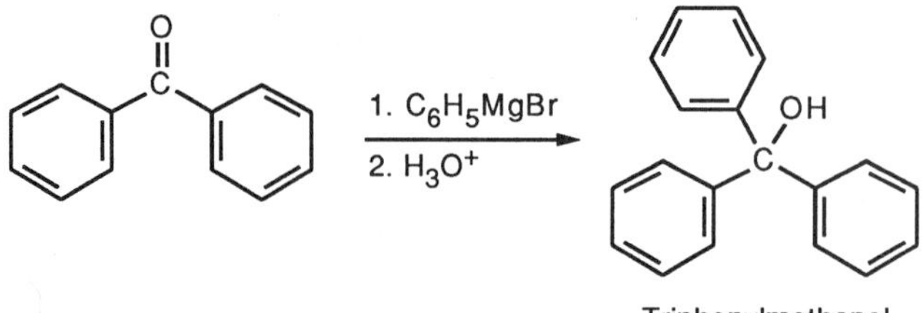

[Benzophenone] $\xrightarrow{\text{1. C}_6\text{H}_5\text{MgBr}}_{\text{2. H}_3\text{O}^+}$ Triphenylmethanol

c)

$\text{CH}_3\text{CH}_2\overset{\overset{\text{O}}{\|}}{\text{C}}\text{CH}_2\text{CH}_3 \xrightarrow{\text{1. C}_6\text{H}_5\text{MgBr}}_{\text{2. H}_3\text{O}^+}$ 3-Phenyl-3-pentanol

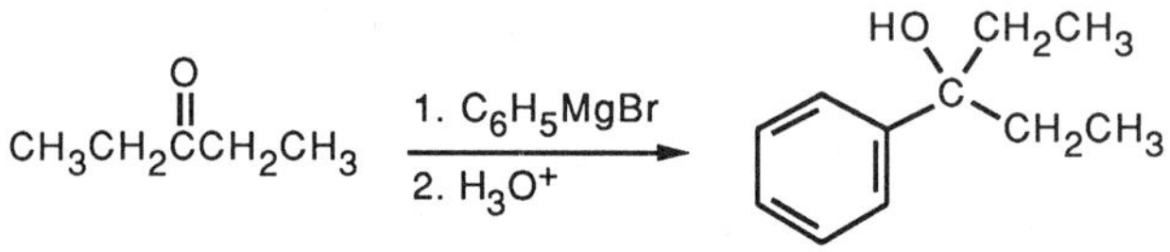

9.36

	Carbonyl compound	Grignard reagent	Product
a)	CH_2O	$\text{CH}_3\text{CHCH}_2\text{CH}_2\text{CH}_2\text{MgBr}$ with CH₃ branch	$\text{CH}_3\text{CHCH}_2\text{CH}_2\text{CH}_2\text{OH}$ with CH₃ branch
b)	cyclohexanecarbaldehyde	CH_3MgBr	cyclohexyl-CH(OH)CH₃
	$\text{CH}_3\overset{\overset{\text{O}}{\|}}{\text{C}}\text{H}$	cyclohexyl-MgBr	
c)	$\text{CH}_3\text{CH}_2\overset{\overset{\text{O}}{\|}}{\text{C}}\text{H}$	$\text{CH}_3\text{CH=CHMgBr}$	$\text{CH}_3\text{CH}_2\overset{\overset{\text{OH}}{|}}{\text{C}}\text{HCH=CHCH}_3$
	$\text{CH}_3\text{CH=CH}\overset{\overset{\text{O}}{\|}}{\text{C}}\text{H}$	$\text{CH}_3\text{CH}_2\text{MgBr}$	

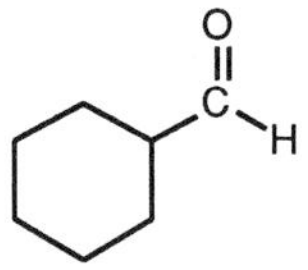

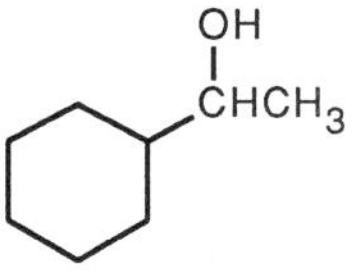

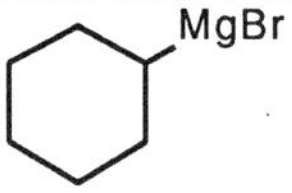

9.37

PhBr $\xrightarrow[\text{3. H}_3\text{O}^+]{\text{1. Mg; 2. CH}_2\text{O}}$ PhCH$_2$OH $\xrightarrow[\text{H}_3\text{O}^+]{\text{CrO}_3}$ PhCOOH

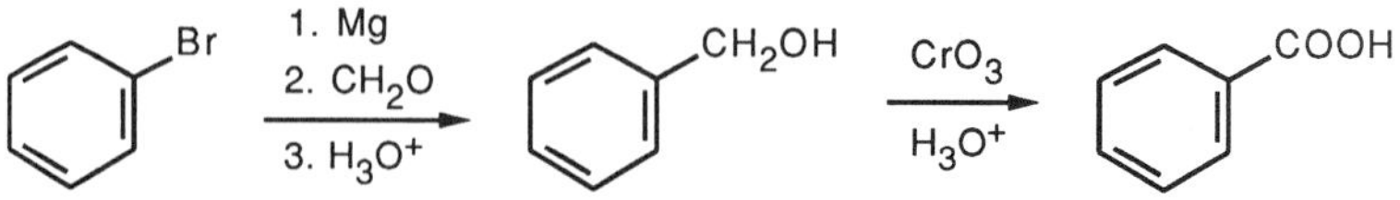

9.38

Carbonyl compound	Alcohol	Hemiacetal	Acetal
a) C₆H₅−C(=O)−CH₃	CH₃CH(OH)CH₃	C₆H₅−C(OH)(OCH(CH₃)₂)−CH₃	C₆H₅−C(OCH(CH₃)₂)₂−CH₃ with (CH₃)₂CHO and OCH(CH₃)₂
b) CH₃CH₂−C(=O)−CH₂CH₃	cyclopentanol	CH₃CH₂−C(OH)(O-cyclopentyl)−CH₂CH₃	CH₃CH₂−C(O-cyclopentyl)₂−CH₂CH₃

9.39

a) CH₃CH₂CH(CH₃)CH₂CHO + 2 CH₃OH $\xrightarrow{\text{H}^+ \text{ catalyst}}$ CH₃CH₂CH(CH₃)CH₂CH(OCH₃)₂

b) C₆H₅−C(=O)−CH₃ + 2 CH₃CH₂OH $\xrightarrow{\text{H}^+ \text{ catalyst}}$ C₆H₅−C(OCH₂CH₃)₂−CH₃

c) cyclopentanone + HOCH₂CH₂OH $\xrightarrow{\text{H}^+ \text{ catalyst}}$ cyclopentanone ethylene ketal

9.40

a) CH₃CH₂CH₂−C(=O)−CH₃ + NH₂OH ⟶ CH₃CH₂CH₂−C(=NOH)−CH₃

b) CH₃CH₂CH₂−C(=O)−CH₃ + 2,4-dinitrophenylhydrazine ⟶ 2,4-dinitrophenylhydrazone of 2-pentanone

c) CH₃CH₂CH₂−C(=O)−CH₃ + H₂NNH₂ $\xrightarrow{\text{KOH}}$ CH₃CH₂CH₂CH₂CH₃ + N₂

9.41

a) Cyclohexanone → (1. CH₃MgBr, 2. H₃O⁺) → 1-methylcyclohexanol (HO, CH₃) → (H₃O⁺) → 1-methylcyclohexene

b) Cyclohexanone → (1. NaBH₄, 2. H₃O⁺) → cyclohexanol (H, OH) → (H₃O⁺) → cyclohexene → (KMnO₄, H₂O, NaOH) → trans-1,2-cyclohexanediol (HO, OH)

c) Cyclohexanone → (1. CH₃MgBr, 2. H₃O⁺) → 1-methylcyclohexanol (HO, CH₃) → (PBr₃) → 1-bromo-1-methylcyclohexane (Br, CH₃)

d) Cyclohexanone → (1. NaBH₄, 2. H₃O⁺) → cyclohexanol (HO, H) → (PBr₃) → bromocyclohexane (Br, H) → (Mg, ether) → cyclohexylmagnesium bromide (BrMg, H)

cyclohexanone + cyclohexylMgBr → 1-cyclohexylcyclohexanol (HO)

9.42 The first equivalent of CH₃MgBr reacts with the hydroxyl hydrogen of 4–hydroxycyclohexanone.

4-hydroxycyclohexanone (O, OH) →(CH₃MgBr)→ 4-(BrMgO)cyclohexanone + CH₄

The second equivalent of CH₃MgBr adds to the ketone in the expected manner to yield 1–methyl-1,4–cyclohexanediol.

BrMgO–cyclohexanone (=O) →(1. CH₃MgBr, 2. H₃O⁺)→ HO–cyclohexane–(CH₃)(OH)

9.43

a) through d): [Reactions of carvone shown]

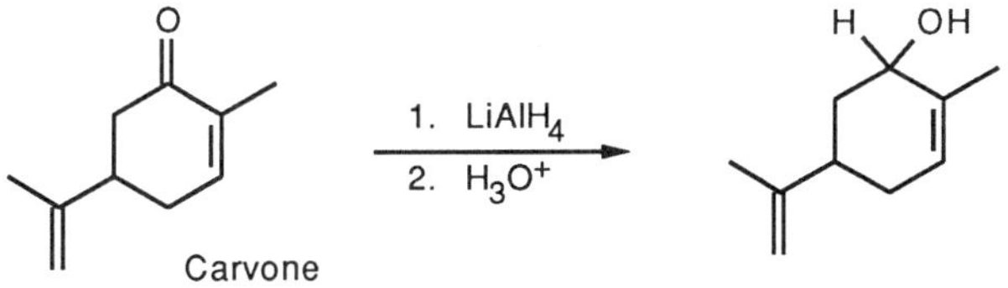

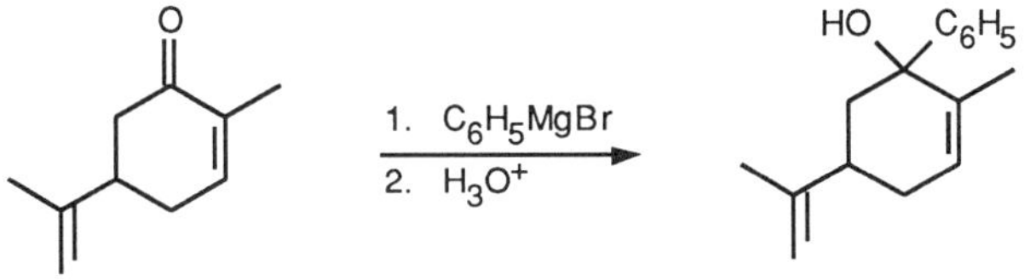

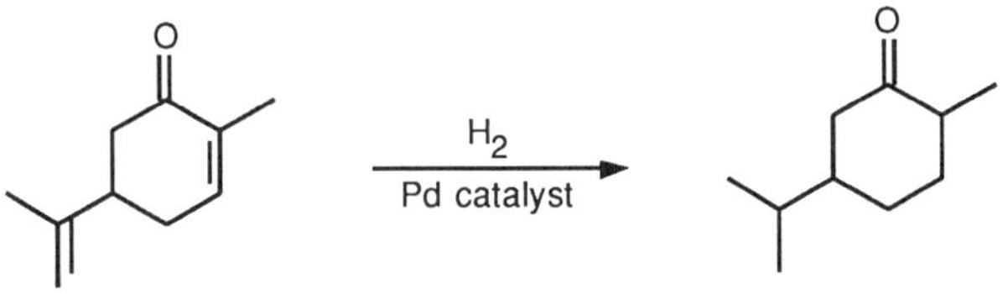

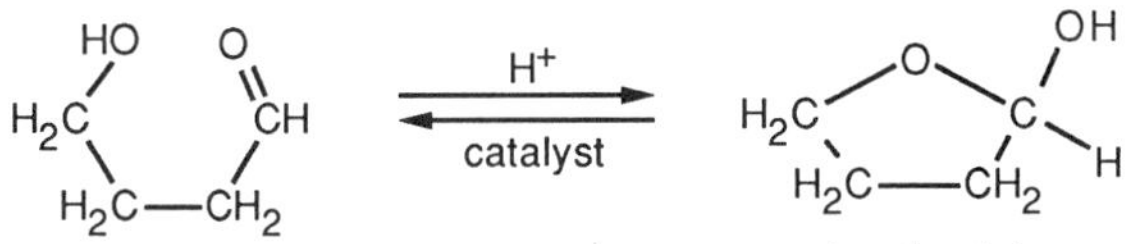

a) Carvone + 1. LiAlH₄, 2. H₃O⁺ → allylic alcohol (H, OH added to carbonyl)

b) Carvone + 1. C₆H₅MgBr, 2. H₃O⁺ → tertiary allylic alcohol (HO, C₆H₅ added to carbonyl)

c) Carvone + H₂, Pd catalyst → fully saturated ketone (isopropyl cyclohexanone derivative)

d) Carvone + CH₃OH, H⁺ catalyst → dimethyl ketal (CH₃O, OCH₃ at former carbonyl carbon)

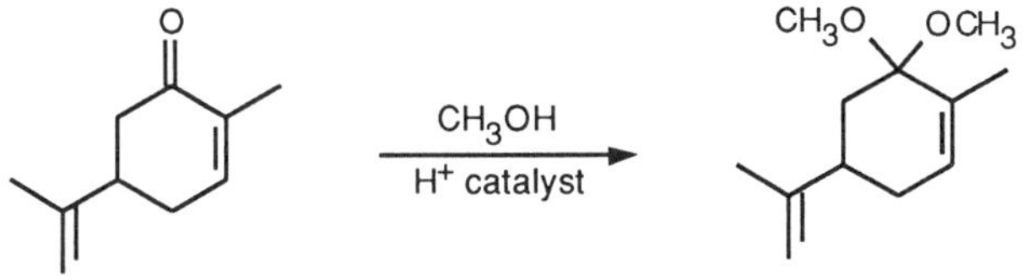

9.44

a) 4-hydroxybutanal ⇌ (H⁺ catalyst) cyclic hemiacetal

b) cyclic hemiacetal + CH₃OH ⇌ (H⁺ catalyst) 2-methoxytetrahydrofuran + H₂O

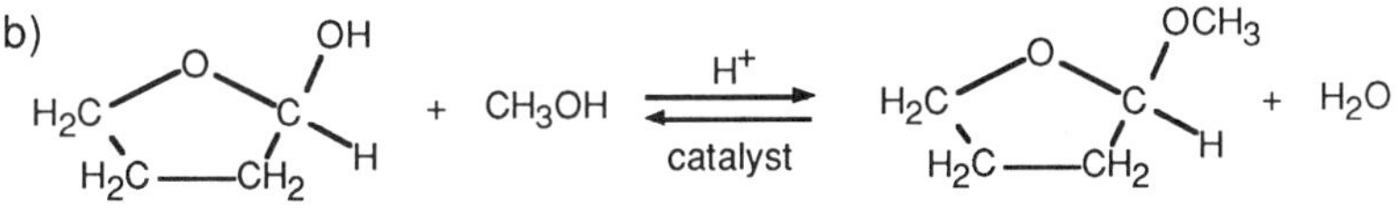

2–Methoxytetrahydrofuran is a cyclic acetal. The hydroxyl oxygen of 4–hydroxybutanal reacts with the aldehyde group to form the cyclic ether linkage.

9.45 In general, ketones are less reactive than aldehydes for both steric (excess crowding) and electronic reasons. If the keto aldehyde in this problem were reduced with *one* equivalent of NaBH₄, the aldehyde functional group would be reduced in preference to the ketone.

For the same reason, reaction of the ketoaldehyde with *one* equivalent of ethylene glycol selectively forms the acetal of the aldehyde functional group. The ketone is then reduced with NaBH₄, and the acetal protecting group is cleaved.

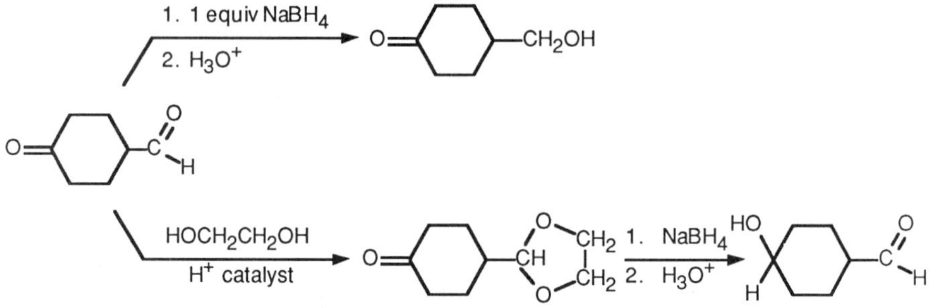

158   Chapter 9

9.46

a) [ketone] + R'SH ⇌ (H⁺ catalyst) hemithioacetal

b) hemithioacetal + R'SH ⇌ (H⁺ catalyst) thioacetal + H₂O

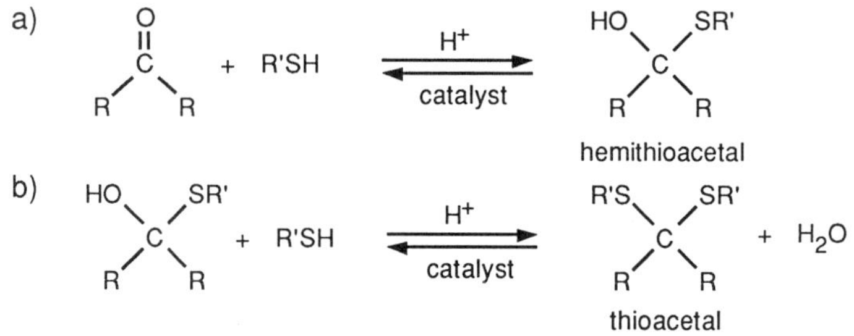

The reaction is analogous to acetal formation that occurs on treatment of a ketone or aldehyde with an alcohol and an acid catalyst.

9.47

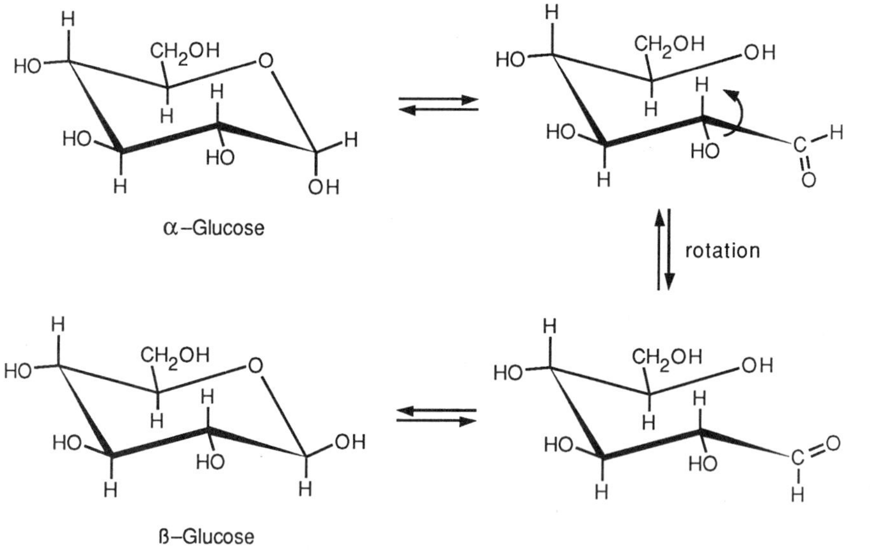

α-Glucose ⇌ [open-chain aldehyde] ⇅ rotation

β-Glucose ⇌ [open-chain aldehyde rotated]

In this series of equilibrium steps, the hemiacetal ring of α–glucose opens to yield the aldehyde. Rotation of the aldehyde group is followed by formation of the cyclic hemiacetal of ß–glucose. The reaction is catalyzed by both acid and base.

9.48

[Mechanism scheme: cyclohexanone + ⁻:CH₂–S⁺(CH₃)₂ → tetrahedral alkoxide intermediate with CH₂–S⁺(CH₃)₂ group → epoxide (1-oxaspiro cyclohexane with CH₂) + :S(CH₃)₂]

Nucleophilic addition to the carbonyl group . . .

. . . is followed by S_N2 displacement of dimethyl sulfide by oxygen.

Chapter Outline

I. Introduction to carbonyl chemistry (Sections 9.1 - 9.3)
   A. Kinds of carbonyl compounds (Section 9.1)
   B. Structure and properties of the carbonyl group (Section 9.2)
   C. Naming aldehydes and ketones (Section 9.3)

II. Preparation of ketones and aldehydes (Sections 9.4 - 9.5)
   A. Aldehydes (Section 9.4)
      1. Oxidation of primary alcohols
      2. Oxidative cleavage of alkenes
   B. Ketones (Section 9.5)
      1. Oxidation of secondary alcohols
      2. Ozonolysis of alkenes
      3. Hydration of terminal alkynes
      4. Friedel-Crafts acylation of arenes

III. Reactions of ketones and aldehydes (Sections 9.6 - 9.13)
   A. Oxidation of aldehydes (Section 9.6)
   B. Nucleophilic addition reactions (Sections 9.7 - 9.12)
      1. Relative reactivity of ketones and aldehydes (Section 9.7)
      2. Reactions (Sections 9.8 - 9.11)
         a. Addition of water - hydration (Section 9.8)
         b. Addition of alcohols - acetal formation (Section 9.9)
         c. Addition of amines - reduction to alkanes (Section 9.10)
         d. Addition of Grignard reagents - alcohol formation (Section 9.11)
      3. Biological nucleophilic reactions (Section 9.12)

Study Skills for Chapter 9

After studying this chapter, you should be able to:

1. Identify carbonyl-containing functional groups (Problems 9.1, 9.20, 9.25).
2. Name aldehydes and ketones (Problems 9.2, 9.23, 9.26).
3. Draw structures of aldehydes and ketones corresponding to give names (Problems 9.3, 9.22, 9.23).
4. Synthesize aldehydes and ketones (Problems 9.4, 9.5, 9.6)
5. Predict the product of reactions of aldehydes and ketones (Problems 9.7, 9.8, 9.11, 9.13, 9.15, 9.16, 9.17, 9.18, 9.19, 9.28, 9.29, 9.30, 9.32, 9.34, 9.35, 9.36, 9.37, 9.38, 9.39, 9.40, 9.41, 9.43, 9.45, 9.46).
6. Formulate the mechanisms of nucleophilic addition reactions (Problems 9.9, 9.12, 9.14, 9.30, 9.31, 9.33, 9.42, 9.44, 9.46, 9.47, 9.48).
7. Compare the relative reactivities of aldehydes and ketones (Problems 9.10, 9.21).

Chapter 10 – Carboxylic Acids and Derivatives

10.1

a) (CH₃)₂CHCH₂COOH
   3-Methylbutanoic acid

b) CH₃CHBrCH₂CH₂COOH
   4-Bromopentanoic acid

c) CH₃CH=CHCH₂CH₂COOH
   4-Hexenoic acid

d) CH₃CH₂CH(COOH)CH₂CH₂CH₃
   2-Ethylpentanoic acid

e) trans-2-Methylcyclohexanecarboxylic acid

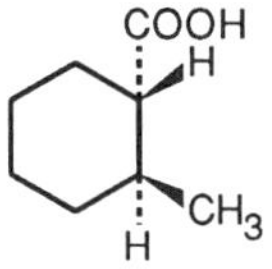

10.2

a) CH₃CH₂CH₂CH(CH₃)CH(CH₃)COOH
   2,3-Dimethylhexanoic acid

b) (CH₃)₂CHCH₂CH₂COOH
   4-Methylpentanoic acid

c) o-Hydroxybenzoic acid

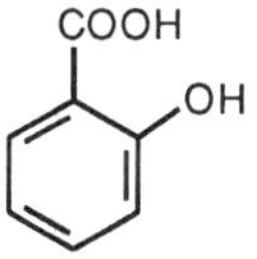

d) trans-1,2-Cyclobutanedicarboxylic acid

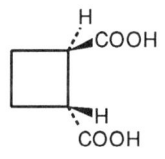

10.3

a) (CH₃)₂CHCH₂CH₂COCl
   4-Methylpentanoyl chloride

b) CH₃CH₂CH(CH₃)CN
   2-Methylbutanenitrile

c) H₂C=CHCH₂CH₂CONH₂
   4-Pentenamide

d) (CH₃CH₂)₂CHCN
   2-Ethylbutanenitrile

e) Cyclopentyl 2-methylpropanoate

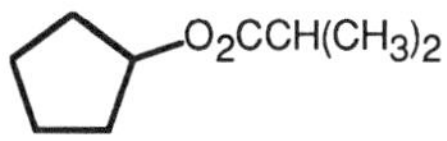

f) 2,3-Dimethyl-2-butenoyl chloride

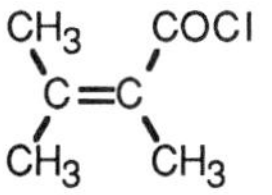

g) 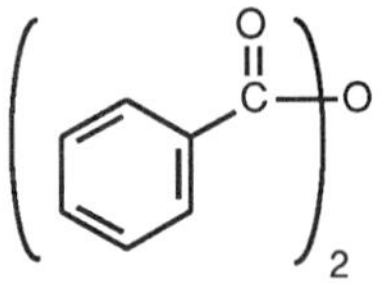 (C₆H₅-CO-O)₂

Benzoic anhydride

h) 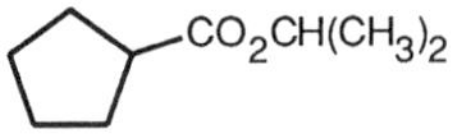 (cyclopentyl)–CO₂CH(CH₃)₂

Isopropyl cyclopentanecarboxylate

10.4

a) (CH₃)₃CCOCl

2,2-Dimethylpropanoyl chloride

b) 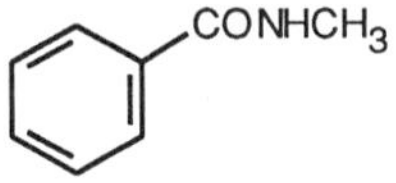 C₆H₅–CONHCH₃

N-Methylbenzamide

c) (CH₃)₃CCH₂CH₂CH₂CN

5,5-Dimethylhexanenitrile

d) CH₃CH₂CH₂COOC(CH₃)₃

tert-Butyl butanoate

e) 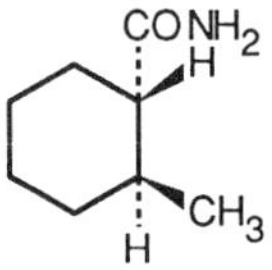 trans-2-Methylcyclohexane-carboxamide (CONH₂, CH₃)

f) 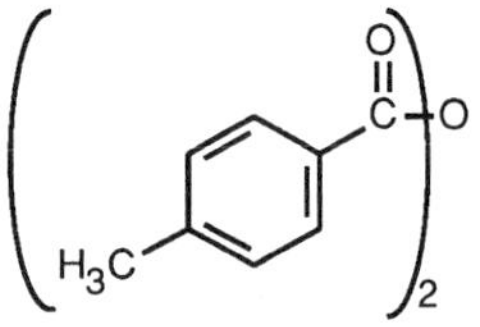 (p-CH₃–C₆H₄–CO–O)₂

p-Methylbenzoic anhydride

g) 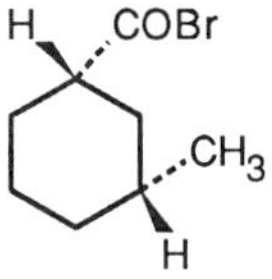 cis-3-Methylcyclohexane-carbonyl bromide (COBr, CH₃)

h) 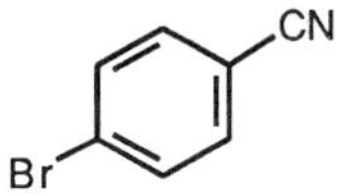 p-Br–C₆H₄–CN

p-Bromobenzonitrile

10.5

a)  C₆H₅COOH + CH₃O⁻ Na⁺ ⟶ C₆H₅COO⁻ Na⁺ + CH₃OH

b) (CH₃)₃CCO–H + KOH ⟶ (CH₃)₃CCO⁻ K⁺ + H₂O

10.6

Least acidic ⟶ Most acidic

Methanol < Phenol < p-Nitrophenol < Acetic acid < Sulfuric acid

Carboxylic Acids and Derivatives   163

10.7 Remember that an electron-withdrawing group increases acidity by stabilizing the carboxylate anion. Also note that the effect of a substituent decreases with distance from the carboxyl group.

Least acidic ⟶ Most acidic

a) CH_3CH_2COOH < $BrCH_2CH_2COOH$ < $BrCH_2COOH$

b) Ethanol < benzoic acid < p–cyanobenzoic acid

10.8

a) Bromobenzene → (1. Mg, ether; 2. CO_2; 3. H_3O^+) → benzoic acid

b) $CH_3CHCH_2CH_2Br$ (with CH_3 on the first CH) → (1. NaCN; 2. NaOH, H_2O; 3. H_3O^+) → $CH_3CHCH_2CH_2COH$ (with CH_3 on the first CH, carbonyl C=O, OH)

10.9

CH_3I —NaCN→ CH_3CN —(1. ⁻OH, H_2O; 2. H_3O^+)→ CH_3COOH

This reaction sequence can't be used to convert iodobenzene to benzoic acid because aryl halides don't undergo S_N2 substitution.

10.10

Iodobenzene —Mg, ether (formation of Grignard reagent)→ PhMgI —(1. CO_2; 2. H_3O^+) carboxylation of Grignard reagent→ PhCOOH

CH_3I —Mg→ CH_3MgI —(1. CO_2; 2. H_3O^+)→ CH_3COOH

Grignard carboxylation can be used to convert both iodobenzene and iodomethane to carboxylic acids.

10.11

	More reactive	Less reactive	Reason
a)	CH_3COCl Acid chloride	CH_3COOCH_3 Ester	Acid chlorides are more reactive (more polar) than esters.
b)	$CH_3CH_2COOCH_3$ Ester	$(CH_3)_2CHCONH_2$ Amide	Esters are more reactive than amides in nucleophilic acyl substitution reactions.

164   Chapter 10

c) CH₃COOCOCH₃   CH₃COOCH₃   Acid anhydrides are more
   Acid anhydride   Ester   reactive than esters in
                                        nucleophilic acyl substi-
                                        tution reactions.

d) CH₃COOCH₃   CH₃CHO   Aldehydes do not undergo
   Ester   Aldehyde   nucleophilic acyl substi-
                                        tution reactions.

10.12

$$F_3C\underset{\delta^-\ \delta^+\ \delta^-}{-\overset{\overset{O^{\delta-}}{\|}}{C}-OCH_3} \qquad H_3C\underset{\delta^+\ \delta^-}{-\overset{\overset{O^{\delta-}}{\|}}{C}-OCH_3}$$

The strongly electron-withdrawing trifluoromethyl group makes the carbonyl carbon more electron-poor and more reactive toward nucleophiles than the methyl acetate carbonyl group. Methyl trifluoroacetate is thus more reactive than methyl acetate in nucleophilic acyl substitution reactions.

10.13

a) PhCOOH + SOCl₂ → PhCOCl

b) PhCOOH + CH₃OH / HCl → PhCOOCH₃ + H₂O

c) PhCOOH 1. LiAlH₄ 2. H₃O⁺ → PhCH₂OH

d) PhCOOH + NaOH → PhCOO⁻ Na⁺ + H₂O

10.14

a) CH₃C(O)—OH + H—OCH₂CH₂CH₂CH₃ ⇌ (H⁺ catalyst) CH₃C(O)—OCH₂CH₂CH₂CH₃ + H₂O
   Acetic acid   Butanol   Butyl acetate

Carboxylic Acids and Derivatives   165

b) $CH_3CH_2CH_2C(=O)OH$ + $H-OCH_3$ $\underset{\text{catalyst}}{\overset{H^+}{\rightleftharpoons}}$ $CH_3CH_2CH_2\overset{O}{\overset{\|}{C}}-OCH_3$ + H_2O

   Butanoic acid   Methanol   Methyl butanoate

10.15

5-Hydroxypentanoic acid $\underset{\text{catalyst}}{\overset{H^+}{\rightleftharpoons}}$ A *lactone* + H_2O

10.16

a) $CH_3CH_2\overset{O}{\overset{\|}{C}}-Cl$ + $H-OCH_3$ $\xrightarrow[\text{solvent}]{\text{pyridine}}$ $CH_3CH_2\overset{O}{\overset{\|}{C}}OCH_3$

   Methyl propanoate

b) $CH_3\overset{O}{\overset{\|}{C}}-Cl$ + $H-OCH_2CH_3$ $\xrightarrow[\text{solvent}]{\text{pyridine}}$ $CH_3\overset{O}{\overset{\|}{C}}OCH_2CH_3$

   Ethyl acetate

c) $CH_3\overset{O}{\overset{\|}{C}}-Cl$ + $H-O-\text{cyclohexyl}$ $\xrightarrow[\text{solvent}]{\text{pyridine}}$ $CH_3\overset{O}{\overset{\|}{C}}O-\text{cyclohexyl}$

   Cyclohexyl acetate

10.17

$(CH_3)_2CH-\overset{O}{\overset{\|}{C}}-Cl$:NH$_3$ $\underset{\text{Nucleophilic addition of ammonia}}{\rightleftharpoons}$ $\left[(CH_3)_2CH-\underset{+NH_3}{\overset{:\ddot{O}:^-}{\overset{|}{C}}}-Cl \underset{\text{Loss of } Cl^-}{\rightleftharpoons} (CH_3)_2CH-\overset{O}{\overset{\|}{C}}-\overset{+}{N}H_3 \right]$

2-Methylpropanoyl chloride

$\downarrow NH_3$

$(CH_3)_2CH-\overset{O}{\overset{\|}{C}}-NH_2$

2-Methylpropanamide

10.18

	Acid chloride	+	Amine	⟶	Amide
a)	CH$_3$CH$_2$COCl	+	NH$_3$	⟶	CH$_3$CH$_2$CONH$_2$
b)	(CH$_3$)$_2$CHCH$_2$COCl	+	CH$_3$NH$_2$	⟶	(CH$_3$)$_2$CHCH$_2$CONHCH$_3$
c)	CH$_3$CH$_2$COCl	+	(CH$_3$)$_2$NH	⟶	CH$_3$CH$_2$CON(CH$_3$)$_2$
d)	C$_6$H$_5$COCl	+	(CH$_3$CH$_2$)$_2$NH	⟶	C$_6$H$_5$CON(CH$_2$CH$_3$)$_2$

10.19

[Mechanism showing acetic anhydride + p-aminophenol reacting through a tetrahedral intermediate, with proton transfer by −OH, followed by loss of acetate to form acetaminophen + CH$_3$COO$^-$]

Acetaminophen

Carboxylic Acids and Derivatives 167

10.20

[cyclic anhydride (phthalic anhydride)] + CH₃OH ⟶ [benzene ring with –COCH₃ and –COH (C=O) groups ortho]

The second half of a cyclic anhydride becomes a carboxylic acid functional group.

10.21

a) CH₃C(=O)—OCH(CH₃)₂ → (1. NaOH, H₂O; 2. H₃O⁺) → CH₃COOH + HOCH(CH₃)₂
                ↑   Acetic acid   Isopropanol
          bond cleaved

b) cyclohexyl–C(=O)—OCH₃ → (1. NaOH, H₂O; 2. H₃O⁺) → cyclohexyl–C(=O)OH + HOCH₃
                    ↑   Cyclohexane-   Methanol
               bond cleaved   carboxylic acid

10.22

RC(=O)–OH + R'O⁻ ⇌ R–C(=O)–O⁻ + R'OH

The principal reaction of a carboxylic acid and an alkoxide is an acid-base reaction, which yields an alcohol and a carboxylate anion. The negative carboxylate group is not reactive toward nucleophiles, and thus the reverse of saponification is unlikely to occur.

10.23

<u>Ester</u>   <u>Alcohols</u>

a) CH₃CH₂CH₂CH(CH₃)COOCH₃ →(1. LiAlH₄; 2. H₃O⁺)→ CH₃CH₂CH₂CH(CH₃)CH₂OH + HOCH₃

b) Ph–C(=O)–O–Ph →(1. LiAlH₄; 2. H₃O⁺)→ Ph–CH₂OH + HO–Ph

10.24

[Butyrolactone: 5-membered ring with O and C=O, H₂C–CH₂–CH₂–C(=O)–O] →(1. LiAlH₄; 2. H₃O⁺)→ HOCH₂CH₂CH₂CH₂OH

Butyrolactone   1,4-Butanediol

10.25

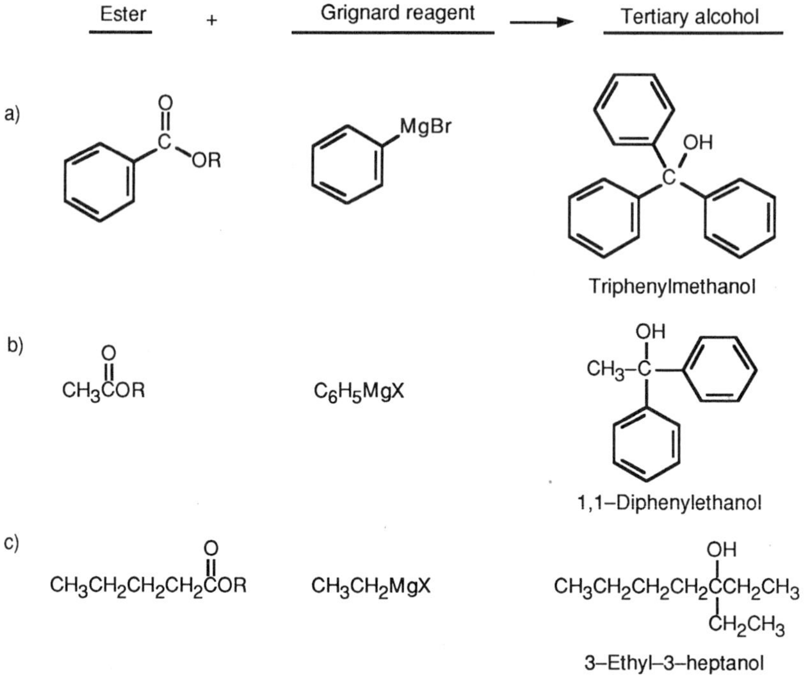

Ester	+	Grignard reagent	→	Tertiary alcohol

a) Benzoate ester (PhCOOR) + PhMgBr → Triphenylmethanol (Ph₃COH)

b) CH₃COR + C₆H₅MgX → 1,1-Diphenylethanol (CH₃-C(OH)(Ph)₂)

c) CH₃CH₂CH₂CH₂COR + CH₃CH₂MgX → CH₃CH₂CH₂CH₂C(OH)(CH₂CH₃)CH₂CH₃ — 3-Ethyl-3-heptanol

10.26

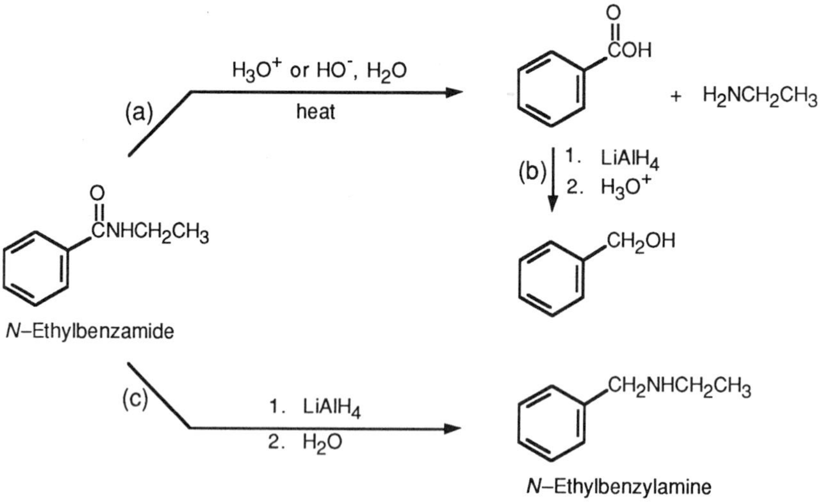

N-Ethylbenzamide (PhCONHCH₂CH₃)

(a) H₃O⁺ or HO⁻, H₂O, heat → PhCOOH + H₂NCH₂CH₃

(b) 1. LiAlH₄; 2. H₃O⁺ → PhCH₂OH

(c) 1. LiAlH₄; 2. H₂O → PhCH₂NHCH₂CH₃ — N-Ethylbenzylamine

10.27

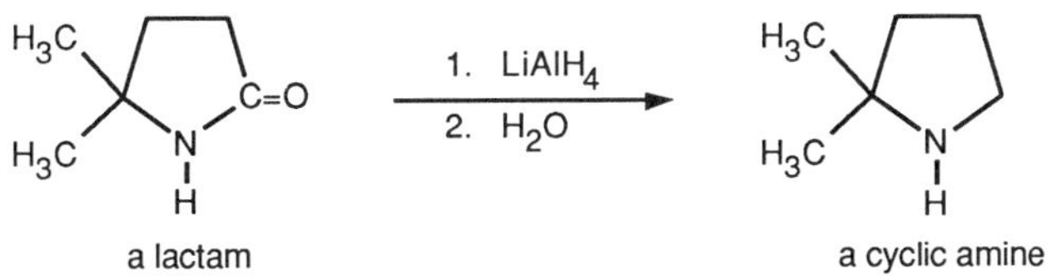

<chemical structure: a lactam (5,5-dimethyl pyrrolidinone with H₃C, H₃C groups, N-H, C=O)> → 1. LiAlH₄, 2. H₂O → <chemical structure: a cyclic amine (5,5-dimethylpyrrolidine)>

a lactam → a cyclic amine

10.28

	Nitrile	+	Grignard Reagent	→	Ketone
a)	CH₃CH₂CN		CH₃CH₂MgX		CH₃CH₂COCH₂CH₃
b)	CH₃CH₂CN		(CH₃)₂CHMgX		CH₃CH₂COCH(CH₃)₂
	or (CH₃)₂CHCN		CH₃CH₂MgX		
c)	C₆H₅CN		CH₃MgX		C₆H₅COCH₃
	or CH₃CN		C₆H₅MgX		
d)	cyclohexyl-CN		cyclohexyl-MgX		dicyclohexyl ketone

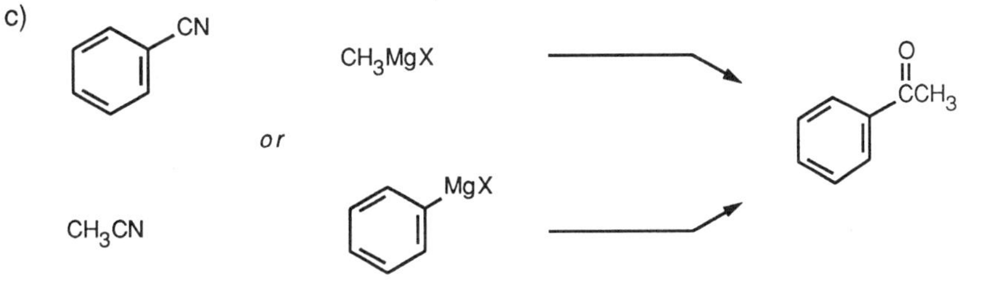

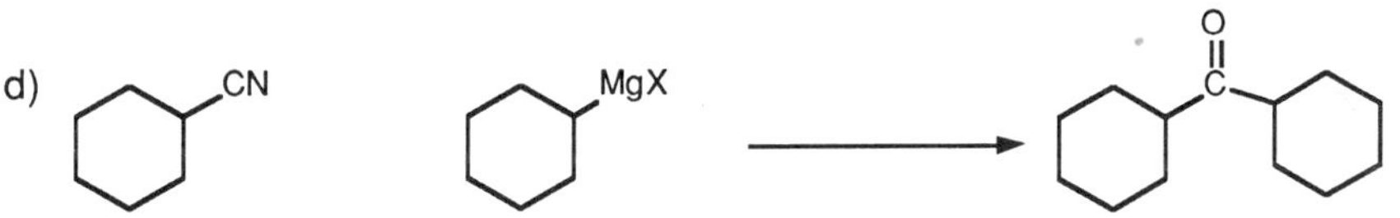

10.29

a) $(CH_3)_2CHCH_2I$ —NaCN→ $(CH_3)_2CHCH_2CN$ —1. LiAlH₄, 2. H₂O→ $(CH_3)_2CHCH_2CH_2NH_2$

b) $C_6H_5CH_2Br$ —NaCN→ $C_6H_5CH_2CN$ —1. CH₃CH₂MgBr, 2. H₃O⁺→ $C_6H_5CH_2COCH_2CH_3$

10.30

HOOC–C₆H₄–COOH + H₂N–C₆H₄–NH₂ → [−CO–C₆H₄–CO–NH–C₆H₄–NH−]ₙ

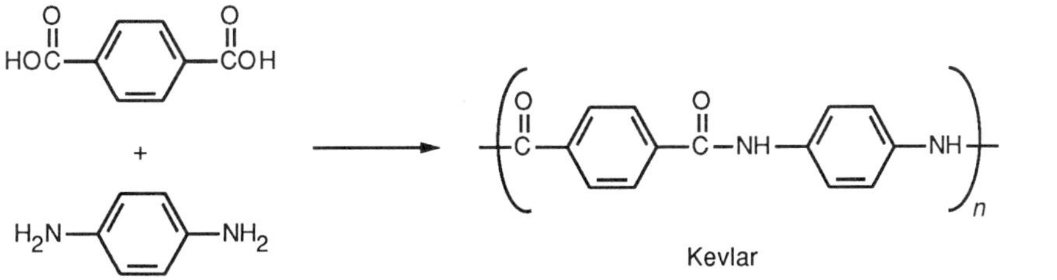

Kevlar

10.31

a) CH₃CHCH₂CH₂CHCH₃ with COOH on C1 and C5
   2,5–Dimethylhexanedioic acid

b) (CH₃)₃CCOOH
   2,2–Dimethylpropanoic acid

c) CH₃CH₂CH₂CH(CH₂CH₂CH₃)CH₂COOH
   3–Propylhexanoic acid

d) *p*-substituted benzene with COOH and NO₂
   p–Nitrobenzoic acid

e) cyclodecene ring with COOH
   1–Cyclodecenecarboxylic acid

f) BrCH₂CHBrCH₂CH₂COOH
   4,5–Dibromopentanoic acid

10.32

a) benzene with CONH₂ and H₃C (para)
   p–Methylbenzamide

b) (CH₃CH₂)₂CHCH=CHCN
   4–Ethyl–2–hexenenitrile

c) CH₃O₂CCH₂CH₂CO₂CH₃
   Dimethyl butanedioate

d) benzene–CH₂CH₂CO₂CH(CH₃)₂
   Isopropyl 3–phenylpropanoate

e) phenyl–C(=O)–O–phenyl
   Phenyl benzoate

f) CH₃CHBrCH₂CONHCH₃
   N–Methyl–3–bromobutanamide

g) 3,5-dibromobenzene with C(=O)–Cl
   3,5–Dibromobenzoyl chloride

h) cyclopentene with CN
   1–Cyclopentenecarbonitrile

10.33

a) CH₃CH₂CH(CH₃)CH(CH₃)CH₂CH₂COOH
4,5-Dimethylheptanoic acid

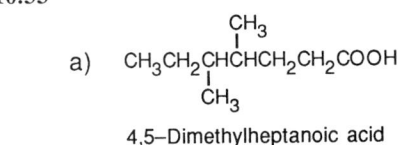

b) cis-1,2-Cyclohexanedicarboxylic acid (cyclohexane with two COOH groups, both H's shown on same face)

c) HOOCCH₂CH₂CH₂CH₂CH₂COOH
Heptanedioic acid

d) (C₆H₅)₃CCOOH
Triphenylacetic acid

e) CH₃CH₂CH₂CH₂C(CH₃)(CH₃)CONH₂
2,2-Dimethylhexanamide

f) C₆H₅CH₂CONH₂
Phenylacetamide

g) 2-Cyclobutenecarbonitrile (cyclobutene ring with CN substituent)

h) Ethyl cyclohexanecarboxylate (cyclohexane ring with COOCH₂CH₃)

10.34 Acetic acid molecules are strongly associated because of hydrogen bonding. Molecules of the ethyl ester are much more weakly associated, and less heat is required to overcome the attractive forces between molecules. Even though the ethyl ester has a greater molecule weight, it boils at a lower temperature than the acid.

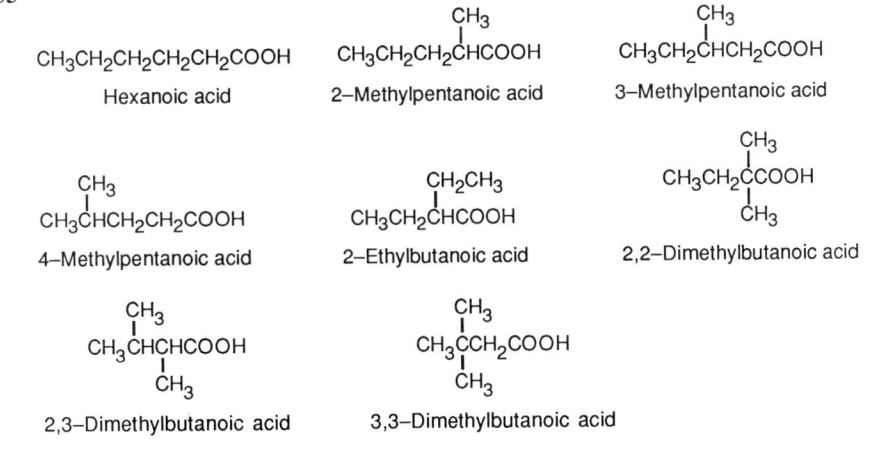

10.35

CH₃CH₂CH₂CH₂CH₂COOH
Hexanoic acid

CH₃CH₂CH₂CH(CH₃)COOH
2-Methylpentanoic acid

CH₃CH₂CH(CH₃)CH₂COOH
3-Methylpentanoic acid

CH₃CH(CH₃)CH₂CH₂COOH
4-Methylpentanoic acid

CH₃CH₂CH(CH₂CH₃)COOH
2-Ethylbutanoic acid

CH₃CH₂C(CH₃)(CH₃)COOH
2,2-Dimethylbutanoic acid

CH₃CH(CH₃)CH(CH₃)COOH
2,3-Dimethylbutanoic acid

CH₃C(CH₃)(CH₃)CH₂COOH
3,3-Dimethylbutanoic acid

172   Chapter 10

10.36 Many other compounds having these formulas can be drawn.

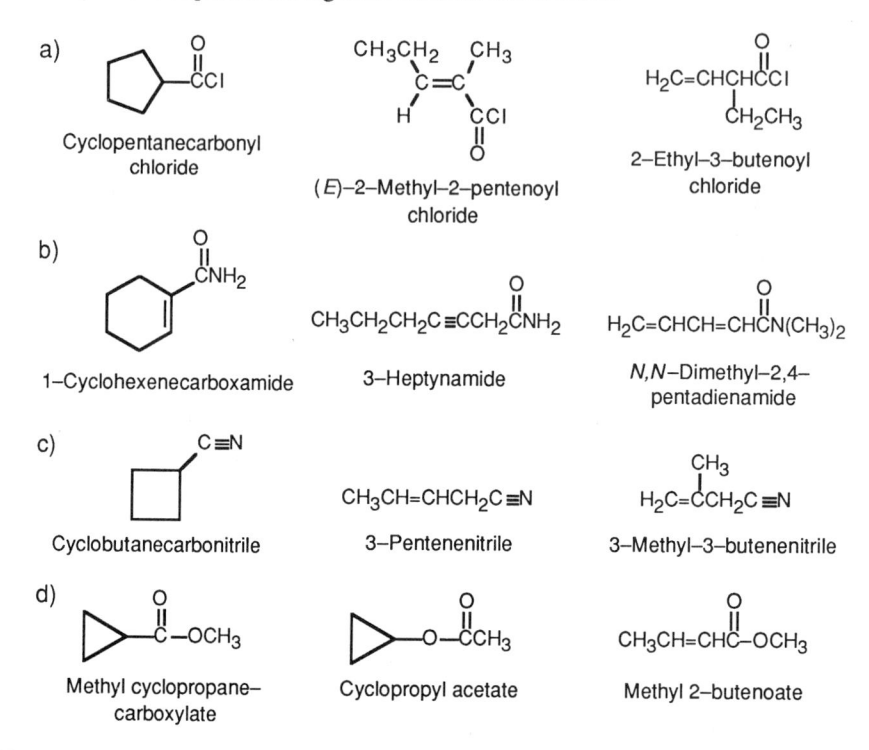

a)

Cyclopentanecarbonyl chloride

(E)-2-Methyl-2-pentenoyl chloride

2-Ethyl-3-butenoyl chloride

b)

1-Cyclohexenecarboxamide

3-Heptynamide

N,N-Dimethyl-2,4-pentadienamide

c)

Cyclobutanecarbonitrile

3-Pentenenitrile

3-Methyl-3-butenenitrile

d)

Methyl cyclopropane-carboxylate

Cyclopropyl acetate

Methyl 2-butenoate

10.37 The reactivity of esters in saponification reactions is influenced by steric factors. Branching in both the acyl and alkyl portions of an ester makes it harder for the hydroxide nucleophile to approach the carbonyl carbon. This effect is less pronounced in the alkyl portion of the ester than in the acyl portion because alkyl branching is one atom farther away from the site of attack. The reactivity order for saponification of alkyl acetates:

Most reactive ⟶ Least reactive

$$CH_3\overset{O}{\underset{\|}{C}}OCH_3 \; > \; CH_3\overset{O}{\underset{\|}{C}}OCH_2CH_3 \; > \; CH_3\overset{O}{\underset{\|}{C}}OCH(CH_3)_2 \; > \; CH_3\overset{O}{\underset{\|}{C}}OC(CH_3)_3$$

10.38 The lower the pK_a, the stronger the acid. Thus, tartaric acid (pK_a = 2.98) is a stronger acid than citric acid (pK_a = 3.14).

10.39

Least acidic ⟶ Most acidic

a) Acetic acid < chloroacetic acid < trifluoroacetic acid
b) Benzoic acid < p–bromobenzoic acid < p–nitrobenzoic acid
c) Cyclohexanol < phenol < acetic acid

10.40 The pK$_a$ of 2-chlorobutanoic acid is lowered by the electronegative chlorine atom, which stabilizes the adjacent carboxylate group. Since this effect decreases with increasing distance form the reaction site, the chlorine atom in 3-chlorobutanoic acid lowers the pK$_a$ much less. The chlorine atom in 4-chlorobutanoic acid is so far from the carboxylic acid group that it has a very small effect on pK$_a$.

10.41

Most reactive ———————————→ Least reactive

CH$_3$COCl > CH$_3$COOCOCH$_3$ > CH$_3$COOCH$_3$ > CH$_3$CONH$_2$

10.42

a) PhCN $\xrightarrow{\text{1. CH}_3\text{MgBr} \\ \text{2. H}_3\text{O}^+}$ PhCOCH$_3$

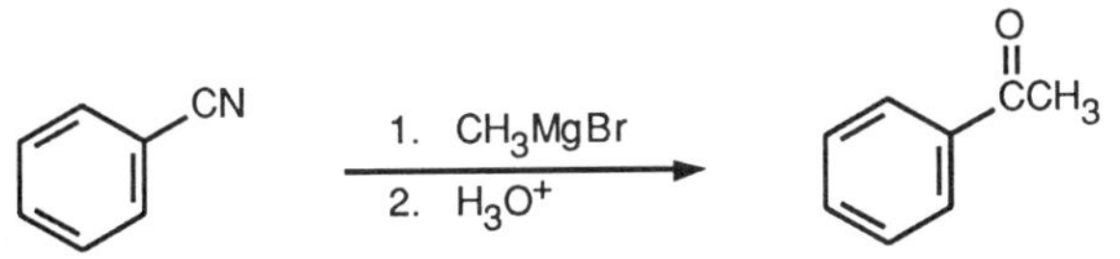

b) PhBr $\xrightarrow{\text{Mg, ether}}$ PhMgBr $\xrightarrow{\text{1. CH}_3\text{CHO} \\ \text{2. H}_3\text{O}^+}$ PhCH(OH)CH$_3$ $\xrightarrow{\text{PCC} \\ \text{CH}_2\text{Cl}_2}$ PhCOCH$_3$

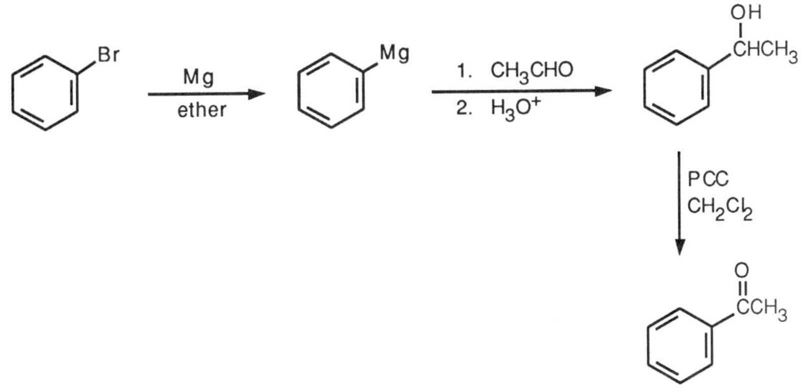

c) PhCOOCH$_3$ $\xrightarrow{\text{1. LiAlH}_4 \\ \text{2. H}_3\text{O}^+}$ PhCH$_2$OH $\xrightarrow{\text{PCC} \\ \text{CH}_2\text{Cl}_2}$ PhCHO $\xrightarrow{\text{1. CH}_3\text{MgBr} \\ \text{2. H}_3\text{O}^+}$ PhCH(OH)CH$_3$ $\xrightarrow{\text{PCC} \\ \text{CH}_2\text{Cl}_2}$ PhCOCH$_3$

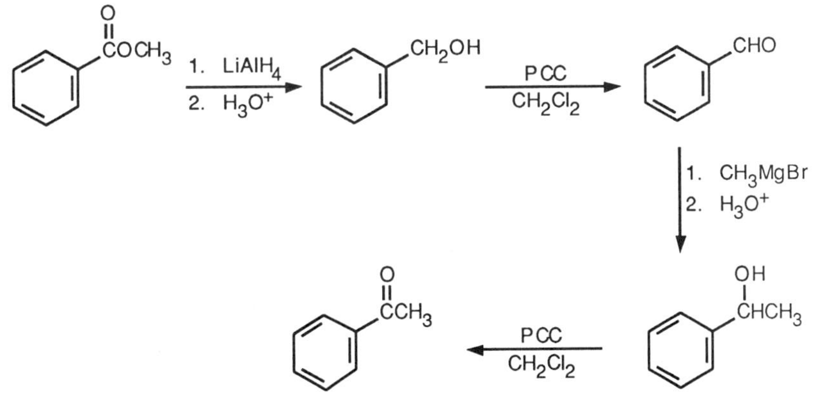

d)

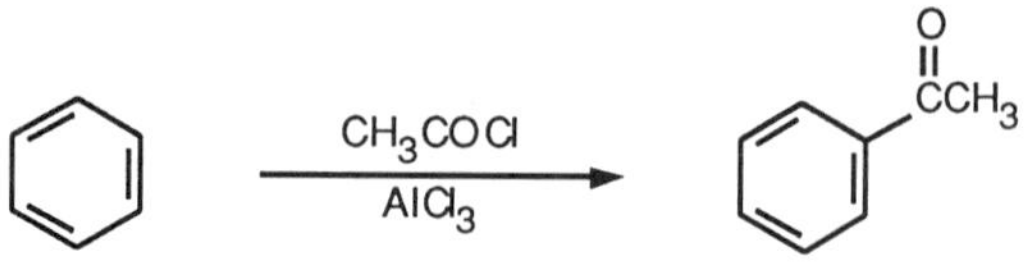

C₆H₆ $\xrightarrow[\text{AlCl}_3]{\text{CH}_3\text{COCl}}$ C₆H₅COCH₃

10.43

a) CH₃CH₂CH₂COOH $\xrightarrow[\text{2. H}_3\text{O}^+]{\text{1. LiAlH}_4}$ CH₃CH₂CH₂CH₂OH

b) CH₃CH₂CH₂CH₂OH (part (a)) $\xrightarrow[\text{CH}_2\text{Cl}_2]{\text{PCC}}$ CH₃CH₂CH₂CHO

c) CH₃CH₂CH₂CH₂OH (part (a)) $\xrightarrow{\text{PBr}_3}$ CH₃CH₂CH₂CH₂Br

d) CH₃CH₂CH₂CH₂Br (part (c)) $\xrightarrow{\text{NaCN}}$ CH₃CH₂CH₂CH₂C≡N

e) CH₃CH₂CH₂CH₂Br (part (c)) $\xrightarrow[\text{ethanol}]{\text{KOH}}$ CH₃CH₂CH=CH₂

f) CH₃CH₂CH₂COOH $\xrightarrow{\text{SOCl}_2}$ CH₃CH₂CH₂COCl $\xrightarrow{2\ \text{NH}_3}$ CH₃CH₂CH₂CONH₂ $\xrightarrow[\text{2. H}_2\text{O}]{\text{1. LiAlH}_4}$ CH₃CH₂CH₂CH₂NH₂

10.44

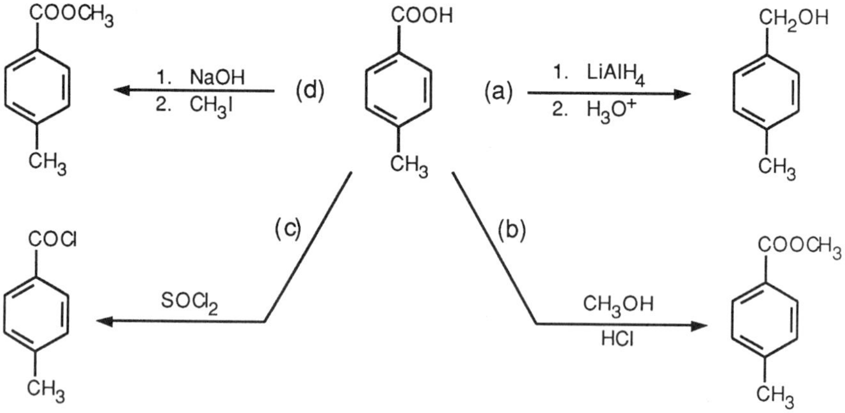

p-CH₃-C₆H₄-COOCH₃ $\xleftarrow[\text{2. CH}_3\text{I}]{\text{1. NaOH}}$ (d) p-CH₃-C₆H₄-COOH $\xrightarrow[\text{2. H}_3\text{O}^+]{\text{1. LiAlH}_4}$ (a) p-CH₃-C₆H₄-CH₂OH

(c) ↙ ↘ (b)

p-CH₃-C₆H₄-COCl $\xleftarrow{\text{SOCl}_2}$ ... $\xrightarrow[\text{HCl}]{\text{CH}_3\text{OH}}$ p-CH₃-C₆H₄-COOCH₃

10.45 Substitution of chloride by cyanide usually proceeds by an S_N2 mechanism. In this case, however, 2–chloro–2–methylpentane, a tertiary chloride, is more likely to undergo elimination to yield 2–methyl–2–pentene. A better route to the desired carboxylic acid is shown below.

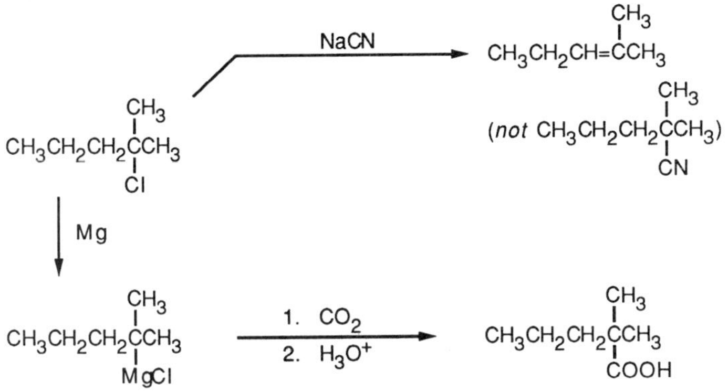

$$CH_3CH_2CH_2C(CH_3)(Cl)CH_3 \xrightarrow{NaCN} CH_3CH_2CH=C(CH_3)CH_3$$

$$(not\ CH_3CH_2CH_2C(CH_3)(CN)CH_3)$$

$$\downarrow Mg$$

$$CH_3CH_2CH_2C(CH_3)(MgCl)CH_3 \xrightarrow[2.\ H_3O^+]{1.\ CO_2} CH_3CH_2CH_2C(CH_3)(COOH)CH_3$$

10.46 a) Grignard carboxylation can't be used to prepare the carboxylic acid because of the acidic hydroxyl group. Use *nitrile hydrolysis*.

b) Either method produces the carboxylic acid in suitable yield, but *Grignard carboxylation* is a better reaction for preparing a carboxylic acid from a secondary bromide.

c) Neither method of acid synthesis yields the desired product. Any Grignard reagent formed will react with the carbonyl functional group present in the starting material. Reaction with cyanide occurs at the carbonyl functional group as well as at halogen. However, if the ketone is first protected by forming an acetal, *either method* can be used for producing a carboxylic acid.

d) Since the hydroxyl proton interferes with formation of the Grignard reagent, *nitrile hydrolysis* must be used to form the carboxylic acid.

10.47

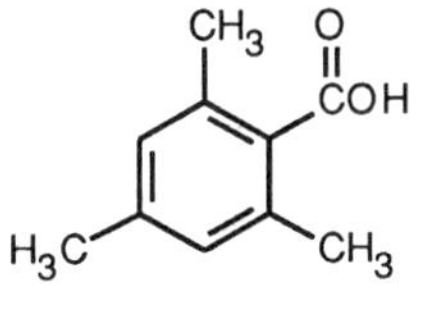

[Structure: 2,4,6-trimethylbenzoic acid — benzene ring with CH_3 at 2, 4, 6 positions and COOH at position 1]

2,4,6–Trimethylbenzoic acid has two methyl groups *ortho* to the carboxylic acid functional group. These bulky methyl groups block the approach of the alcohol and prevent esterification from occurring under Fischer esterification conditions.

10.48

a) 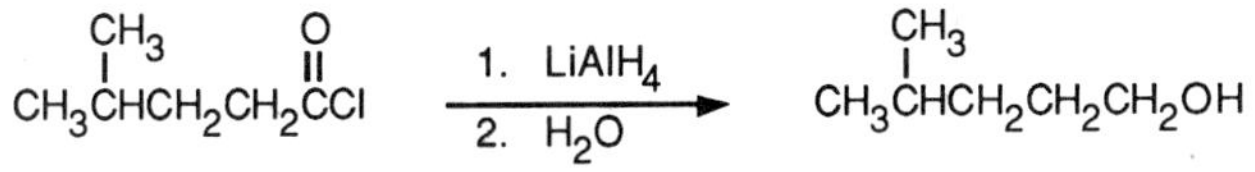 $$CH_3CH(CH_3)CH_2CH_2COCl \xrightarrow[2.\ H_2O]{1.\ LiAlH_4} CH_3CH(CH_3)CH_2CH_2CH_2OH$$

176   Chapter 10

b) 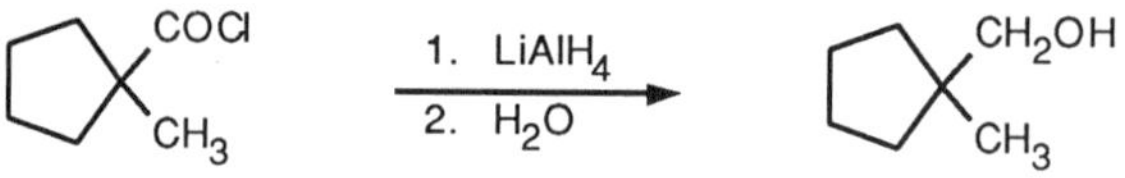 [cyclopentane with COCl and CH₃ substituents] →(1. LiAlH₄; 2. H₂O)→ [cyclopentane with CH₂OH and CH₃ substituents]

10.49

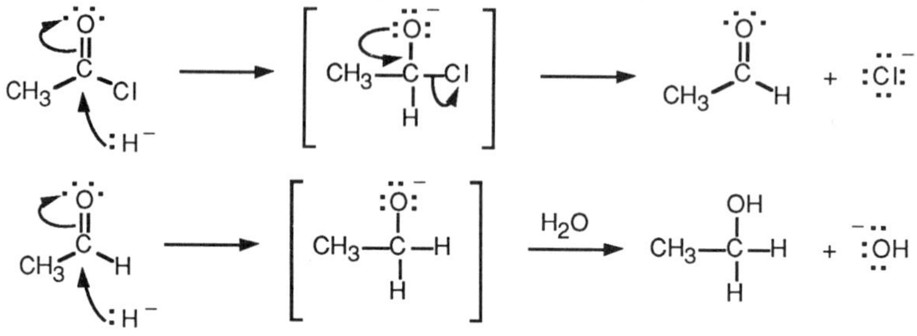

CH₃–C(=O)–Cl with :H⁻ attacking → [CH₃–C(O⁻)(H)–Cl intermediate] → CH₃–C(=O)–H + :Cl:⁻

CH₃–C(=O)–H with :H⁻ attacking → [CH₃–C(O⁻)(H)–H intermediate] →(H₂O)→ CH₃–CH(OH)–H + :OH⁻

10.50

R–C(=O)–Cl with :R'⁻MgBr attacking → [R–C(O⁻)(R')–Cl intermediate] → R–C(=O)–R'

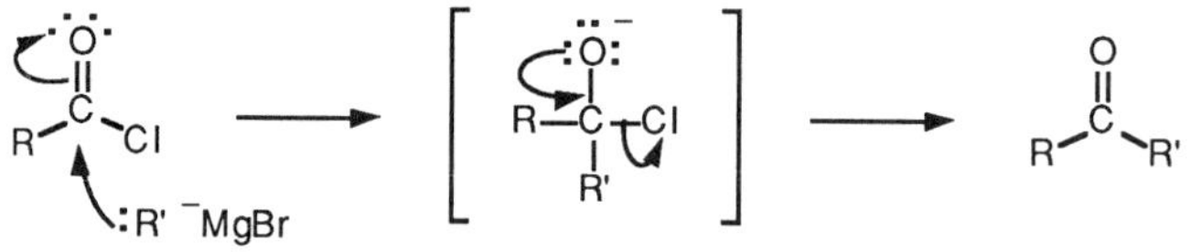

10.51 a) The first step of the reaction of a Grignard reagent with an acid chloride is pictured in Problem 10.50. A second molecule of Grignard reagent adds to the ketone to yield a tertiary alcohol.

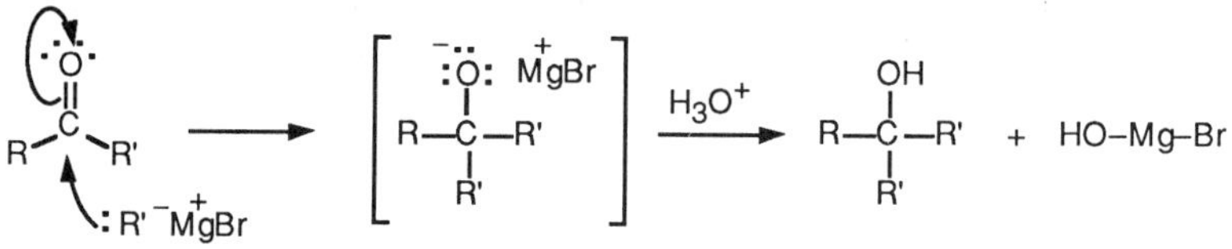

R–C(=O)–R' with :R'⁻MgBr attacking → [R–C(O⁻MgBr⁺)(R')(R') intermediate] →(H₃O⁺)→ R–C(OH)(R')–R' + HO–Mg–Br

b) 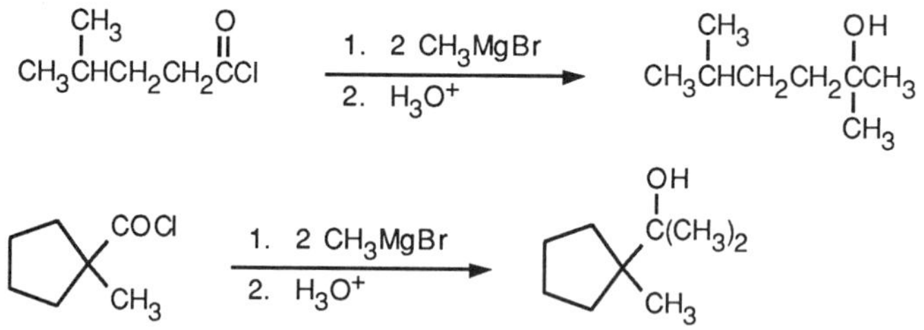 CH₃CH(CH₃)CH₂CH₂–C(=O)Cl →(1. 2 CH₃MgBr; 2. H₃O⁺)→ CH₃CH(CH₃)CH₂CH₂–C(OH)(CH₃)CH₃

[cyclopentane with COCl and CH₃] →(1. 2 CH₃MgBr; 2. H₃O⁺)→ [cyclopentane with C(OH)(CH₃)₂ and CH₃]

Carboxylic Acids and Derivatives   177

10.52 Dimethyl carbonate is a diester. Use your knowledge of the Grignard reaction to work your way through this problem.

[Mechanism scheme showing dimethyl carbonate reacting with PhMgBr through successive addition-elimination steps, eliminating $^-OCH_3$, forming benzophenone intermediate, then adding a third PhMgBr, followed by H_3O^+ workup to give Triphenylmethanol]

10.53

a) $CH_3CH_2\overset{O}{\underset{\|}{C}}Cl \xrightarrow[\text{2. }H_3O^+]{\text{1. 2 }CH_3MgBr} CH_3CH_2\underset{CH_3}{\underset{|}{\overset{OH}{\underset{|}{C}}}}CH_3$

b) $CH_3CH_2\overset{O}{\underset{\|}{C}}Cl \xrightarrow[H_2O]{NaOH} CH_3CH_2\overset{O}{\underset{\|}{C}}O^-Na^+$

c) $CH_3CH_2\overset{O}{\underset{\|}{C}}Cl \xrightarrow{2\ CH_3NH_2} CH_3CH_2\overset{O}{\underset{\|}{C}}NHCH_3$

d) $CH_3CH_2\overset{O}{\underset{\|}{C}}Cl \xrightarrow[\text{2. }H_3O^+]{\text{1. LiAlH}_4} CH_3CH_2CH_2OH$

178   Chapter 10

e) $CH_3CH_2\overset{\overset{O}{\|}}{C}Cl \xrightarrow[\text{pyridine}]{C_6H_{11}OH} CH_3CH_2\overset{\overset{O}{\|}}{C}O-\bigcirc$

f) $CH_3CH_2\overset{\overset{O}{\|}}{C}Cl \xrightarrow{CH_3COO^-Na^+} CH_3CH_2\overset{\overset{O}{\|}}{C}O\overset{\overset{O}{\|}}{C}CH_3$

10.54

a) $CH_3CH_2\overset{\overset{O}{\|}}{C}OCH_3 \xrightarrow[\text{2. } H_3O^+]{\text{1. 2 } CH_3MgBr} CH_3CH_2\underset{\underset{CH_3}{|}}{\overset{\overset{OH}{|}}{C}}CH_3 + CH_3OH$

b) $CH_3CH_2\overset{\overset{O}{\|}}{C}OCH_3 \xrightarrow[H_2O]{NaOH} CH_3CH_2\overset{\overset{O}{\|}}{C}O^-Na^+ + CH_3OH$

c) $CH_3CH_2\overset{\overset{O}{\|}}{C}OCH_3 \xrightarrow{2\ CH_3NH_2} CH_3CH_2\overset{\overset{O}{\|}}{C}NHCH_3 + CH_3OH$

d) $CH_3CH_2\overset{\overset{O}{\|}}{C}OCH_3 \xrightarrow[\text{2. } H_3O^+]{\text{1. LiAlH}_4} CH_3CH_2CH_2OH + CH_3OH$

e) $CH_3CH_2\overset{\overset{O}{\|}}{C}OCH_3 \xrightarrow[\text{pyridine}]{C_6H_{11}OH}$ no reaction

f) $CH_3CH_2\overset{\overset{O}{\|}}{C}OCH_3 \xrightarrow{CH_3COO^-Na^+}$ no reaction

In general, esters are less reactive than acid chlorides (Problem 10.53).

10.55

| Ester | Grignard reagent | → | Alcohol |

a) $CH_3CH_2CH_2\overset{\overset{O}{\|}}{C}OR$ + CH_3MgBr → $CH_3CH_2CH_2\underset{\underset{CH_3}{|}}{\overset{\overset{OH}{|}}{C}}CH_3$

b) $CH_3\overset{\overset{O}{\|}}{C}OR$ + $\bigcirc\!-MgBr$ → $\underset{}{\overset{HO\ \ CH_3}{\underset{}{C}}}(\text{with two phenyl groups})$

Remember that the Grignard reagent contributes the two identical groups to the tertiary alcohol.

Carboxylic Acids and Derivatives 179

10.56

a) $CH_3CHCH_2CH_2COH + HOCH_2CH_3 \xrightarrow{HCl} CH_3CHCH_2CH_2COCH_2CH_3$ (with CH₃ substituent on the chain)

$CH_3CHCH_2CH_2CCl + HOCH_2CH_3 \xrightarrow{pyridine} CH_3CHCH_2CH_2COCH_2CH_3$ (with CH₃ substituent on the chain)

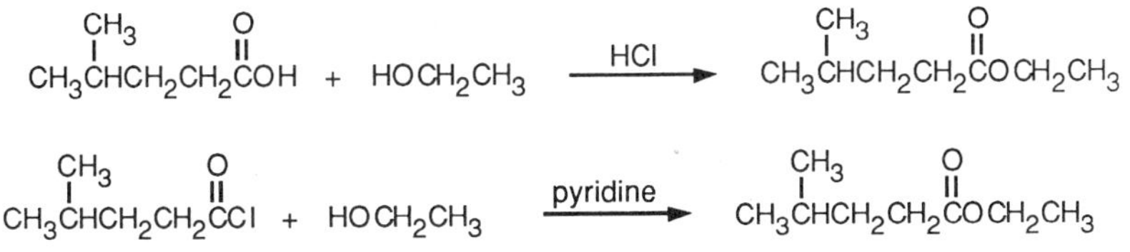

b) cyclopentyl–CH₂OH + HOCCH₃ $\xrightarrow{HCl}$ cyclopentyl–CH₂OCCH₃

cyclopentyl–CH₂OH + ClCCH₃ $\xrightarrow{pyridine}$ cyclopentyl–CH₂OCCH₃

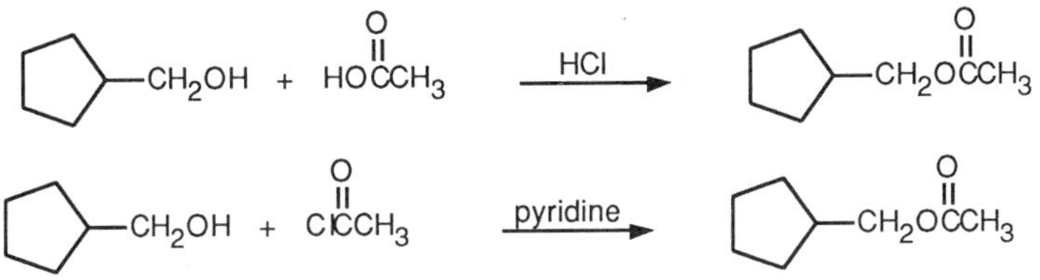

10.57

a) 4-Br-C₆H₄–C(=O)–OCH(CH₃)₂ + NaOH/H₂O → 4-Br-C₆H₄–C(=O)–O⁻Na⁺ + HOCH(CH₃)(CH₃)

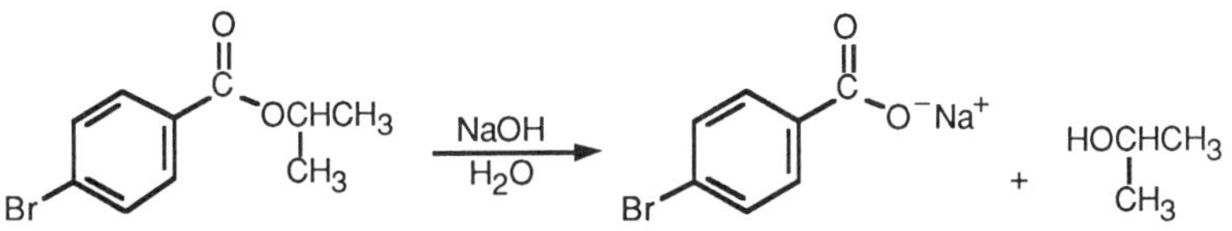

b) CH₃CH₂C(=O)–O–cyclohexyl + NaOH/H₂O → CH₃CH₂C(=O)–O⁻Na⁺ + HO–cyclohexyl

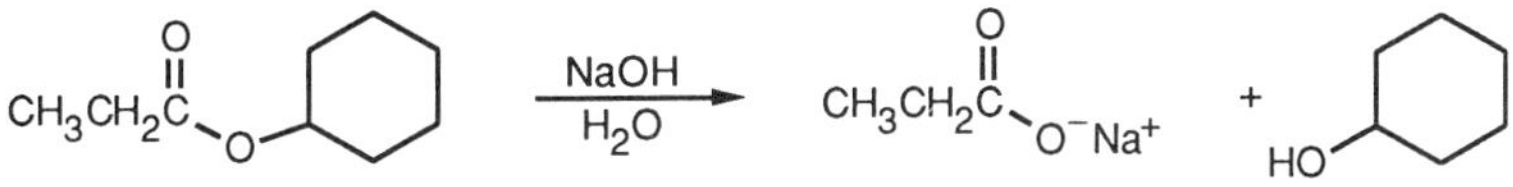

10.58

$CH_3C(=O)-OCH_3 \;\xrightleftharpoons{H^+}\; \left[CH_3C(^+OH)-OCH_3 \;\xrightarrow[HOCH_2CH_3]{}\; CH_3C(OH)(OCH_3)(HOCH_2CH_3)^+ \right]$

$\Updownarrow$

$CH_3COCH_2CH_3 \;\xleftharpoons{-H^+}\; \left[CH_3C(=^+OH)(OCH_2CH_3) + HOCH_3 \;\xrightleftharpoons\; CH_3C(OH)(OCH_3)(H)(OCH_2CH_3)^+ \right]$

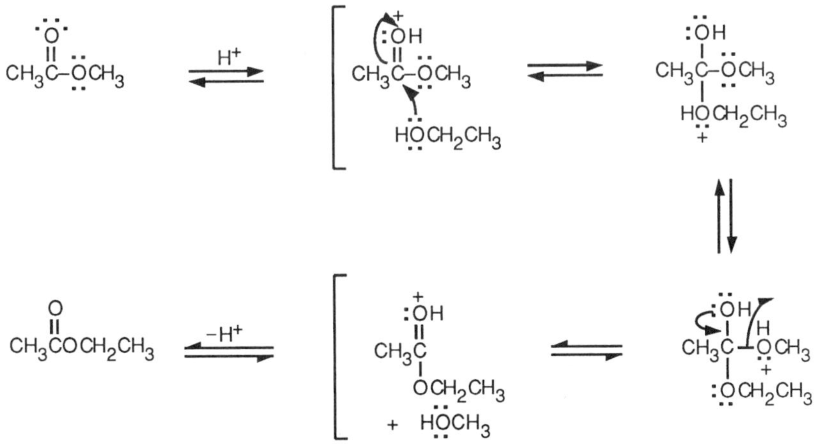

The above mechanism is very similar to the mechanism of Fischer esterification, illustrated in Section 10.6. Conversion of the methyl ester to the ethyl ester occurs because of the large excess of the solvent, ethanol.

10.59

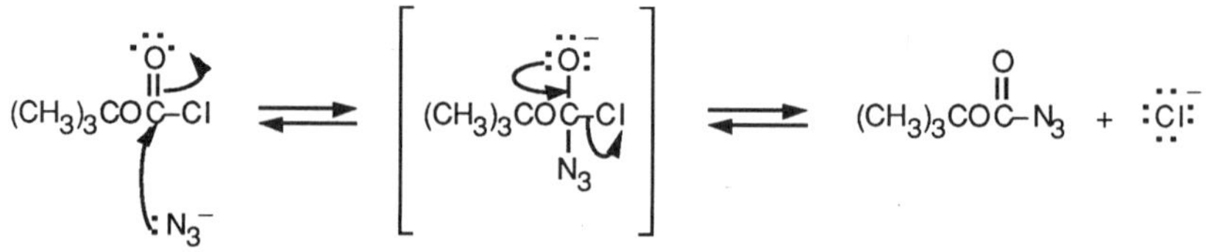

$$(CH_3)_3COC(=O)-Cl + :N_3^- \rightleftharpoons [(CH_3)_3CO-C(O^-)(Cl)(N_3)] \rightleftharpoons (CH_3)_3COC(=O)-N_3 + :Cl:^-$$

10.60

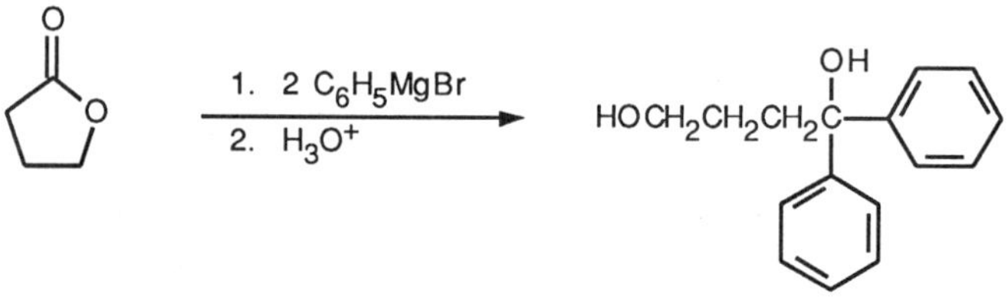

γ-butyrolactone + 1. 2 C$_6$H$_5$MgBr; 2. H$_3$O$^+$ → HOCH$_2$CH$_2$CH$_2$C(OH)(C$_6$H$_5$)$_2$

10.61

m-bromotoluene —Mg→ m-CH$_3$C$_6$H$_4$MgBr —1. CO$_2$; 2. H$_3$O$^+$→ m-toluic acid (COOH) —SOCl$_2$/CHCl$_3$→ m-toluoyl chloride —HN(CH$_2$CH$_3$)$_2$, NaOH→ N,N-diethyl-m-toluamide

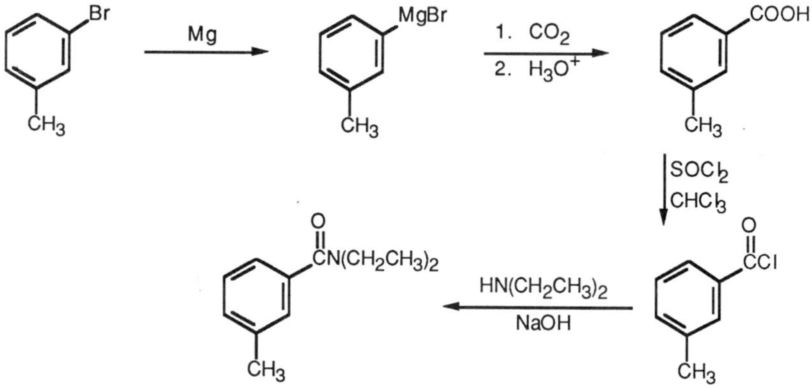

N,N-Diethyl-*m*-toluamide (DEET)

10.62

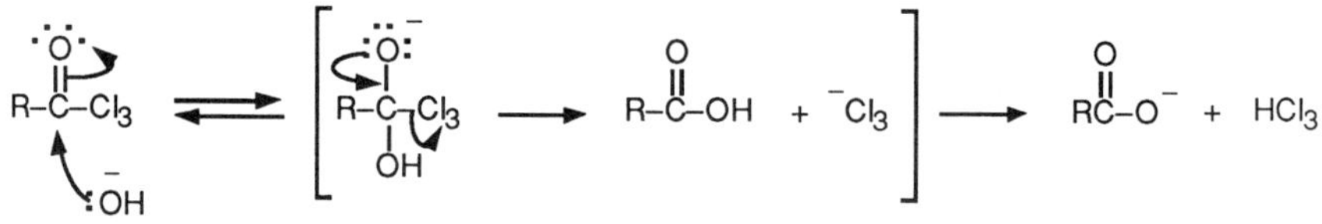

$$R-C(=O)-Cl_3 + {}^-OH \rightleftharpoons [R-C(O^-)(Cl_3)(OH)] \longrightarrow R-C(=O)-OH + {}^-Cl_3 \longrightarrow RC(=O)-O^- + HCl_3$$

10.63

$$BrCH_2C(=O)-H + H_2O \underset{}{\overset{K_a}{\rightleftharpoons}} BrCH_2C(=O)O^- + H_3O^+$$

$$K_a = \frac{[BrCH_2COO^-][H_3O^+]}{[BrCH_2COOH]} = 10^{-3}$$

	Initial molarity	Molarity after dissociation
BrCH₂COOH	0.1	0.1 − x
BrCH₂COO⁻	0	x
H₃O⁺	~0	x

$$K_a = \frac{x \cdot x}{(0.1 - x)} = 10^{-3}$$

Using the quadratic formula to solve for x, we find that
x = 0.01 − .0005 ~ 0.01.

Percent dissociation = $\frac{0.01}{0.1}$ × 100% = 10%

Chapter Outline

I. Naming carboxylic acids and derivatives (Section 10.1)
II. Carboxylic acids (Sections 10.2 - 10.4)
    A. Occurrence, structure, and properties (Section 10.2)
    B. Acidity (Sections 10.3)
        1. Resonance stabilization of carboxylate anion
        2. Substituent effects
    C. Synthesis of carboxylic acids (Section 10.4)
        1. Oxidation of alkylbenzenes
        2. Oxidation of primary alcohols and aldehydes
        3. Hydrolysis of nitriles
        4. Reaction of Grignard reagents with CO_2
III. Nucleophilic acyl substitution reactions (Sections 10.5 - 10.12)
    A. Relative reactivity of carboxylic acid derivatives (Section 10.5)
    B. Reactions of carboxylic acids (Section 10.6)
        1. Conversion to acid chlorides
        2. Conversion to anhydrides
        3. Conversion to esters - Fischer esterification
        4. Conversion to amides
    C. Acid halides (Section 10.7)
        1. Preparation from carboxylic acids
        2. Reactions of acid halides
            a. Hydrolysis to give carboxylic acids
            b. Alcoholysis to give esters
            c. Aminolysis to give amides

- D. Acid anhydrides (Section 10.8)
    1. Preparation from acid halides
    2. Reactions (same as acid halides)
- E. Esters (Section 10.9)
    1. Preparation from acids and acid halides
    2. Reactions
        a. Hydrolysis to give acids
        b. Aminolysis to give amides
        c. Reduction with LiAlH$_4$ to give primary alcohols
        d. Reaction with Grignard reagents to give tertiary alcohols
- F. Amides (Section 10.10)
    1. Preparation from acid chlorides
    2. Reactions
        a. Hydrolysis to give carboxylic acids
        b. Reduction with LiAlH$_4$ to give amines
- G. Nitriles (Section 10.11)
    1. Preparation from alkyl halides
    2. Reactions
        a. Hydrolysis to give carboxylic acids
        b. Reduction with LiAlH$_4$ to give amines
        c. Reaction with Grignard reagents to give ketones

IV. Nylon and polyester: step-growth polymers (Section 10.12)

Study Skills for Chapter 10

After studying this chapter, you should be able to:
1. Name carboxylic acids and derivatives (Problems 10.1, 10.3, 10.31, 10.32, 10.35, 10.36).
2. Draw structures of carboxylic acids and derivatives from given names (Problems 10.2, 10.4, 10.33, 10.35, 10.36).
3. Rank compounds in order of increasing acidity (Problems 10.6, 10.7, 10.38, 10.39).
4. Use electronegativity and resonance arguments to predict the acidity of carboxylic acids (Problem 10.40).
5. Synthesize carboxylic acids (Problems 10.8, 10.9, 10.10, 10.46).
6. Predict the relative reactivity of carboxylic acid derivatives (Problems 10.11, 10.12, 10.37, 10.41, 10.47).
7. Predict the products of reactions for:
   Carboxylic acids (Problems 10.5, 10.13, 10.14, 10.15, 10.43, 10.44, 10.56)
   Acid halides (Problems 10.16, 10.18, 10.48, 10.53)
   Acid anhydrides (Problem 10.20)
   Esters (Problems 10.21, 10.23, 10.24, 10.25, 10.54, 10.55, 10.57, 10.60)
   Amides (Problems 10.26, 10.27)
   Nitriles (Problem 10.28)
8. Use nucleophilic acyl substitution reactions in synthesis (Problems 10.29, 10.42, 10.61)
9. Formulate mechanisms of simple nucleophilic acyl substitution reactions (Problems 10.17, 10.19, 10.45, 10.49, 10.50, 10.51, 10.52, 10.58, 10.59, 10.62).

Chapter 11 – Carbonyl alpha-Substitution Reactions and Condensation Reactions

11.1-11.2 Acidic hydrogens in the keto form of each of these compounds are underlined.

	Keto Form	Enol Form	Number of Acidic Hydrogens
a)	cyclopentanone with 4 α-H underlined	cyclopentenol (OH)	4
b)	H-C(H)(H)-CCl=O (α-H's underlined)	H-C=C(OH)-Cl	3
c)	H-C(H)(H)-CO-OCH$_2$CH$_3$	H-C=C(OH)-OCH$_2$CH$_3$	3
d)	H-C(H)(H)-C(=O)-OH	H-C=C(OH)-OH	4
e)	H-C(H)(H)-C(=O)-Ph	H-C=C(OH)-Ph	3

11.3

2-methylcyclohexan-1-ol (enol, double bond toward C6) ⇌ 2-methylcyclohexanone ⇌ 2-methylcyclohex-1-en-1-ol (enol, double bond toward C2)

Enolization toward carbon 6   Enolization toward carbon 2

Enolization can occur in either direction from the carbonyl group. Two different enols are formed because carbon 2 has a methyl substituent and carbon 6 does not.

Carbonyl alpha-Substitution Reactions   185

11.4 As suggested in Practice Problem 11.2, locate the acidic protons and replace one of them with a halogen in order to carry out the alpha-substitution reaction.

a)
$$CH_3\underset{CH_3}{\overset{H}{\underset{|}{C}}}-\overset{O}{\underset{\|}{C}}-\underset{CH_3}{\overset{H}{\underset{|}{C}}}CH_3 + Cl_2 \xrightarrow[\text{solvent}]{CH_3COOH} CH_3\underset{CH_3}{\overset{H}{\underset{|}{C}}}-\overset{O}{\underset{\|}{C}}-\underset{CH_3}{\overset{Cl}{\underset{|}{C}}}CH_3 + HCl$$

b) [cyclohexanone with two CH₃ groups and two α-H's] + Br₂ $\xrightarrow[\text{solvent}]{CH_3COOH}$ [α-bromo cyclohexanone with one H replaced by Br] + HBr

11.5 *Step 1.* Treat 3–pentanone with Br₂ in acetic acid to form the alpha-bromo ketone.

$$CH_3CH_2\overset{O}{\underset{\|}{C}}CH_2CH_3 \xrightarrow[CH_3COOH]{Br_2} CH_3CH_2\overset{O}{\underset{\|}{C}}\underset{Br}{\overset{|}{C}}HCH_3$$

Step 2. Heat the alpha-bromo ketone in pyridine to form 1–penten–3–one.

$$CH_3CH_2\overset{O}{\underset{\|}{C}}\underset{Br}{\overset{|}{C}}HCH_3 \xrightarrow[\Delta]{C_5H_5N} CH_3CH_2\overset{O}{\underset{\|}{C}}CH=CH_2$$

11.6

[Stereochemistry scheme: (R)-3-phenyl-2-butanone ⇌ (via H₃O⁺) enol (C₆H₅, CH₃, OH, CH₃) ⇌ (via H₃O⁺) (S)-3-phenyl-2-butanone]

Loss of the proton at carbon 3 during enolization results in a loss of stereochemical configuration. Reattachment of a proton at carbon 3 can occur from either side of the sp^2 carbon, producing racemic 3–phenyl–2–butanone.

11.7

a) $CH_3CH_2\overset{O}{\underset{\|}{C}}H$ — weakly acidic

b) $(CH_3)_3C\overset{O}{\underset{\|}{C}}CH_3$ — weakly acidic

c) $CH_3\overset{O}{\underset{\|}{C}}OH$ — most acidic / weakly acidic

d) $CH_3CH_2CH_2C\equiv N$ — weakly acidic

186   Chapter 11

e)

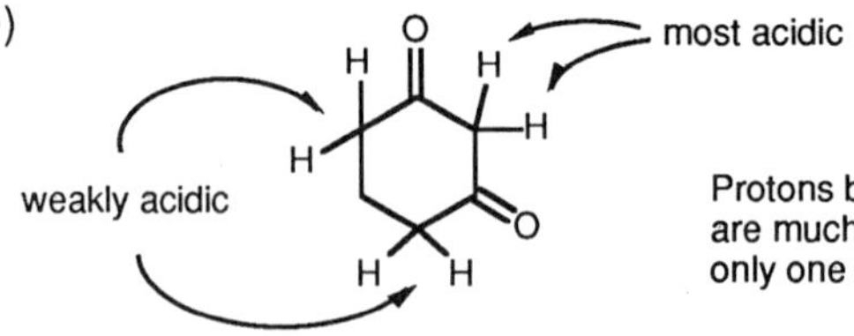

most acidic — Protons between the two carbonyl groups are much more acidic than protons next to only one carbonyl group.

weakly acidic

11.8

a) CH₃CH₂CH(H)–C(=O)H →(Base) CH₃CH₂C⁻(H)–C(=O)H ↔ CH₃CH₂CH=C(H)–O:⁻

b) CH₃–C(=O)–CH₂–H →(Base) or

- CH₃–C(=O)–C⁻H–CH₃ ↔ CH₃–C(O⁻)=CH–CH₃
- CH₃CH₂–C(=O)–C⁻H–H ↔ CH₃CH₂–C(O⁻)=CH₂

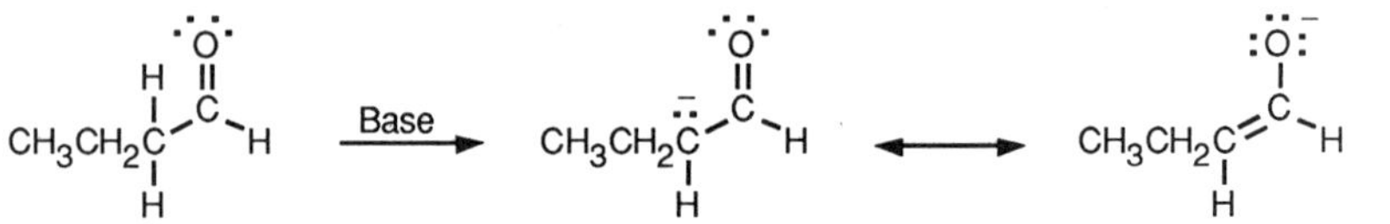

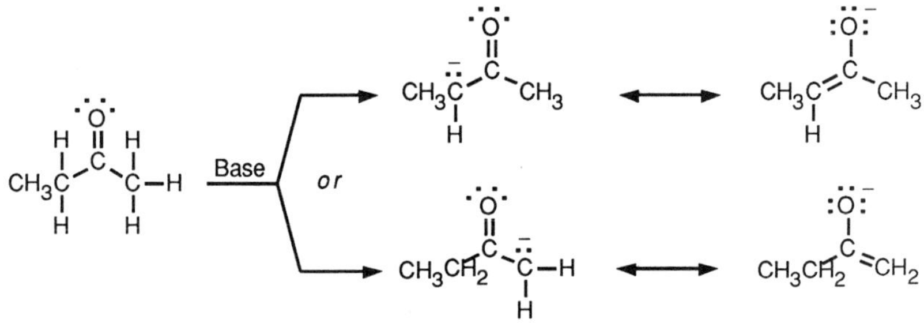

c) (2-methylcyclohexanone) →(Base) or

- enolate at C between C=O and CH₃ ↔ enol form
- enolate at other α-carbon ↔ enol form

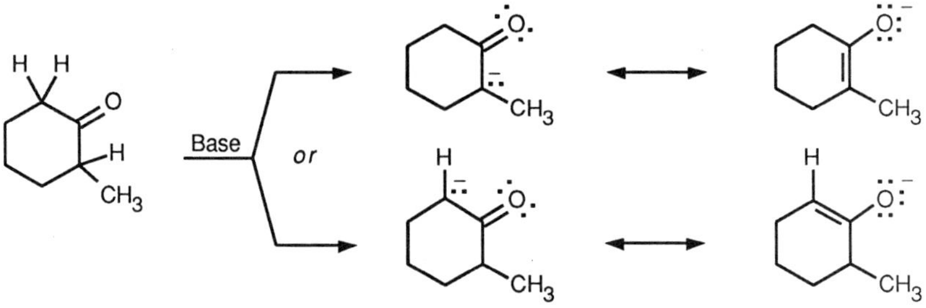

In parts (b) and (c), two different enolate ions can form.

11.9

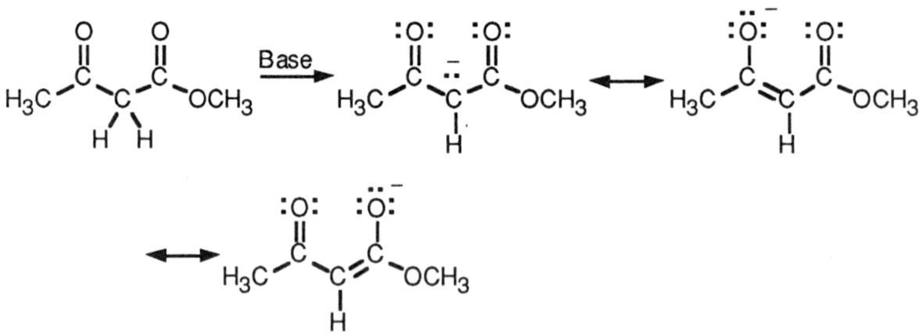

H₃C–C(=O)–CH₂–C(=O)–OCH₃ →(Base) H₃C–C(=O)–C⁻H–C(=O)–OCH₃ ↔ H₃C–C(O⁻)=CH–C(=O)–OCH₃

↔ H₃C–C(=O)–CH=C(O⁻)–OCH₃

The most stable enolate is formed when one of the acidic hydrogens from carbon 2 is abstracted. The enolate resulting from abstraction of a hydrogen at carbon 4 is much less stable.

11.10 The malonic ester synthesis produces substituted acetic acid compounds. Look for the acetic acid component of the compound you want to synthesize. The remainder of the molecule comes from the alkyl halide. Remember that the alkyl halide should be primary or methyl.

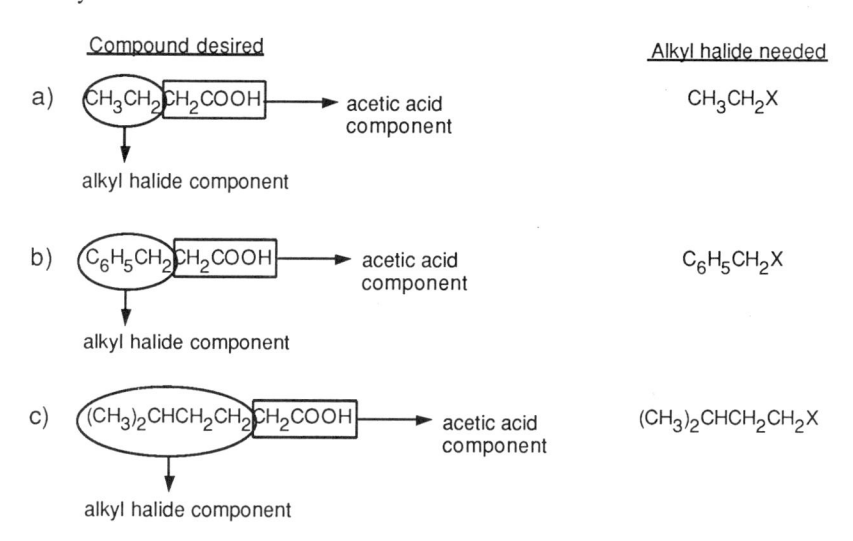

        Compound desired   Alkyl halide needed

a) (CH$_3$CH$_2$)CH$_2$COOH ⟶ acetic acid component   CH$_3$CH$_2$X

    ↓ alkyl halide component

b) (C$_6$H$_5$CH$_2$)CH$_2$COOH ⟶ acetic acid component   C$_6$H$_5$CH$_2$X

    ↓ alkyl halide component

c) ((CH$_3$)$_2$CHCH$_2$CH$_2$)CH$_2$COOH ⟶ acetic acid component   (CH$_3$)$_2$CHCH$_2$CH$_2$X

    ↓ alkyl halide component

11.11 As in Problem 11.10, locate the acetic acid component of the target molecule. This fragment comes from malonic ester. The rest of the molecule comes from the alkyl halide or halides.

a) ((CH$_3$)$_2$CHCH$_2$)CH$_2$COOH ⟶ acetic acid component

    ↓ alkyl halide component

CH$_2$(COOCH$_2$CH$_3$)$_2$ $\xrightarrow{\text{Na}^+ \; ^-\text{OCH}_2\text{CH}_3}$ Na$^+$:CH(COOCH$_2$CH$_3$)$_2$

                                                            ↓ (CH$_3$)$_2$CHCH$_2$Br

(CH$_3$)$_2$CHCH$_2$CH$_2$COOH $\xleftarrow[\Delta]{\text{H}_3\text{O}^+}$ (CH$_3$)$_2$CHCH$_2$CH(COOCH$_2$CH$_3$)$_2$ + NaBr

4–Methylpentanoic acid

+ 2 CH$_3$CH$_2$OH + CO$_2$

188   Chapter 11

b)

CH₃CH₂CH₂CHCOOH → acetic acid component
        |
        CH₃

(alkyl halide components: CH₃CH₂CH₂— circled, CH₃— circled)

alkyl halide components

CH₂(COOCH₂CH₃)₂ →[Na⁺ ⁻OCH₂CH₃]→ Na⁺ :C̄H(COOCH₂CH₃)₂

↓ CH₃CH₂CH₂Br

Na⁺ :C̄(COOCH₂CH₃)₂ ←[Na⁺ ⁻OCH₂CH₃]— CH₃CH₂CH₂CH(COOCH₂CH₃)₂ + NaBr
     |
     CH₂CH₂CH₃

↓ CH₃Br

            CH₃   CH₃
            | |
CH₃CH₂CH₂C(COOCH₂CH₃)₂ →[H₃O⁺, Δ]→ CH₃CH₂CH₂CHCOOH
+ NaBr   2-Methylpentanoic acid
                                    + 2 CH₃CH₂OH + CO₂

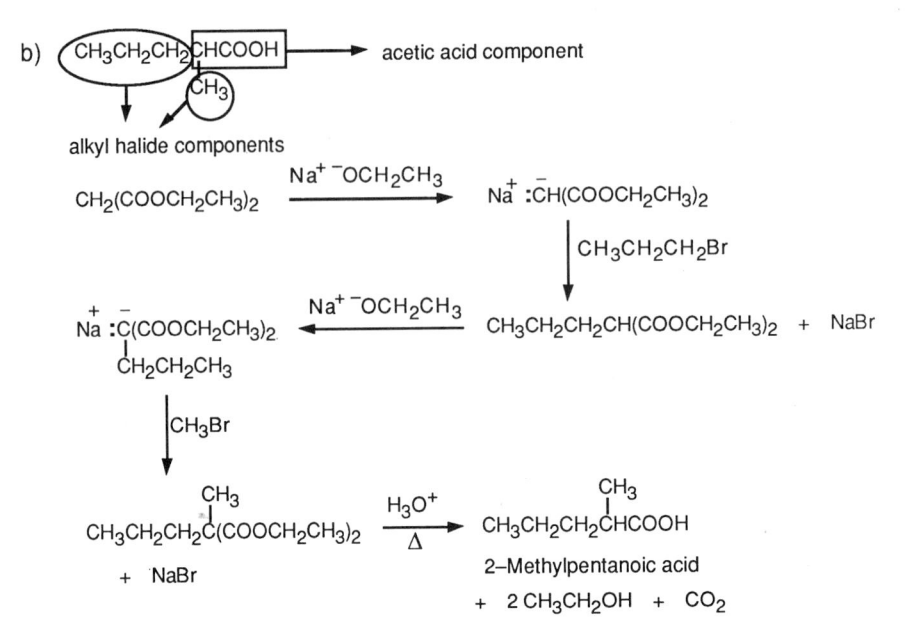

11.12 Compounds that can undergo the aldol reaction must have acidic α–hydrogen atoms. Cyclohexanone is the only compound in this problem capable of undergoing the aldol reaction.

2 (cyclohexanone with α-H's highlighted) →[NaOH]⇌ (aldol product: cyclohexanone with cyclohexanol substituent, OH, H)

No α-hydrogens: (2,2,6,6-tetramethylcyclohexanone with H₃C, H₃C, CH₃, CH₃)   (benzaldehyde, PhCHO)   H₂CO

11.13 It is easier to draw the product of an aldol condensation if the two components are written so that the new bond connects them.

a) CH₃CH₂CH₂CH(=O) + CH₂CH(=O)   →[NaOH]⇌   CH₃CH₂CH₂CH(OH)—CHCH(=O)
                           | |
                           CH₂CH₃   CH₂CH₃

Carbonyl alpha-Substitution Reactions   189

b) 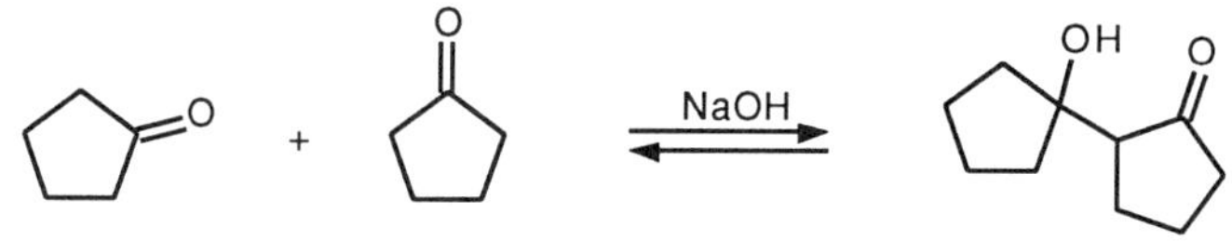 cyclopentanone + cyclopentanone ⇌ (NaOH) 1-(1-hydroxycyclopentyl)-2-cyclopentanone (aldol adduct)

c) 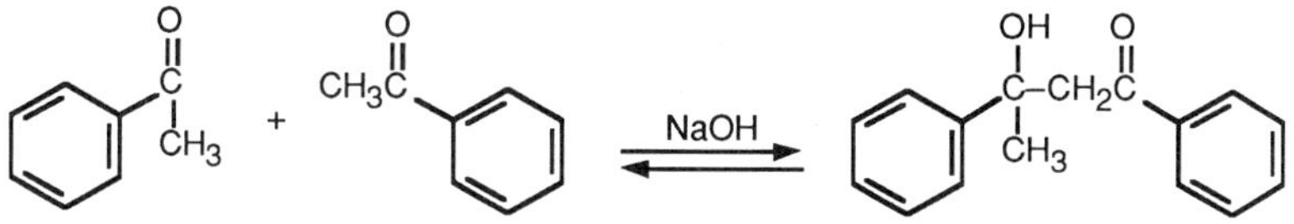 PhC(=O)CH(CH₃) + CH₃C(=O)Ph ⇌ (NaOH) Ph-C(OH)(CH₃)-CH₂-C(=O)-Ph

11.14

a) 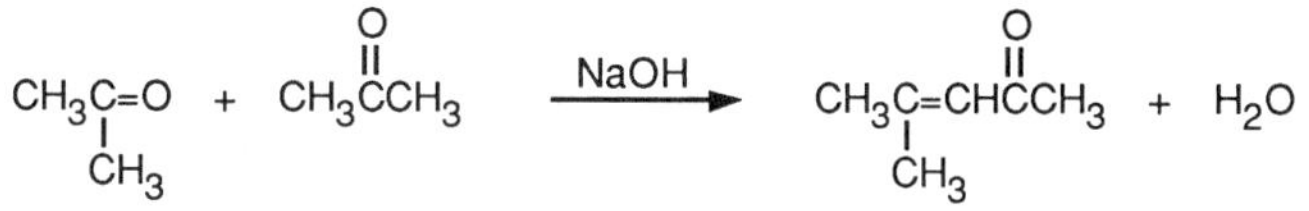 CH₃C(=O)CH₃ (acetone with CH₃ branch shown) + CH₃C(=O)CH₃ →(NaOH)→ CH₃C(CH₃)=CHC(=O)CH₃ + H₂O

b) 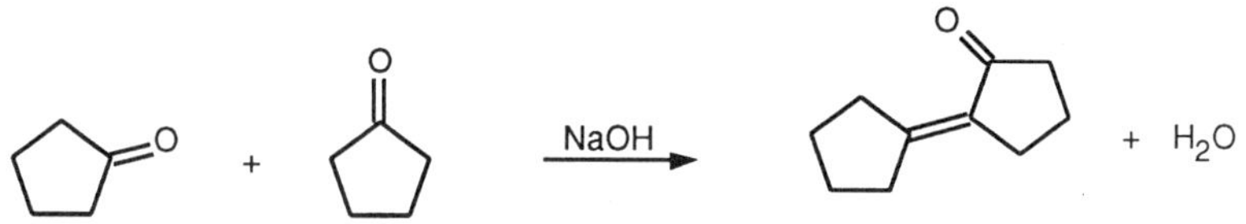 cyclopentanone + cyclopentanone →(NaOH)→ cyclopentylidenecyclopentanone + H₂O

c) 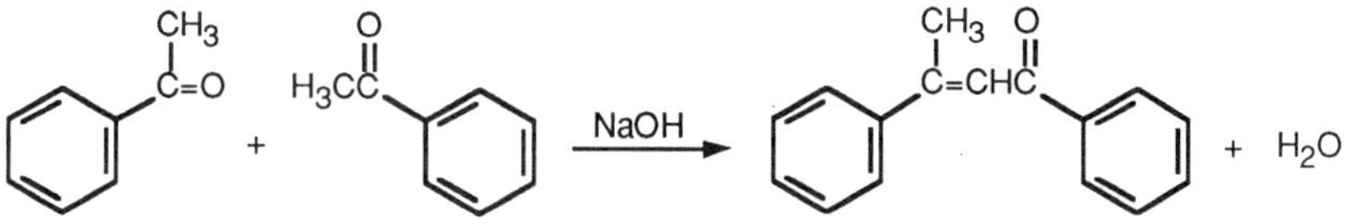 PhC(CH₃)=O + H₃CC(=O)Ph →(NaOH)→ Ph-C(CH₃)=CH-C(=O)-Ph + H₂O

d) 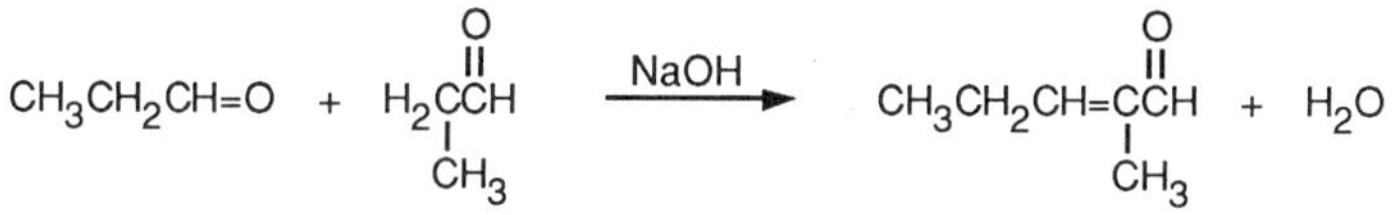 CH₃CH₂CH=O + H₂C(CH₃)CH=O →(NaOH)→ CH₃CH₂CH=C(CH₃)CH=O + H₂O

11.15 Two different enolate anions can be formed from 2–butanone. Each enolate can react with a second molecule of 2–butanone to yield two different enones.

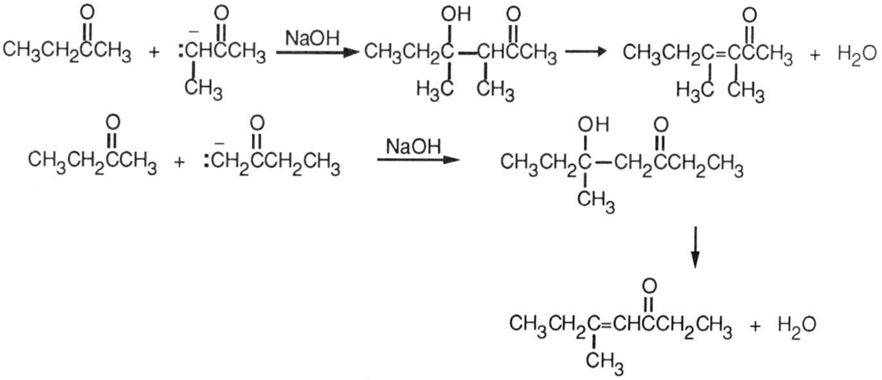

CH₃CH₂C(=O)CH₃ + ⁻:CH(CH₃)C(=O)CH₃ →(NaOH)→ CH₃CH₂C(OH)(CH₃)—CH(CH₃)C(=O)CH₃ → CH₃CH₂C(CH₃)=C(CH₃)C(=O)CH₃ + H₂O

CH₃CH₂C(=O)CH₃ + ⁻:CH₂C(=O)CH₂CH₃ →(NaOH)→ CH₃CH₂C(OH)(CH₃)—CH₂C(=O)CH₂CH₃

↓

CH₃CH₂C(CH₃)=CHC(=O)CH₂CH₃ + H₂O

190   Chapter 11

11.16 As in the aldol condensation, compounds that undergo Claisen condensation reactions must have acidic alpha-hydrogens. Only methyl propanoate (c) yields a Claisen condensation product.

a)
$$H-\overset{\overset{O}{\|}}{C}-OCH_3$$
no alpha hydrogens

b)
$$H_2C=CH-\overset{\overset{O}{\|}}{C}-OCH_3$$
no acidic alpha hydrogens

c) $CH_3CH_2\overset{\overset{O}{\|}}{C}OCH_3 \ + \ \underset{\underset{CH_3}{|}}{CH_2}\overset{\overset{O}{\|}}{C}OCH_3 \ \underset{}{\overset{Na^+ \ ^-OCH_3}{\rightleftharpoons}} \ CH_3CH_2\overset{\overset{O}{\|}}{C}CH_2\overset{\overset{O}{\|}}{C}OCH_3 \ + \ CH_3OH$

11.17 As in the aldol condensation, writing the two Claisen components in the correct orientation makes it easier to predict the product.

a) $(CH_3)_2CHCH_2\overset{\overset{O}{\|}}{C}OCH_3 \ + \ \underset{\underset{CH(CH_3)_2}{|}}{CH_2}\overset{\overset{O}{\|}}{C}OCH_3 \ \underset{2. \ H_3O^+}{\overset{1. \ NaOCH_3}{\longrightarrow}} \ (CH_3)_2CHCH_2\overset{\overset{O}{\|}}{C}-\underset{\underset{CH(CH_3)_2}{|}}{CH}\overset{\overset{O}{\|}}{C}OCH_3 \ + \ CH_3OH$

b) PhCH₂CO₂CH₃ + PhCH(CH₂CO₂CH₃) → (via 1. NaOCH₃, 2. H₃O⁺) → PhCH(CH₂COCHCO₂CH₃·Ph) + CH₃OH

c) (cyclohexyl)CH₂CO₂CH₃ + (cyclohexyl)CH(CH₂CO₂CH₃) → (1. NaOCH₃, 2. H₃O⁺) → (cyclohexyl)CH(CH₂COCHCO₂CH₃·cyclohexyl) + CH₃OH

11.18

a) $HOCH_2\overset{\overset{O}{\|}}{C}CH_3$

All hydrogens are acidic. The alcohol hydrogen is more acidic than the other hydrogens.

b) $HOCH_2CH_2\overset{\overset{O}{\|}}{C}C(CH_3)_3$
       ↑   ↑
    acidic   acidic

Carbonyl alpha-Substitution Reactions 191

c) 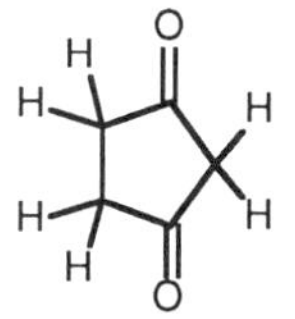 [cyclopentane-1,3-dione with all H labeled] All hydrogens are acidic. The two hydrogens between the two carbonyl groups are much more acidic than the others.

d) 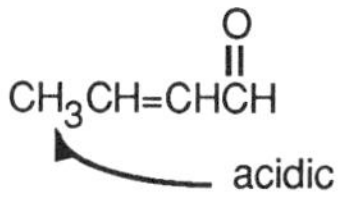 CH₃CH=CHCH=O
      ↖ acidic

11.19

[four enol forms of 1,3-cyclohexanedione shown in equilibrium]

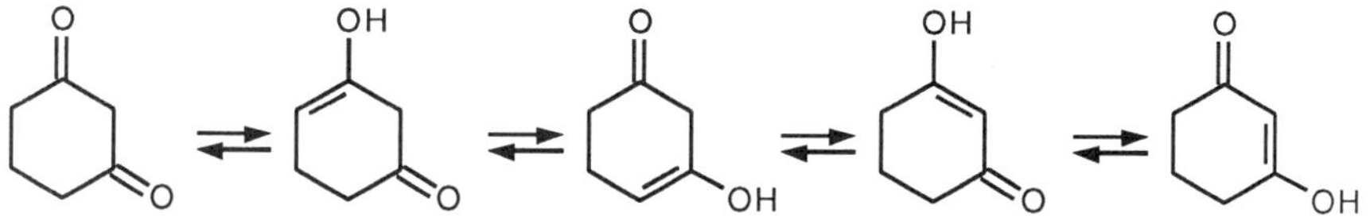

Of the four possible enol forms of 1,3–cyclohexanedione, the last two are the most stable because the enol double bond is conjugated with the second ketone group.

11.20

Least acidic ⟶ Most acidic

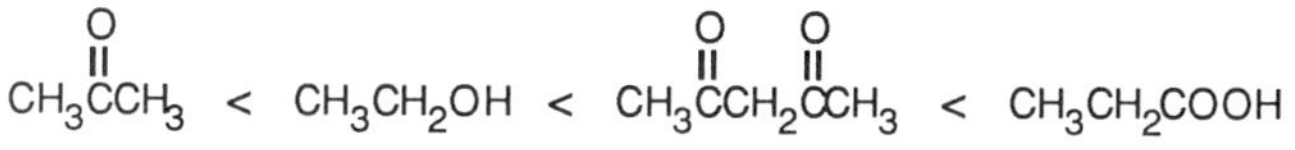

$$CH_3COCH_3 < CH_3CH_2OH < CH_3COCH_2COCH_3 < CH_3CH_2COOH$$

11.21 2,4–Pentanedione is enolized to a greater extent than acetone because its enol form is more stable than the enol form of acetone. This enol stability is due to two factors: (1) conjugation of the enol double bond with the second carbonyl group, and (2) hydrogen bonding between the enol hydrogen and the second carbonyl group.

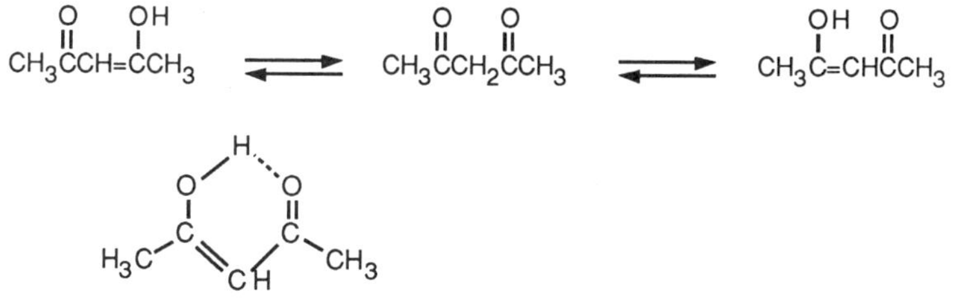

$$CH_3C(OH)=CHCOCH_3 \rightleftharpoons CH_3COCH_2COCH_3 \rightleftharpoons CH_3C(OH)=CHCOCH_3$$

[hydrogen-bonded enol structure shown with six-membered ring containing O–H···O]

11.22

a) 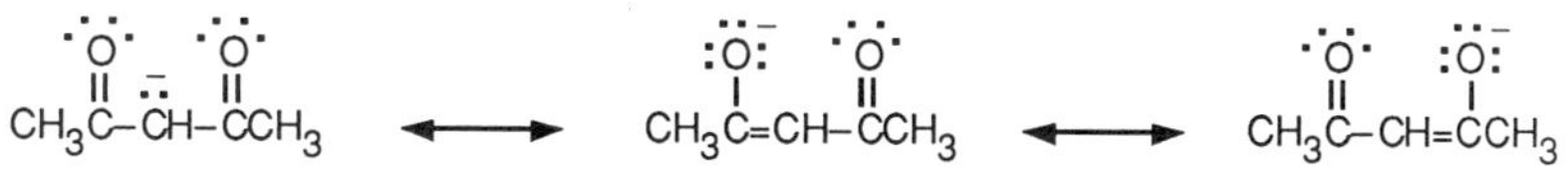 CH₃C(O)–CH⁻–C(O)CH₃ ⟷ CH₃C(O⁻)=CH–C(O)CH₃ ⟷ CH₃C(O)–CH=C(O⁻)CH₃

b) :CH₂–C≡N: ⟷ CH₂=C=N:⁻

c) CH₃CH=CH–CH⁻–C(O)CH₃ ⟷ CH₃CH=CH–CH=C(O⁻)CH₃ ⟷ CH₃CH⁻–CH=CH–C(O)CH₃

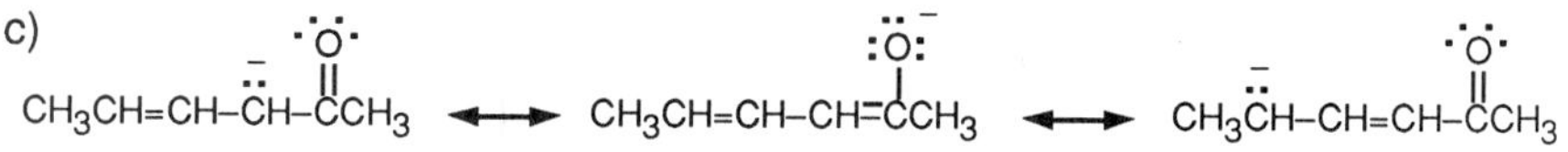

11.23 d)

$:N\equiv C-\overset{..}{\underset{..}{C}}H-\overset{\overset{\overset{..}{O}\cdot}{\|}}{C}OCH_2CH_3 \longleftrightarrow :N\equiv C-CH=\overset{\overset{:\overset{..}{O}:^-}{|}}{C}OCH_2CH_3 \longleftrightarrow {}^-:N=C=CH-\overset{\overset{\overset{..}{O}\cdot}{\|}}{C}OCH_2CH_3$

$\underset{H_3C}{\overset{\overset{\overset{..}{O}\cdot}{\|}}{C}}\diagdown_{CH_3} + D_3O^+ \rightleftarrows \left[\underset{H_2C}{\overset{\overset{\overset{+}{:\overset{..}{O}D}}{|}}{\underset{H}{C}}}\diagdown_{CH_3} \right] \rightleftarrows \underset{H_2C}{\overset{:\overset{..}{O}D}{C}}\diagdown_{CH_3} + D_2\overset{+}{O}H$

enol

$\underset{H_2C}{\overset{:\overset{..}{O}D}{C}}\diagdown_{CH_3} + D_3O^+ \rightleftarrows \left[D_2O + \underset{H_2\overset{|}{C}}{\overset{\overset{\overset{+}{:\overset{..}{O}D}}{|}}{C}}\diagdown_{CH_3} \right] \rightleftarrows \underset{H_2\overset{|}{C}}{\overset{\overset{\overset{\overset{..}{O}\cdot}{\|}}{D}}{C}}\diagdown_{CH_3} + D_3O^+$

11.24 An enolate is generally more reactive than an enol because an enolate is negatively charged and is thus more nucleophilic than an enol.

11.25 In *acid-catalyzed enolization*, protonation of the carbonyl carbon is followed by the loss of a proton on the carbon alpha to the carbonyl group to give the enol. In *base-catalyzed enolization*, abstraction of a proton on the alpha carbon is followed by protonation of the negatively charged oxygen to give the enol.

11.26

[Cyclohexanone with CH₃ at C2 (chiral)] + H₃O⁺ or ⁻:OH ⇌ [enol: cyclohexenol with CH₃] + H₂O ⇌ [Cyclohexanone with CH₃ at C2 (racemized)] + H₃O⁺ or ⁻:OH

Carbon 2 loses its chirality in the step in which the enol double bond is formed. Protonation occurs with equal probability from either side of sp^2-hybridized carbon 2, resulting in racemic product.

11.27

[(R)-3-methylcyclohexanone] + H₃O⁺ or ⁻:OH ⇌ [enol form] + H₂O ⇌ [(R)-3-methylcyclohexanone] + H₃O⁺ or ⁻:OH

(R)–3–methylcyclohexanone is not racemized by acid or base because its stereogenic center is not involved in the enolization reaction.

Carbonyl alpha-Substitution Reactions   193

11.28 Malonic ester has two acidic protons and can be alkylated twice, to yield a compound of the structure $R_2C(COOC_2H_5)_2$. Decarboxylation then gives $R_2CHCOOH$. The alpha proton is only weakly acidic and is no longer activated by two adjacent ester groups. Thus, trialkylation does not occur.

11.29

a) $(CH_3)_3CCHO$

c) benzophenone (diphenyl ketone)

Neither of these compounds undergoes aldol condensation because neither one has acidic protons alpha to the carbonyl group.

b) cyclobutanone + cyclobutanone $\xrightarrow{NaOH}$ β-hydroxy aldol product $\xrightarrow{NaOH}$ α,β-unsaturated ketone

d)
$$CH_3(CH_2)_7CH_2\overset{O}{\overset{\|}{C}}H \;+\; \underset{CH_2(CH_2)_6CH_3}{CH_2\overset{O}{\overset{\|}{C}}H} \;\underset{}{\overset{NaOH}{\rightleftharpoons}}\; CH_3(CH_2)_7CH_2\underset{}{\overset{OH}{\overset{|}{C}}}H-\underset{CH_2(CH_2)_6CH_3}{\overset{O}{\overset{\|}{C}}H}$$

$\updownarrow$ NaOH

$$CH_3(CH_2)_7CH_2CH=\underset{CH_2(CH_2)_6CH_3}{\overset{O}{\overset{\|}{C}}CH}$$

11.30 First, convert malonic ester into its sodium salt.

$$CH_2(COOCH_2CH_3)_2 \;\xrightarrow{\overset{+\;-}{Na}OCH_2CH_3}\; \overset{+}{Na}\; :\overset{-}{C}H(COOCH_2CH_3)_2 \;+\; CH_3CH_2OH$$

a)

$$\overset{+}{Na}\; :\overset{-}{C}H(COOCH_2CH_3)_2 \;+\; CH_3CH_2CH_2Br \;\longrightarrow\; CH_3CH_2CH_2CH(COOCH_2CH_3)_2 \;+\; NaBr$$

$\downarrow H_3O^+, \Delta$

$$CH_3CH_2CH_2CH_2COOCH_2CH_3 \;\underset{H^+}{\overset{CH_3CH_2OH}{\longleftarrow}}\; CH_3CH_2CH_2CH_2COOH \;+\; 2\;CH_3CH_2OH \;+\; CO_2$$

Ethyl pentanoate

b) $CH_3\underset{CH_3}{\overset{|}{C}}HCH_2COOCH_2CH_3$

Ethyl 3-methylbutanoate

194   Chapter 11

This ester would be difficult to synthesize by the malonic ester route because the necessary alkyl halide, 2–bromopropane, is a secondary bromide and would undergo elimination as well as substitution.

c) Na$^+$:CH(COOCH$_2$CH$_3$)$_2$ + CH$_3$Br ⟶ CH$_3$CH(COOCH$_2$CH$_3$)$_2$ + NaBr

$\Big\downarrow$ Na$^+$$^-OCH_2CH_3$

CH$_3$CH$_2$C(COOCH$_2$CH$_3$)$_2$ + NaBr ⟵ CH$_3$CH$_2$Br Na$^+$:C(COOCH$_2$CH$_3$)$_2$
     | |
     CH$_3$ CH$_3$

$\Big\downarrow$ H$_3$O$^+$, Δ

CH$_3$CH$_2$CHCOOH + 2 CH$_3$CH$_2$OH + CO$_2$ $\xrightarrow{\text{CH}_3\text{CH}_2\text{OH}, \text{H}^+}$ CH$_3$CH$_2$CHCOOCH$_2$CH$_3$
     | |
     CH$_3$ CH$_3$

Ethyl 2–methylbutanoate

d)   (CH$_3$)$_3$CCOOCH$_2$CH$_3$
     Ethyl 2,2–dimethylpropanoate

This ester cannot be prepared by the malonic ester route because it is trisubstituted at the alpha carbon.

11.31

2,6-Heptanedione $\xrightarrow{\text{NaOH, ethanol}}$ 3-Methyl-2-cyclohexenone

11.32

3-Cyclohexenone ⇌$^{\text{H}^+}$ enol ⇌$^{\text{H}^+}$ 2-Cyclohexenone

3–Cyclohexenone and 2–cyclohexenone can be interconverted because they are both in equilibrium with the same enol under acidic conditions.

Carbonyl alpha-Substitution Reactions   195

11.33

3–Cyclohexenone and 2–cyclohexenone form the same enolate anion on base treatment and can thus be interconverted.

11.34

As in Problem 11.33, the isomers are in equilibrium through their common enolate.

1.35 Two of the products result from Claisen *self*-condensation.

$$\text{CH}_3\overset{\text{O}}{\underset{\text{OCH}_2\text{CH}_3}{\text{C}}} + \text{CH}_3\overset{\text{O}}{\text{C}}\text{OCH}_2\text{CH}_3 \xrightarrow[\text{2. H}_3\text{O}^+]{\text{1. Na}^+\bar{\text{O}}\text{CH}_2\text{CH}_3} \text{CH}_3\overset{\text{O}}{\text{C}}-\text{CH}_2\overset{\text{O}}{\text{C}}\text{OCH}_2\text{CH}_3 + \text{CH}_3\text{CH}_2\text{OH}$$

Ethyl acetate   Ethyl acetate

$$\text{CH}_3\text{CH}_2\overset{\text{O}}{\underset{\text{OCH}_2\text{CH}_3}{\text{C}}} + \text{CH}_2\overset{\text{O}}{\underset{\text{CH}_3}{\text{C}}}\text{OCH}_2\text{CH}_3 \xrightarrow[\text{2. H}_3\text{O}^+]{\text{1. Na}^+\bar{\text{O}}\text{CH}_2\text{CH}_3} \text{CH}_3\text{CH}_2\overset{\text{O}}{\text{C}}-\overset{\text{O}}{\underset{\text{CH}_3}{\text{CH}}}\text{COCH}_2\text{CH}_3 + \text{CH}_3\text{CH}_2\text{OH}$$

Ethyl propanoate Ethyl propanoate

The other two products are formed from *mixed* Claisen condensations.

$$\text{CH}_3\overset{\text{O}}{\underset{\text{OCH}_2\text{CH}_3}{\text{C}}} + \text{CH}_2\overset{\text{O}}{\underset{\text{CH}_3}{\text{C}}}\text{OCH}_2\text{CH}_3 \xrightarrow[\text{2. H}_3\text{O}^+]{\text{1. Na}^+\bar{\text{O}}\text{CH}_2\text{CH}_3} \text{CH}_3\overset{\text{O}}{\text{C}}-\overset{\text{O}}{\underset{\text{CH}_3}{\text{CH}}}\text{COCH}_2\text{CH}_3 + \text{CH}_3\text{CH}_2\text{OH}$$

Ethyl acetate   Ethyl propanoate

$$\text{CH}_3\text{CH}_2\overset{\text{O}}{\underset{\text{OCH}_2\text{CH}_3}{\text{C}}} + \text{CH}_3\overset{\text{O}}{\text{C}}\text{OCH}_2\text{CH}_3 \xrightarrow[\text{2. H}_3\text{O}^+]{\text{1. Na}^+\bar{\text{O}}\text{CH}_2\text{CH}_3} \text{CH}_3\text{CH}_2\overset{\text{O}}{\text{C}}-\text{CH}_2\overset{\text{O}}{\text{C}}\text{OCH}_2\text{CH}_3 + \text{CH}_3\text{CH}_2\text{OH}$$

Ethyl propanoate   Ethyl acetate

11.36 As in the previous problem, a Claisen *self* condensation between two molecules of ethyl acetate yields one product.

$$CH_3\overset{O}{\underset{OCH_2CH_3}{C}} + CH_3\overset{O}{C}OCH_2CH_3 \xrightarrow[2.\ H_3O^+]{1.\ Na^+\bar{O}CH_2CH_3} CH_3\overset{O}{C}-CH_2\overset{O}{C}OCH_2CH_3 + CH_3CH_2OH$$

The other product results from a *mixed* Claisen condensation.

$$C_6H_5\overset{O}{C}-OCH_2CH_3 + CH_3\overset{O}{C}OCH_2CH_3 \xrightarrow[2.\ H_3O^+]{1.\ Na^+\bar{O}CH_2CH_3} C_6H_5\overset{O}{C}-CH_2\overset{O}{C}OCH_2CH_3 + CH_3CH_2OH$$

Ethyl benzoate does not undergo Claisen self-condensation because it has no alpha protons.

11.37

Aldol self-condensation reaction:

$$CH_3\overset{O}{C}H + CH_3\overset{O}{C}H \xrightarrow{NaOH} CH_3CH=CH\overset{O}{C}H + H_2O$$
$$\text{2-Butenal}$$

Mixed aldol reaction:

$$C_6H_5\overset{O}{C}H + CH_3\overset{O}{C}H \xrightarrow{NaOH} C_6H_5CH=CH\overset{O}{C}H + H_2O$$
$$\text{Cinnamaldehyde}$$

11.38 Try to think backwards from the product to recognize the aldol components. In these two products, the double bond is the new bond formed.

a) $C_6H_5\overset{CH_3}{\underset{}{C}}=\overset{O}{\underset{}{CHCC_6H_5}} + H_2O$

$\uparrow$ NaOH

$C_6H_5\overset{CH_3}{\underset{}{C}}=O + CH_3\overset{O}{C}C_6H_5$

Acetophenone

b) $CH_3\overset{CH_3}{\underset{}{C}}=\overset{O}{\underset{}{CHCCH_3}} + H_2O$

$\uparrow$ NaOH

$CH_3\overset{CH_3}{\underset{}{C}}=O + CH_3\overset{O}{C}CH_3$

Acetone

11.39

$$CH_3CHO + CH_3CHO \xrightarrow{NaOH} CH_3CH=CHCHO \xrightarrow[\text{2. }H_3O^+]{\text{1. LiAlH}_4} CH_3CH=CHCH_2OH$$

$$\downarrow H_2, \text{ Pd catalyst}$$

$$CH_3CH_2CH_2CH_2OH$$
1-Butanol

11.40

$$CH_2(COOCH_2CH_3)_2 + \overset{+}{Na}\overset{-}{O}CH_2CH_3 \rightleftharpoons \overset{+}{Na} :\overset{-}{C}H(COOCH_2CH_3)_2$$

$$\overset{+}{Na} :\overset{-}{C}H(COOCH_2CH_3)_2 + BrCH_2CH_2CH_2CH_2Br \longrightarrow BrCH_2CH_2CH_2CH_2CH(COOCH_2CH_3)_2 + NaBr$$

$$\downarrow \overset{+}{Na}\overset{-}{O}CH_2CH_3$$

cyclopentane-CHCOOH + 2 CH$_3$CH$_2$OH + CO$_2$ $\xleftarrow[\Delta]{H_3O^+}$ cyclopentane-C(COOCH$_2$CH$_3$)$_2$ + NaBr

11.41

[cyclopentanone with CH$_2$-CHO side chain] $\xrightarrow{NaOH}$ [cyclopentenone condensation product] + H$_2$O

11.42

Treatment of either the *cis* or *trans* isomer with base causes enolization alpha to the carbonyl group and results in loss of configuration at the *a*-position. Reattachment of the proton at carbon 2 produces either of the diastereomeric 4–*tert*–butyl–2–methyl-cyclohexanones. In both diastereomers the *tert*–butyl group of carbon 4 occupies the equatorial position for steric reasons. The methyl group of the *cis* isomer is also equatorial; the methyl group of the *trans* isomer is axial. The *trans* isomer is less stable because of interactions of the axial methyl group with the ring protons.

cis ⇌ :OH⁻ ⇌ trans

198   Chapter 11

11.43 First, treat geraniol with PBr$_3$ to form geranyl bromide:

$(CH_3)_2C=CHCH_2CH_2C(CH_3)=CHCH_2Br$.

$CH_2(CO_2CH_2CH_3)_2$ $\xrightarrow{\text{1. Na}^+ {}^-OCH_2CH_3}_{\text{2. Geranyl bromide}}$ $CH_3\underset{|}{\overset{CH_3}{C}}=CHCH_2CH_2\underset{|}{\overset{CH_3}{C}}=CHCH_2CH(CO_2CH_2CH_3)_2$

$\downarrow H_3O^+, \Delta$

$2\ CH_3CH_2OH\ +\ CO_2\ +\ CH_3\underset{|}{\overset{CH_3}{C}}=CHCH_2CH_2\underset{|}{\overset{CH_3}{C}}=CHCH_2CH_2\overset{O}{\overset{\|}{C}}OH$

$\downarrow \begin{array}{c}CH_3CH_2OH\\H^+\end{array}$

$CH_3\underset{|}{\overset{CH_3}{C}}=CHCH_2CH_2\underset{|}{\overset{CH_3}{C}}=CHCH_2CH_2\overset{O}{\overset{\|}{C}}OCH_2CH_3$

Ethyl geranylacetate

11.44 As in the malonic acid synthesis, you should identify the structural fragments of the target compound. The "acetone component" comes from acetoacetic ester; the other component comes from a primary alkyl halide.

a)

$(\underbrace{C_6H_5CH_2}_{}\underbrace{CH_2\overset{O}{\overset{\|}{C}}CH_3}_{})$ ⟶ acetone component

↓

alkyl halide component

$CH_3\overset{O}{\overset{\|}{C}}CH_2COOCH_2CH_3$ $\xrightarrow{\text{Na}^+{}^-O\ CH_2CH_3}$ $\overset{+}{Na}\ :\overset{-}{C}H\overset{O}{\overset{\|}{C}}CH_3$
$\quad\quad\quad\quad\quad\quad\quad\quad\quad\quad\quad\quad\quad\quad\quad\quad\quad\quad\quad\quad\quad COOCH_2CH_3$

$\searrow C_6H_5CH_2Br$

$C_6H_5CH_2CH_2\overset{O}{\overset{\|}{C}}CH_3\ +\ CH_3CH_2OH\ +\ CO_2\ \xleftarrow{H_3O^+}_{\Delta}\ C_6H_5CH_2\underset{|}{CH}\overset{O}{\overset{\|}{C}}CH_3\ +\ NaBr$
4–Phenyl–2–butanone $\quad\quad\quad\quad\quad\quad\quad\quad\quad\quad\quad\quad\quad\quad\quad\quad COOCH_2CH_3$

Carbonyl alpha-Substitution Reactions   199

b) (CH₃)₂CHCH₂―CH₂CCH₃ (with C=O) → acetone component

(CH₃)₂CHCH₂ circled → alkyl halide component

$$CH_3CCH_2COOCH_2CH_3 \xrightarrow{Na^+ \ ^-O \ CH_2CH_3} Na^+ \ :\overset{-}{C}HCCH_3$$
$$\phantom{CH_3CCH_2COOCH_2CH_3 \xrightarrow{Na^+ \ ^-O \ CH_2CH_3}} COOCH_2CH_3$$

$$\downarrow (CH_3)_2CHCH_2Br$$

$$(CH_3)_2CHCH_2CH_2CCH_3 \xleftarrow{H_3O^+, \ \Delta} (CH_3)_2CHCH_2CHCCH_3 + NaBr$$
5-Methyl-2-hexanone   COOCH₂CH₃

+ CH₃CH₂OH + CO₂

c) CH₃CH₂CH₂―CHCCH₃ (with C=O and CH₃) → acetone component

CH₃CH₂CH₂ and CH₃ circled → alkyl halide components

$$CH_3CCH_2COOCH_2CH_3 \xrightarrow{Na^+ \ ^-OCH_2CH_3} Na^+ \ :\overset{-}{C}HCCH_3$$
$$ COOCH_2CH_3$$

$$\downarrow CH_3CH_2CH_2Br$$

$$\underset{\substack{Na^+ \ O \\ CH_3CH_2CH_2CCCH_3 \\ COOCH_2CH_3}}{} \xleftarrow{Na^+ \ ^-OCH_2CH_3} CH_3CH_2CH_2CHCCH_3 + NaBr$$
$$ COOCH_2CH_3$$

$$\downarrow CH_3Br$$

$$\underset{\substack{H_3C \ O \\ CH_3CH_2CH_2C-C-CH_3 \\ COOCH_2CH_3 \\ + \ NaBr}}{} \xrightarrow{H_3O^+, \ \Delta} CH_3CH_2CH_2CHCCH_3 + CH_3CH_2OH + CO_2$$
$$ CH_3$$
$$ \text{3-Methyl-2-hexanone}$$

11.45 The acetoacetic ester synthesis can be used only if the desired compounds fill certain qualifications.

    1) Three carbons must come from acetoacetic ester. In other words, compounds of the form RCOCH$_3$ can't be synthesized by the reaction of RX with acetoacetic ester.

    2) Alkyl halides must be primary or methyl since the acetoacetic ester synthesis is an S$_N$2 reaction.

    3) Trisubstitution at the alpha position can't be achieved by an acetoacetic ester synthesis.

a) 2–Butanone is produced by the reaction of sodium acetoacetate with CH$_3$Br.

b) Phenylacetone can't be produced by an acetoacetic acid synthesis. The necessary halide component, bromobenzene, does not enter into S$_N$2 reactions [see (2) above].

c) Acetophenone can't be produced by an acetoacetic acid synthesis [see (1) above].

d) 3,3–Dimethyl–2–butanone can't be produced by an acetoacetic acid synthesis because it is trisubstituted at the alpha position [see (3) above].

11.46

Chapter Outline

I. Alpha-substitution reactions (Sections 11.1 - 11.6)
   A. Keto-enol tautomerism (Section 11.1)
   B. Reactions involving enols (Sections 11.2 - 11.3)
      1. Mechanisms of alpha-substitution reactions (Section 11.2)
      2. Alpha halogenation of ketones and aldehydes (Section 11.3)
   C. Reactions involving enolate ions (Sections 11.4 - 11.6)
      1. Acidity of alpha hydrogens (Section 11.4)
      2. Reactivity of enolate ions (Section 11.5)
      3. Reactions of enolate ions - Malonic ester synthesis (Section 11.6)
II. Carbonyl condensation reactions (Sections 11.7 - 11.10)
   A. General features (Section 11.7)
   B. Aldol condensation reaction (Section 11.8 - 11.9)
      1. Mechanism of aldol reaction (Section 11.8)
      2. Dehydration of aldol products to yield enones (Section 11.9)
   C. Claisen condensation reaction of esters (Section 11.10)

Study Skills for Chapter 11

After studying this chapter, you should be able to:

1. Draw keto and enol tautomers of carbonyl compounds (Problems 11.1, 11.3, 11.19).
2. Identify acidic hydrogens (Problems 11.2, 11.7, 11.18, 11.20).
3. Draw resonance forms of enolate anions (Problems 11.8, 11.9, 11.22).
4. Formulate mechanisms of enolization (Problems 11.6, 11.23, 11.25, 11.26, 11.27, 11.32, 11.33, 11.34, 11.42, 11.46).
5. Use alpha-halogenation reactions in synthesis (Problems 11.4, 11.5).
6. Synthesize alpha-substituted carboxylic acids by the malonic ester route (Problems 11.10, 11.11, 11.28, 11.30, 11.40, 11.43).
7. Synthesize beta-hydroxy ketones, beta-hydroxy aldehydes, and enones using the aldol condensation (Problems 11.12, 11.13, 11.14, 11.15, 11.29, 11.31, 11.37, 11.38, 11.39, 11.41).
8. Synthesize beta-keto esters using the Claisen condensation (Problem 11.16, 11.17, 11.35, 11.36).

Chapter 12 – Amines

12.1

a) (CH$_3$)$_2$CHNH$_2$ — primary amine

b) (CH$_3$CH$_2$)$_2$NH — secondary amine

c) 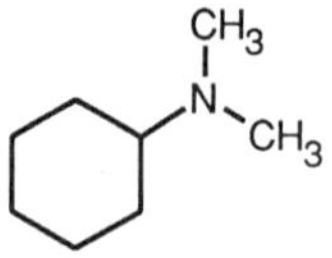 cyclohexyl-N(CH$_3$)$_2$ — tertiary amine

d) 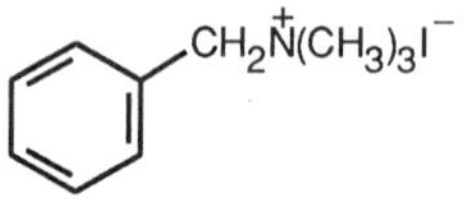 C$_6$H$_5$CH$_2$N$^+$(CH$_3$)$_3$ I$^-$ — quaternary ammonium salt

12.2

a) 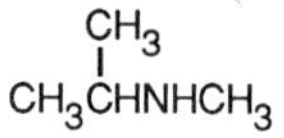 CH$_3$CHNHCH$_3$ with CH$_3$ substituent

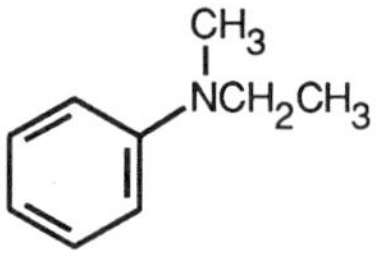

b) C$_6$H$_5$N(CH$_3$)CH$_2$CH$_3$

c) cyclohexyl-N$^+$(CH$_3$)(CH$_2$CH$_2$CH$_3$)(CH$_2$CH$_3$) I$^-$

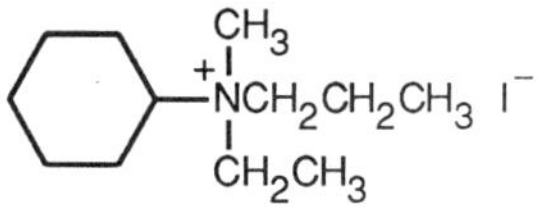

12.3

a) CH$_3$NHCH$_2$CH$_3$ — N-Methylethylamine

b) 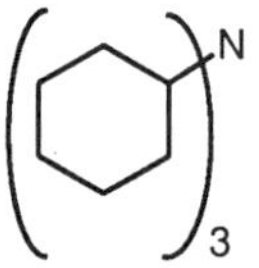 Tricyclohexylamine

c) 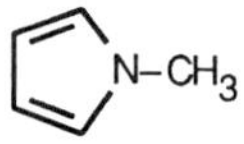 N-Methylpyrrole

d) 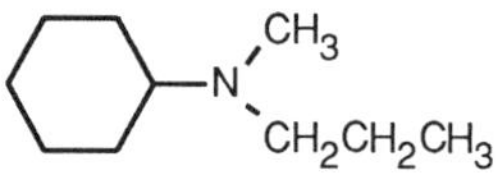 N-Methyl-N-propylcyclohexylamine

e) 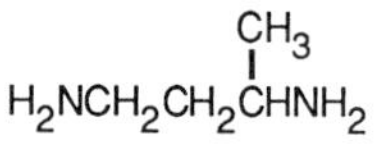 H$_2$NCH$_2$CH$_2$CHNH$_2$ with CH$_3$ substituent — 1,3-Butanediamine

12.4

a) (CH$_3$CH$_2$)$_3$N — Triethylamine

b) 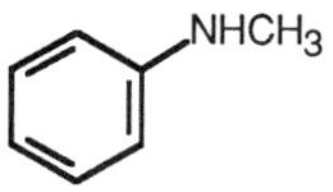 C$_6$H$_5$NHCH$_3$ — N-Methylaniline

c) (CH$_3$CH$_2$)$_4$N$^+$Br$^-$ — Tetraethylammonium bromide

d) 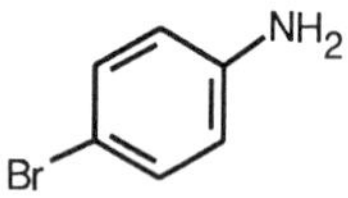 p-Bromoaniline

e) 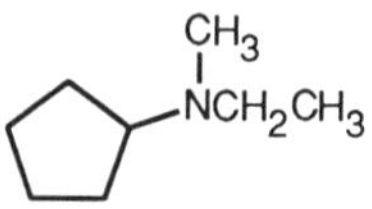 N-Ethyl-N-methylcyclopentylamine

12.5 Because a methyl group is electron-donating, it makes the –NH₂ group of p–methylaniline more electron-rich than the –NH₂ group of aniline. p–Methylaniline is thus the stronger base.

12.6

cyclopentyl-N(CH₃)(H) + HBr ⟶ cyclopentyl-N⁺H₂(CH₃) Br⁻

12.7

	More basic	Less basic
a)	CH₃CH₂NH₂ amine	CH₃CH₂CONH₂ amide
b)	NaOH hydroxide	C₆H₅NH₂ arylamine
c)	CH₃NHCH₃ alkylamine	CH₃NHC₆H₅ arylamine
d)	(CH₃)₃N amine	CH₃OCH₃ ether

The basicity order for the above compounds:

hydroxide > alkylamine > arylamine > amide, ether.

12.8

(R)-lactic acid (OH, COOH, H, CH₃) + (S)-2-butanol (CH₃, HO, H, CH₂CH₃) $\xrightarrow{H^+ \text{ catalyst}}$ ester with R center (OH, COO, H, CH₃) and S center (CH₃, H, CH₂CH₃)

(S)-lactic acid (CH₃, COOH, H, HO) + (S)-2-butanol (CH₃, HO, H, CH₂CH₃) $\xrightarrow{H^+ \text{ catalyst}}$ ester with S center (CH₃, COO, H, HO) and S center (CH₃, H, CH₂CH₃)

The product esters are diastereomers.

12.9 As shown in the previous problem, reaction of (±)–lactic acid with (S)–2-butanol yields a mixture of two diastereomers. Since diastereomers (unlike enantiomers) differ in physical properties and chemical behavior, it should be possible to separate them by a technique such as distillation, fractional crystallization or chromatography. After separation, each ester can be saponified to yield pure (R)– or (S)– lactic acid and (S) 2–butanol.

12.10

a) 3 CH₃CH₂Br $\xrightarrow[\text{2. NaOH}]{\text{1. NH}_3}$ (CH₃CH₂)₃N

Some quaternary ammonium salt is also formed.

Amines   203

204   Chapter 12

12.11 b)   4 CH$_3$Br $\xrightarrow[\text{2. NaOH}]{\text{1. NH}_3}$ (CH$_3$)$_4$N$^+$ Br$^-$

a)   CH$_3$CH$_2$C(=O)NH$_2$ $\xrightarrow[\text{2. H}_2\text{O}]{\text{1. LiAlH}_4}$ CH$_3$CH$_2$CH$_2$NH$_2$
Propanamide

b)   CH$_3$CH$_2$C(=O)NHCH$_2$CH$_2$CH$_3$ $\xrightarrow[\text{2. H}_2\text{O}]{\text{1. LiAlH}_4}$ CH$_3$CH$_2$CH$_2$NHCH$_2$CH$_2$CH$_3$
N–Propylpropanamide

c) 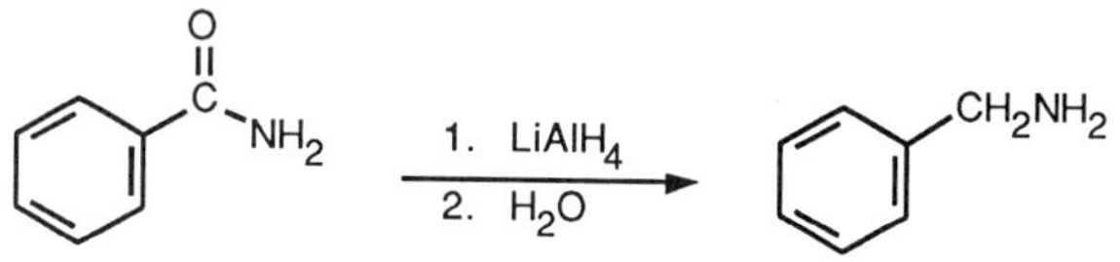   C$_6$H$_5$C(=O)NH$_2$ $\xrightarrow[\text{2. H}_2\text{O}]{\text{1. LiAlH}_4}$ C$_6$H$_5$CH$_2$NH$_2$
Benzamide

12.12

a)   CH$_3$CH(CH$_3$)CH$_2$C≡N $\xrightarrow[\text{2. H}_2\text{O}]{\text{1. LiAlH}_4}$ CH$_3$CH(CH$_3$)CH$_2$CH$_2$NH$_2$
3–Methylbutanenitrile

b) 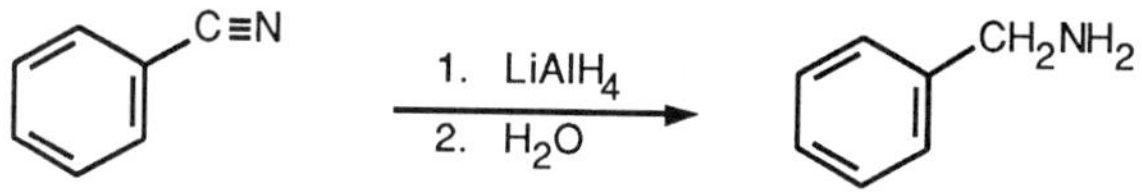   C$_6$H$_5$C≡N $\xrightarrow[\text{2. H}_2\text{O}]{\text{1. LiAlH}_4}$ C$_6$H$_5$CH$_2$NH$_2$
Benzonitrile

12.13

a) 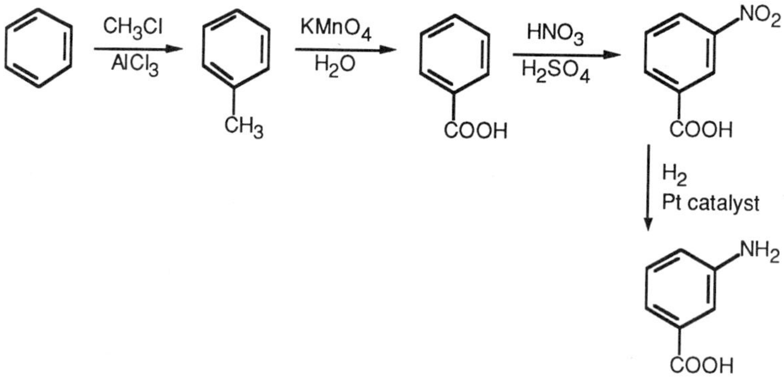   C$_6$H$_6$ $\xrightarrow[\text{AlCl}_3]{\text{CH}_3\text{Cl}}$ C$_6$H$_5$CH$_3$ $\xrightarrow[\text{H}_2\text{O}]{\text{KMnO}_4}$ C$_6$H$_5$COOH $\xrightarrow[\text{H}_2\text{SO}_4]{\text{HNO}_3}$ m-NO$_2$-C$_6$H$_4$-COOH $\xrightarrow{\text{H}_2,\ \text{Pt catalyst}}$ m-NH$_2$-C$_6$H$_4$-COOH

m–Aminobenzoic acid

Amines 205

b)

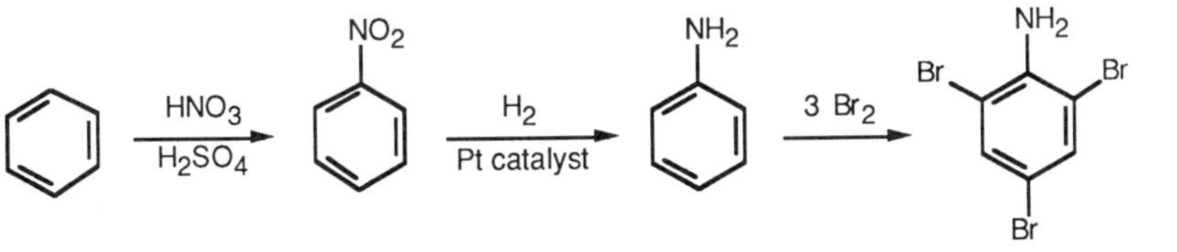

$$\text{benzene} \xrightarrow[\text{H}_2\text{SO}_4]{\text{HNO}_3} \text{nitrobenzene (NO}_2\text{)} \xrightarrow[\text{Pt catalyst}]{\text{H}_2} \text{aniline (NH}_2\text{)} \xrightarrow{3\ \text{Br}_2} \text{2,4,6-Tribromoaniline}$$

Bromination of the reactive aniline ring requires no catalyst.

12.14

$$\text{CH}_3\overset{\text{O}}{\text{C}}\text{Cl} + \text{HN(CH}_2\text{CH}_3)_2 \xrightarrow{\text{NaOH}} \text{CH}_3\overset{\text{O}}{\text{C}}\text{N(CH}_2\text{CH}_3)_2$$

12.15

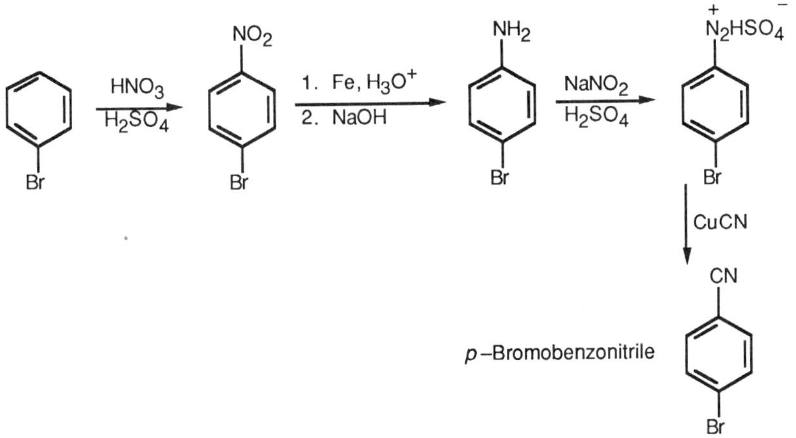

Bromobenzene $\xrightarrow[\text{H}_2\text{SO}_4]{\text{HNO}_3}$ p-bromonitrobenzene $\xrightarrow[\text{2. NaOH}]{\text{1. Fe, H}_3\text{O}^+}$ p-bromoaniline $\xrightarrow[\text{H}_2\text{SO}_4]{\text{NaNO}_2}$ p-bromo diazonium (N_2HSO_4) $\xrightarrow{\text{CuCN}}$ p-Bromobenzonitrile

12.16

a) benzene $\xrightarrow[\text{AlCl}_3]{\text{CH}_3\text{Cl}}$ toluene $\xrightarrow[\text{H}_2\text{O}]{\text{KMnO}_4}$ benzoic acid $\xrightarrow[\text{FeBr}_3]{\text{Br}_2}$ m-Bromobenzoic acid

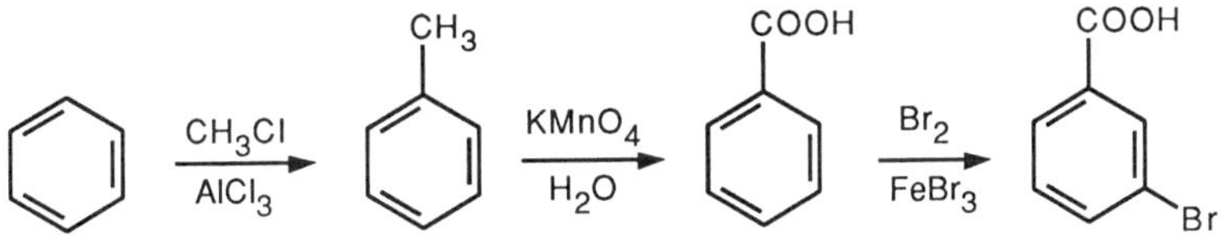

12.17

12.18

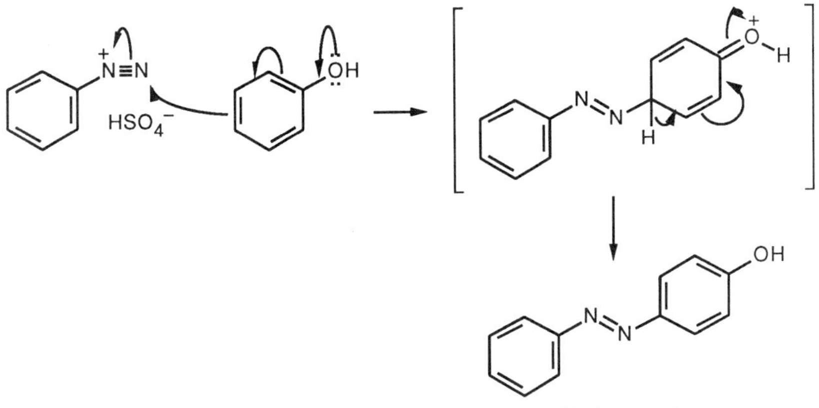
N≡N
HSO4−
OH
N=N
O
H
H
N=N
OH

p–Hydroxyazobenzene

12.19

a)

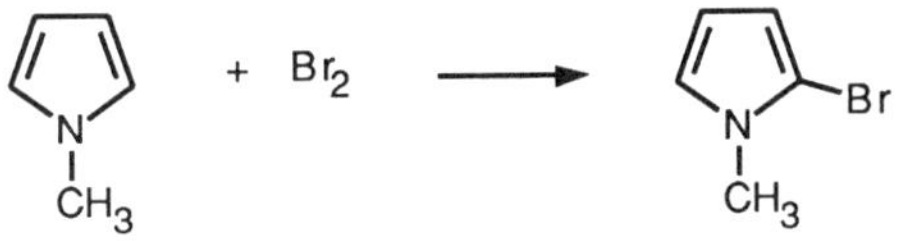
N
CH3
+ Br2
Br

b)

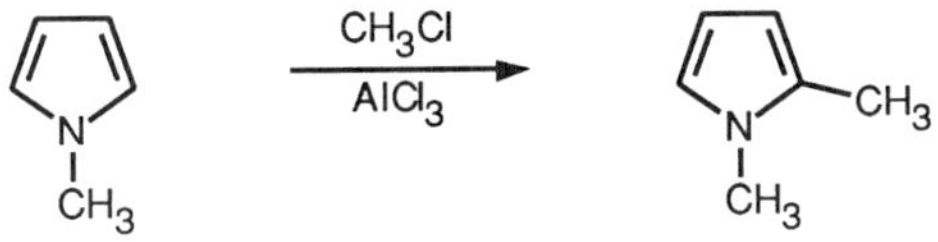
N
CH3
CH3Cl
AlCl3

c)

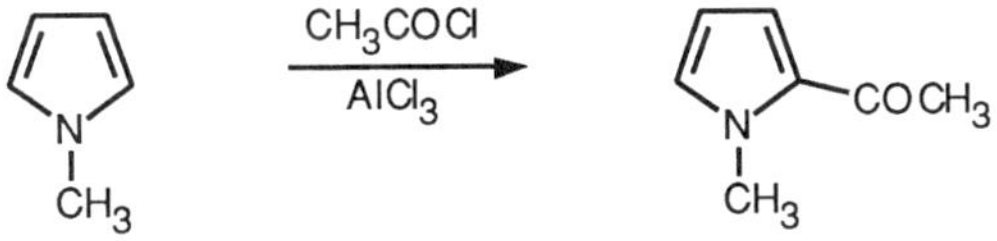
N
CH3
CH3COCl
AlCl3
COCH3

12.20

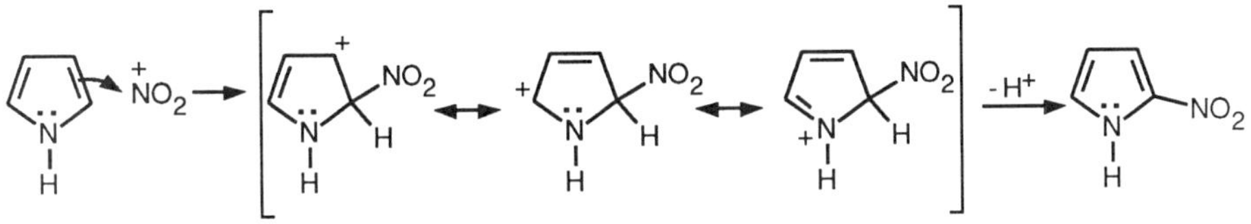
N
H
NO2
- H+

12.21

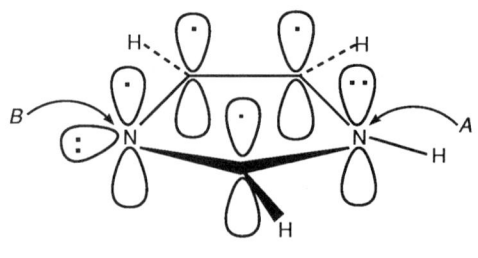

Imidazole

Nitrogen atom *B* is more basic (more pyridine-like) because its lone pair of electrons lies in an sp^2 orbital and is more available for donation to a Lewis acid. The lone pair of electrons of nitrogen *A*, which lies in a *p* orbital that is part of the ring pi system, is more "pyrrole-like."

12.22

Dextromethorphan

(1) aromatic ring
(2) quaternary **carbon**
(3) two carbons
(4) tertiary amine

12.23

a) Lysergic acid diethylamide — tertiary amine, secondary amine

b) Caffeine — tertiary amine

12.24

a) *N,N*–Dimethylaniline

b) *N*–Methylcyclohexylamine

c) (Cyclohexylmethyl)amine

Amines 209

d) (2-Methylcyclohexyl)amine — cyclohexane with NH₂ and CH₃

e) (CH₃)₂NCH₂CH₂COOH
3-(N,N-Dimethylamino)propanoic acid

12.25

a) 2,4-Dibromoaniline (benzene with NH₂, 2-Br, 4-Br)

b) Cyclopentyl-CH₂CH₂NH₂
(2-Cyclopentylethyl)amine

c) Cyclopentyl-NHCH₂CH₃
N-Ethylcyclopentylamine

d) Cyclopentyl-N(CH₃)₂
N,N-Dimethylcyclopentylamine

e) Pyrrolidine-N-CH₂CH₂CH₃
N-Propylpyrrolidine

f) H₂NCH₂CH₂CH₂CN
4-Aminobutanenitrile

12.26

Dimethylamine (showing hydrogen bonding between two (CH₃)₂NH molecules)

Trimethylamine (CH₃)₃N:

Even though dimethylamine has a lower molecular weight than trimethylamine, it boils at a higher temperature. Liquid dimethylamine forms hydrogen bonds that must be broken in the boiling process. Since extra energy must be added to break these hydrogen bonds, dimethylamine has a higher boiling point than trimethylamine, which does not form hydrogen bonds.

12.27

2-(3,4,5-Trimethoxyphenyl)ethylamine

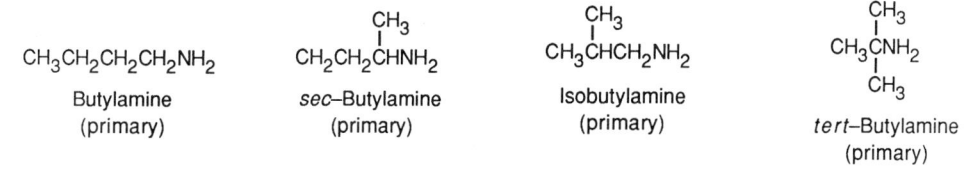

12.28

CH₃CH₂CH₂CH₂NH₂
Butylamine
(primary)

CH₂CH₂CH(CH₃)NH₂
sec-Butylamine
(primary)

(CH₃)₂CHCH₂NH₂
Isobutylamine
(primary)

(CH₃)₃CNH₂
tert-Butylamine
(primary)

210   Chapter 12

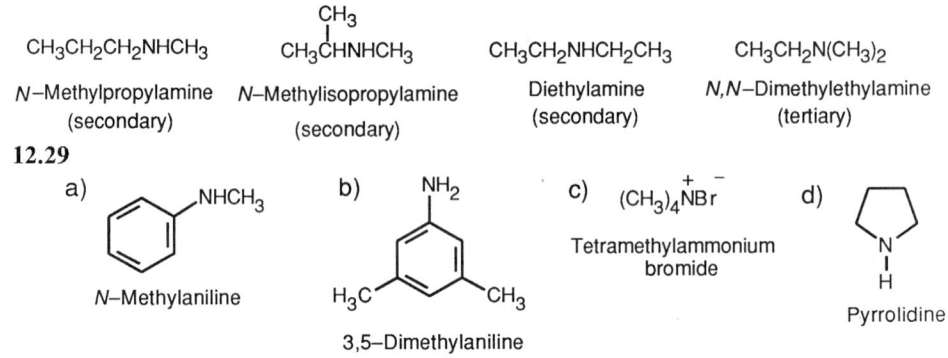

CH₃CH₂CH₂NHCH₃   CH₃CH(CH₃)NHCH₃   CH₃CH₂NHCH₂CH₃   CH₃CH₂N(CH₃)₂

N-Methylpropylamine   N-Methylisopropylamine   Diethylamine   N,N-Dimethylethylamine
(secondary)   (secondary)   (secondary)   (tertiary)

12.29

a) C₆H₅—NHCH₃   b) 3,5-(CH₃)₂C₆H₃—NH₂   c) (CH₃)₄N⁺ Br⁻   d) Pyrrolidine (N-H)

N-Methylaniline   3,5-Dimethylaniline   Tetramethylammonium
                                                bromide

12.30

a) CH₃CH₂CH₂NH₂ + CH₃Br ⟶ CH₃CH₂CH₂NHCH₃
   Polyalkylated products are also formed.

b) C₆H₁₁—NH₂ + HBr ⟶ C₆H₁₁—NH₃⁺ Br⁻

c) CH₃CH₂C(=O)NH₂ $\xrightarrow{1.\ LiAlH_4;\ 2.\ H_2O}$ CH₃CH₂CH₂NH₂

d) C₆H₅—C≡N $\xrightarrow{1.\ LiAlH_4;\ 2.\ H_2O}$ C₆H₅—CH₂NH₂

12.31 In these reactions, polyalkylated products are also formed.

a) CH₃CH₂CH₂CH₂CH₂CH₂Br $\xrightarrow{1.\ NH_3;\ 2.\ NaOH}$ CH₃CH₂CH₂CH₂CH₂CH₂NH₂
   Hexylamine

b) C₆H₅—CH₂Br $\xrightarrow{1.\ NH_3;\ 2.\ NaOH}$ C₆H₅—CH₂NH₂
   Benzylamine

c) 4 CH₃I + NH₃ ⟶ (CH₃)₄N⁺ I⁻
   Tetramethylammonium iodide

d) C₆H₁₁—Br $\xrightarrow{1.\ NH_3;\ 2.\ NaOH}$ C₆H₁₁—NH₂ $\xrightarrow{1.\ CH_3I;\ 2.\ NaOH}$ C₆H₁₁—NHCH₃
   N-Methylcyclohexylamine

12.32 In parts (a) and (b), overalkylation products are also formed.

a) $CH_3CH_2CH_2CH_2Br$ $\xrightarrow[\text{2. NaOH}]{\text{1. excess NH}_3}$ $CH_3CH_2CH_2CH_2NH_2$
 Butylamine

b) $CH_3CH_2CH_2CH_2Br$ $\xrightarrow[\text{2. NaOH}]{\text{1. Butylamine (from a.)}}$ $(CH_3CH_2CH_2CH_2)_2NH$
 Dibutylamine

c) $CH_3CH_2CH_2CH_2Br$ $\xrightarrow{\text{NaCN}}$ $CH_3CH_2CH_2CH_2CN$ $\xrightarrow[\text{2. H}_2\text{O}]{\text{1. LiAlH}_4}$ $CH_3CH_2CH_2CH_2CH_2NH_2$
 Pentylamine

12.33

a) $CH_3CH_2CH_2CH_2OH$ $\xrightarrow[\text{H}_3\text{O}^+]{\text{CrO}_3}$ $CH_3CH_2CH_2COOH$ $\xrightarrow{\text{SOCl}_2}$ $CH_3CH_2CH_2\overset{\overset{O}{\|}}{C}Cl$

$\downarrow$ 2 NH$_3$

$CH_3CH_2CH_2CH_2NH_2$ $\xleftarrow[\text{2. H}_2\text{O}]{\text{1. LiAlH}_4}$ $CH_3CH_2CH_2\overset{\overset{O}{\|}}{C}NH_2$
Butylamine

b) $CH_3CH_2CH_2CH_2NH_2$ [from (a)]
 +
 $CH_3CH_2CH_2\overset{\overset{O}{\|}}{C}Cl$ [from (a)] $\xrightarrow{\text{NaOH}}$ $CH_3CH_2CH_2CH_2NH\overset{\overset{O}{\|}}{C}CH_2CH_2CH_3$

$\downarrow$ 1. LiAlH$_4$ 2. H$_2$O

$(CH_3CH_2CH_2CH_2)_2NH$
Dibutylamine

c) $CH_3CH_2CH_2CH_2OH$ $\xrightarrow{\text{PBr}_3}$ $CH_3CH_2CH_2CH_2Br$ $\xrightarrow{\text{NaCN}}$ $CH_3CH_2CH_2CH_2CN$

$\downarrow$ 1. LiAlH$_4$ 2. H$_2$O

$CH_3CH_2CH_2CH_2CH_2NH_2$
Pentylamine

12.34

a) 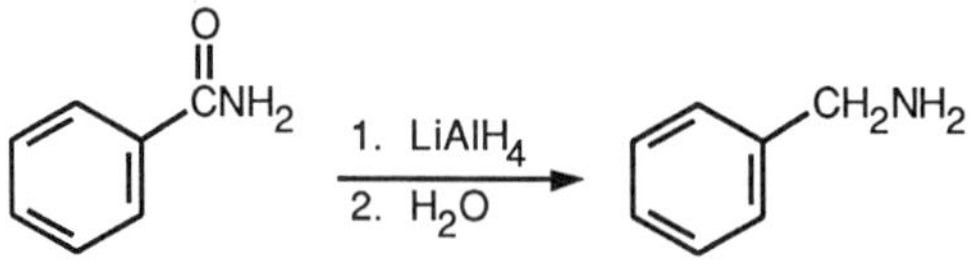 PhC(O)NH₂ —1. LiAlH₄; 2. H₂O→ PhCH₂NH₂

b) 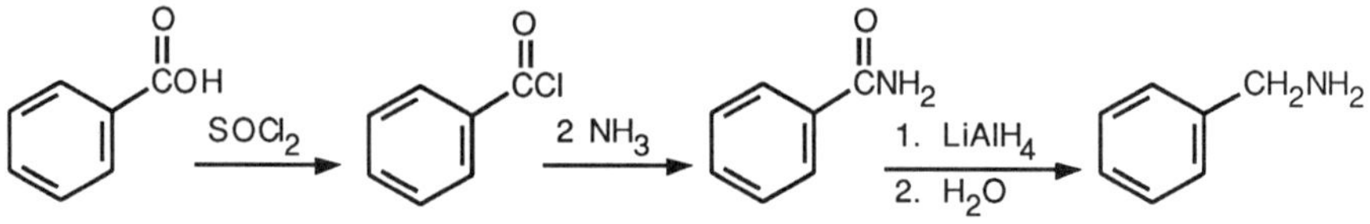 PhCOOH —SOCl₂→ PhCOCl —2 NH₃→ PhC(O)NH₂ —1. LiAlH₄; 2. H₂O→ PhCH₂NH₂

c) 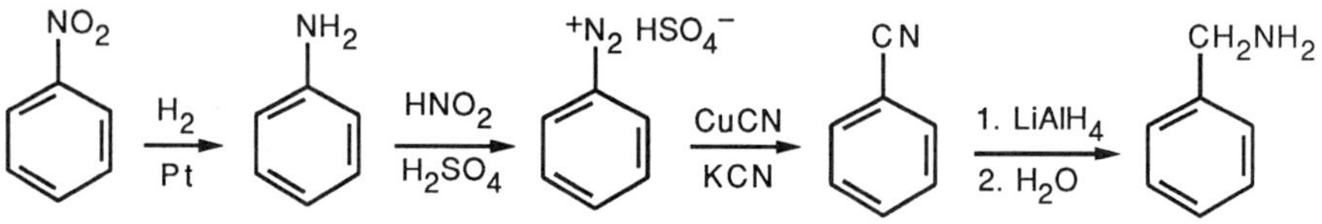 PhNO₂ —H₂, Pt→ PhNH₂ —HNO₂, H₂SO₄→ PhN₂⁺ HSO₄⁻ —CuCN, KCN→ PhCN —1. LiAlH₄; 2. H₂O→ PhCH₂NH₂

d) 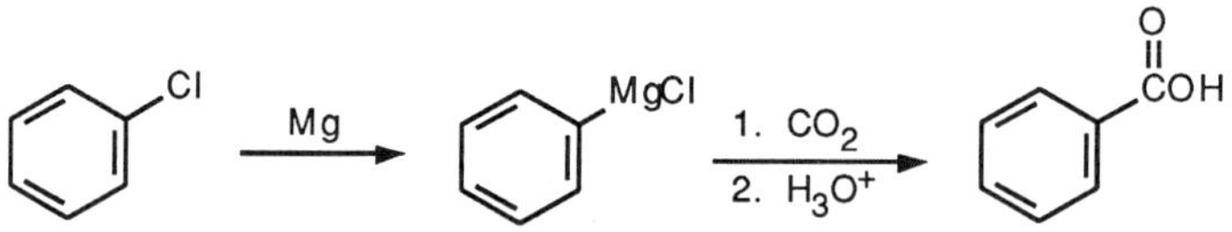 PhCl —Mg→ PhMgCl —1. CO₂; 2. H₃O⁺→ PhCOOH [see (b)]

12.35

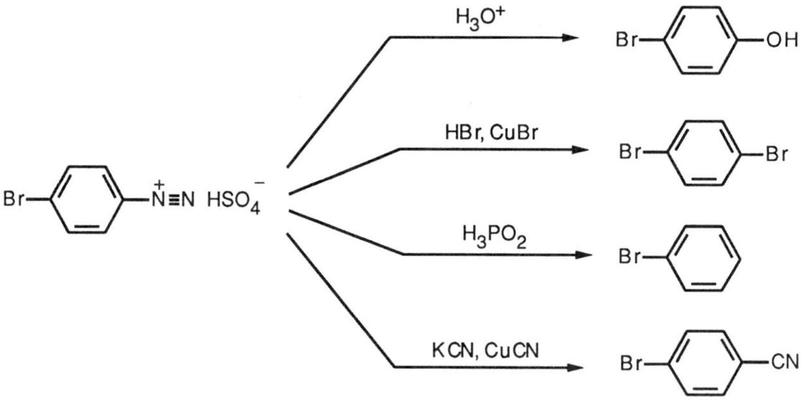

Br–C₆H₄–N≡N⁺ HSO₄⁻:
- H₃O⁺ → Br–C₆H₄–OH
- HBr, CuBr → Br–C₆H₄–Br
- H₃PO₂ → Br–C₆H₅
- KCN, CuCN → Br–C₆H₄–CN

12.36

PhNH₂ —HNO₂, H₂SO₄→ PhN₂⁺ HSO₄⁻ —NaCN→ PhCN —1. NaOH, H₂O; 2. H₃O⁺→ PhCOOH

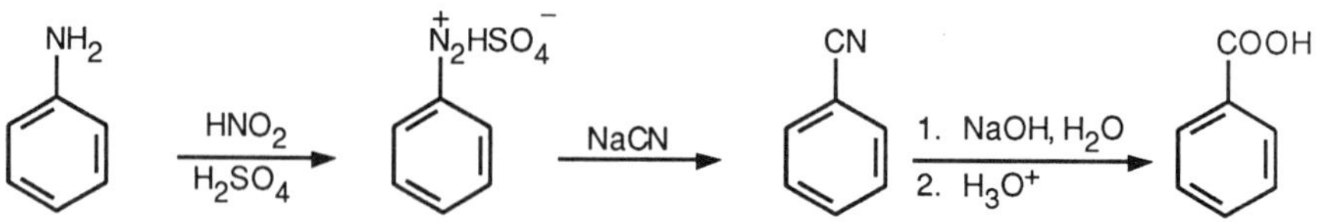

12.37

a) $CH_3CH_2CH_2CH_2\overset{O}{\overset{\|}{C}}NH_2 \xrightarrow[\text{2. }H_2O]{\text{1. LiAlH}_4} CH_3CH_2CH_2CH_2CH_2NH_2$

b) $CH_3CH_2CH_2CH_2CN \xrightarrow[\text{2. }H_2O]{\text{1. LiAlH}_4} CH_3CH_2CH_2CH_2CH_2NH_2$

c) $CH_3CH_2CH_2CH_2COOH \xrightarrow{SOCl_2} CH_3CH_2CH_2CH_2\overset{O}{\overset{\|}{C}}Cl$

$\downarrow$ 2 NH$_3$

$CH_3CH_2CH_2CH_2CH_2NH_2 \xleftarrow[\text{2. }H_2O]{\text{1. LiAlH}_4} CH_3CH_2CH_2CH_2\overset{O}{\overset{\|}{C}}NH_2$

12.38 $CH_3CH_2NH_2$ is more basic than $CF_3CH_2NH_2$ because the electron-withdrawing fluorine atoms make the nitrogen of $CF_3CH_2NH_2$ more electron-poor and less basic.

12.39 The aldehyde group, which is electron-withdrawing, makes *p*–aminobenzaldehyde less basic than aniline.

12.40 Triethylamine is more basic than aniline because the lone-pair electrons of the aniline nitrogen are delocalized by orbital overlap with the aromatic ring pi electrons and are less available for donation to an acid. Thus, the reaction of triethylammonium chloride with aniline does not occur.

12.41

$H_2C=CHCH=CH_2 \xrightarrow{Cl_2} ClCH_2CH=CHCH_2Cl \xrightarrow{\text{2 NaCN}} NCCH_2CH=CHCH_2CN$

$\downarrow$ 1. LiAlH$_4$
$\quad$ 2. H$_2$O

$H_2NCH_2CH_2CH_2CH_2CH_2CH_2NH_2 \xleftarrow[\text{Pd catalyst}]{H_2} H_2NCH_2CH_2CH=CHCH_2CH_2NH_2$

1,6–Hexanediamine

12.42

$HO\overset{O}{\overset{\|}{C}}CH_2CH_2CH_2CH_2\overset{O}{\overset{\|}{C}}OH \xrightarrow{SOCl_2} Cl\overset{O}{\overset{\|}{C}}CH_2CH_2CH_2CH_2\overset{O}{\overset{\|}{C}}Cl$

$\downarrow$ 4 NH$_3$

$H_2NCH_2CH_2CH_2CH_2CH_2CH_2NH_2 \xleftarrow[\text{2. }H_2O]{\text{1. LiAlH}_4} H_2N\overset{O}{\overset{\|}{C}}CH_2CH_2CH_2CH_2\overset{O}{\overset{\|}{C}}NH_2$

12.43

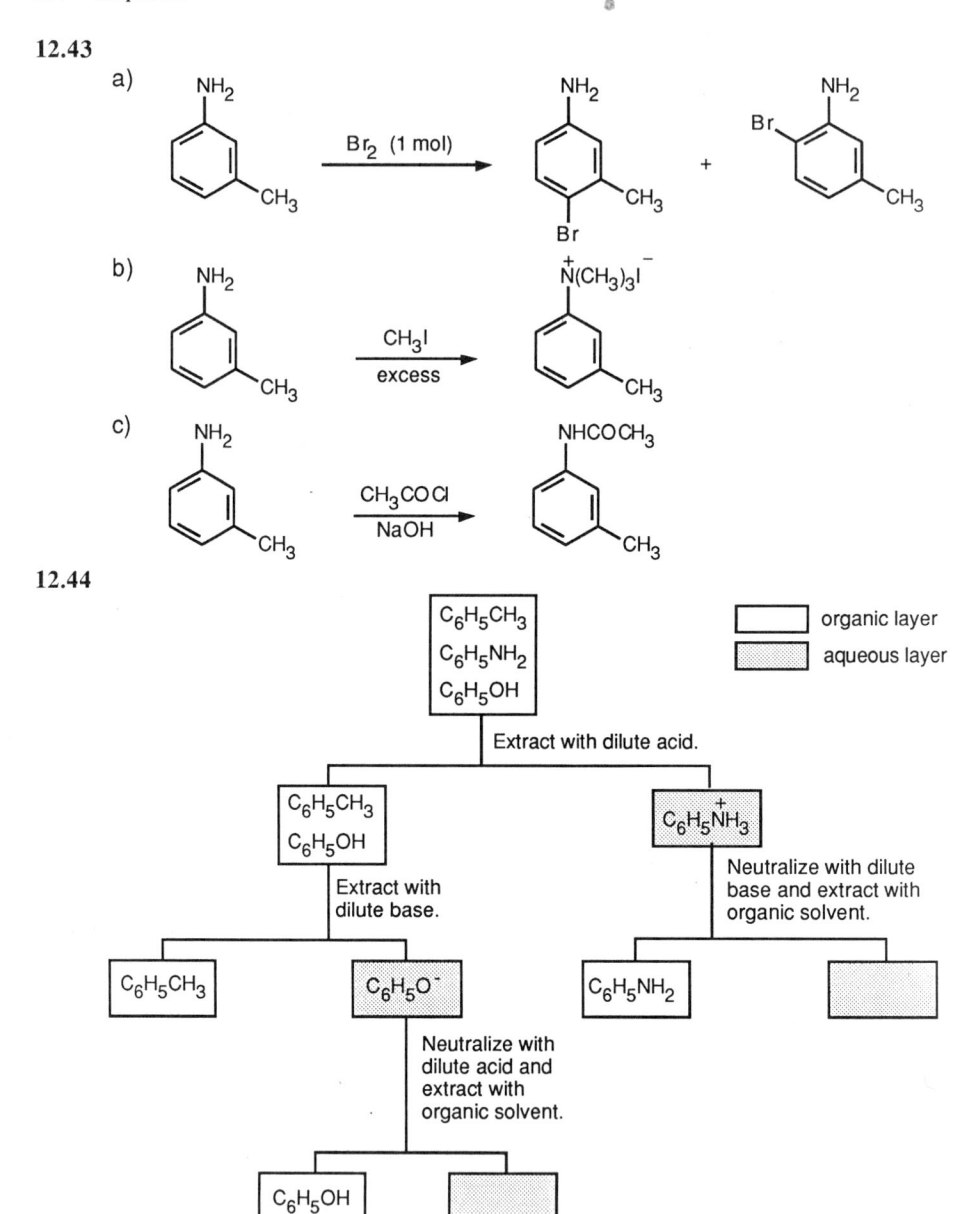

a) 3-methylaniline + Br₂ (1 mol) → 4-bromo-3-methylaniline (NH₂ with CH₃ and Br para to NH₂) + 2-bromo-5-methylaniline (Br ortho to NH₂)

b) 3-methylaniline + CH₃I (excess) → 3-methylphenyltrimethylammonium iodide, N⁺(CH₃)₃ I⁻

c) 3-methylaniline + CH₃COCl / NaOH → N-(3-methylphenyl)acetamide (NHCOCH₃)

12.44

Mixture: C₆H₅CH₃, C₆H₅NH₂, C₆H₅OH

Extract with dilute acid.
- Organic layer: C₆H₅CH₃, C₆H₅OH
  - Extract with dilute base.
    - Organic layer: C₆H₅CH₃
    - Aqueous layer: C₆H₅O⁻
      - Neutralize with dilute acid and extract with organic solvent.
        - Organic layer: C₆H₅OH
        - Aqueous layer
- Aqueous layer: C₆H₅NH₃⁺
  - Neutralize with dilute base and extract with organic solvent.
    - Organic layer: C₆H₅NH₂
    - Aqueous layer

□ organic layer
▓ aqueous layer

12.45 Diphenylamine is less basic than aniline. The nitrogen lone-pair electrons of diphenylamine can overlap with the pi electron system of both aromatic rings, causing even greater electron delocalization than occurs for aniline.

12.46

a) 2-Ethylpyrrole

b) 2,3-Dimethylaniline

c) 3-Methylindole

12.47

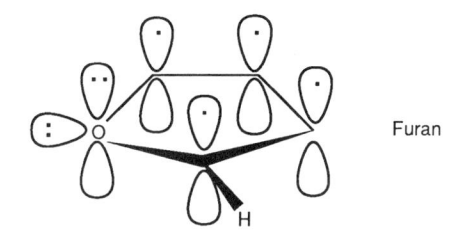

Furan

One oxygen lone pair is in a *p* orbital that is part of the pi electron system of furan. The other oxygen lone pair is in an sp^2 orbital that lies in the plane of the furan ring.

12.48

Furan + Br_2 → 2-bromofuran

12.49

Benzene → (HNO$_3$/H$_2$SO$_4$) → nitrobenzene → (H$_2$, Pd catalyst) → aniline → (3 Br$_2$) → 2,4,6-tribromoaniline → (HNO$_2$, H$_2$SO$_4$) → diazonium salt → (H$_3$PO$_2$) → 1,3,5-Tribromobenzene

12.50

Three resonance forms contribute to the stability of the imide anion.

216   Chapter 12

12.51

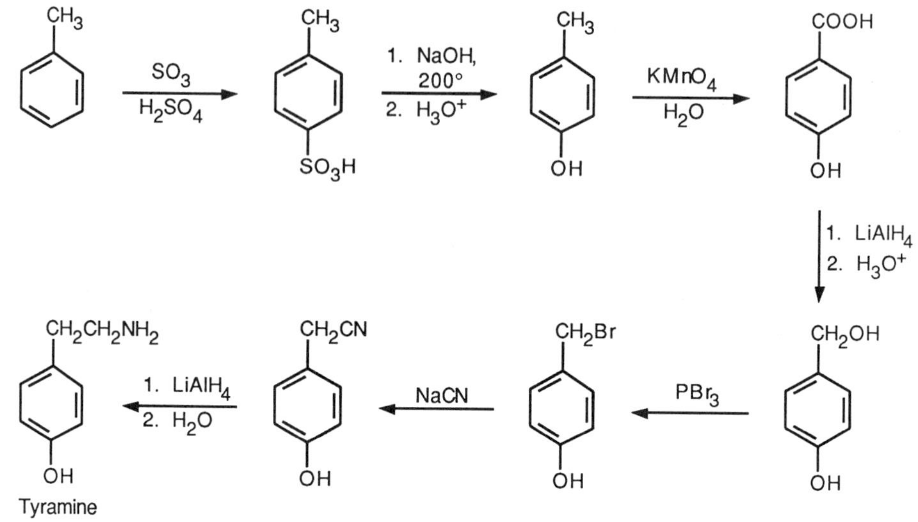

[Synthesis scheme: Toluene → (SO₃/H₂SO₄) → p-toluenesulfonic acid → (1. NaOH, 200°; 2. H₃O⁺) → p-cresol → (KMnO₄/H₂O) → p-hydroxybenzoic acid → (1. LiAlH₄; 2. H₃O⁺) → p-hydroxybenzyl alcohol → (PBr₃) → p-hydroxybenzyl bromide → (NaCN) → p-hydroxybenzyl cyanide → (1. LiAlH₄; 2. H₂O) → Tyramine (p-HOC₆H₄CH₂CH₂NH₂)]

12.52

Atropine [bicyclic N-CH₃ structure with OC(=O)-CH(CH₂OH)C₆H₅ ester] →(⁻OH, H₂O)→ bicyclic N-CH₃ alcohol + C₆H₅CH(CH₂OH)COOH (Tropic acid) →(H₂SO₄)→ Tropidene

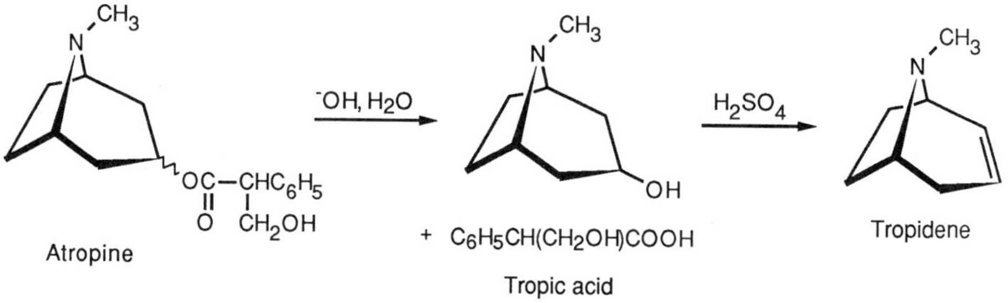

12.53 The reaction of trimethylamine with ethylene oxide is an S_N2 reaction.

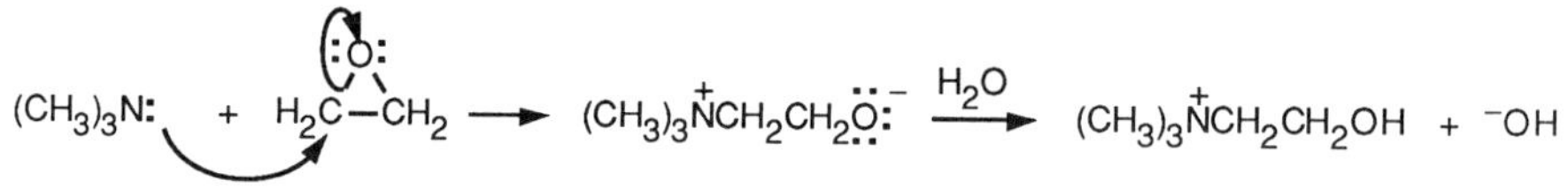

$(CH_3)_3N: \; + \; H_2C\overset{\overset{\displaystyle :\ddot{O}:}{\diagdown}}{-}CH_2 \; \longrightarrow \; (CH_3)_3\overset{+}{N}CH_2CH_2\ddot{O}:^- \; \xrightarrow{H_2O} \; (CH_3)_3\overset{+}{N}CH_2CH_2OH \; + \; {}^-OH$

12.54

Chapter Outline

I. Characteristics of amines (Sections 12.1 - 12.4)
   A. Naming amines (Section 12.1)
   B. Structure and properties (Section 12.2)
   C. Basicity of amines (Section 12.3)
      1. pK_b of amines
      2. Basicity of arylamines
      3. Basicity of amides
   D. Resolution of enantiomers via amine salts (Section 12.4)

II. Synthesis of amines (Section 12.5)
   A. S_N2 reactions of alkyl halides
   B. Reduction of amides and nitriles
   C. Reduction of nitroarenes

III. Reactions of amines (Section 12.6)
    A. Alkylations and acylation
    B. Sandmeyer reaction
        1. Conversion of arylamines into haloarenes
        2. Conversion of arylamines into arenenitriles
        3. Conversion of arylamines into phenols
        4. Conversion of arylamines into arenes
    C. Diazonium coupling reactions
IV. Heterocyclic amines (Section 12.7)
    A. Pyrrole
    B. Pyridine
    C. Fused-ring heterocycles
V. Naturally occurring amines: morphine alkaloids (Section 12.8)

Study Skills for Chapter 12

After studying this chapter, you should be able to:

1. Classify amines as primary, secondary, tertiary, or quaternary (Problems 12.1, 12.2, 12.23, 12.28, 12.29).
2. Name amines (Problems 12.3, 12.25, 12.28).
3. Draw structures of amines corresponding to given names (Problems 12.4, 12.24, 12.27, 12.28, 12.46).
4. Predict the basicity of amines (Problems 12.5, 12.7, 12.38, 12.39, 12.40, 12.45, 12.50).
5. Describe the use of amines to resolve enantiomers (Problems 12.8, 12.9).
6. Synthesize amines (Problems 12.10, 12.11, 12.12, 12.13, 12.31, 12.32, 12.33, 12.34, 12.37, 12.41, 12.42, 12.51, 12.53).
7. Predict the products of reactions of amines (Problems 12.6, 12.14, 12.30, 12.35, 12.36, 12.43).
8. Use amines in synthetic sequences (Problems 12.15, 12.16, 12.17, 12.18, 12.49, 12.54).
9. Be familiar with the properties and chemistry of simple heterocyclic amines (Problems 12.19, 12.20, 12.21, 12.47, 12.48).

Chapter 13 – Structure Determination

13.1

$$\varepsilon = h\nu = \frac{hc}{\lambda}$$

where $h = 6.62 \times 10^{-34}$ J·s
$c = 3 \times 10^{10}$ cm/s
λ = wavelength in centimeters

For infrared radiation ($\lambda = 10^{-4}$ cm):

$$\varepsilon = \frac{(6.62 \times 10^{-34} \text{ J·s})(3 \times 10^{10} \frac{\text{cm}}{\text{s}})}{10^{-4} \text{ cm}} = 2 \times 10^{-19} \text{ J}$$

For an X ray with $\lambda = 3 \times 10^{-7}$ cm:

$$\varepsilon = \frac{(6.62 \times 10^{-34} \text{ J·s})(3 \times 10^{10} \frac{\text{cm}}{\text{s}})}{3 \times 10^{-7} \text{ cm}} = 7 \times 10^{-17} \text{ J}$$

Thus, an X ray is of higher energy than infrared radiation.

13.2 First, convert radiation in cm to radiation in Hz by the equation:

$$\nu = \frac{c}{\lambda} = \frac{3 \times 10^{10} \frac{\text{cm}}{\text{s}}}{9 \times 10^{-4} \text{ cm}} = 3 \times 10^{13} \text{ Hz}$$

The equation $\varepsilon = h\nu$ says that the greater the value of ν, the greater the energy. Thus, radiation with $\nu = 3 \times 10^{13}$ Hz ($\lambda = 9 \times 10^{-4}$ cm) is higher in energy than radiation with $\nu = 4 \times 10^{9}$ Hz.

13.3

IR absorption	Due to:
a) 1715 cm^{-1}	ketone
b) 1540 cm^{-1}	nitro group
c) 2210 cm^{-1}	nitrile group or alkyne
d) 1720 cm^{-1} 2500-3100 cm^{-1}	carboxylic acid
e) 3500 cm^{-1}	alcohol
1735 cm^{-1}	ester

13.4 To use IR to distinguish between isomers, find a strong IR absorption present in one isomer that is absent in the other isomer.

a) CH₃CH₂OH

Strong hydroxyl band at 3400–3640 cm⁻¹

CH₃OCH₃

No band in the region 3400–3640 cm⁻¹

b) CH₃CH₂CH₂CH₂CH=CH₂

Alkene bands at 3020-3100 cm⁻¹ and 1650–1670 cm⁻¹

(benzene ring)

No bands in alkene region

c) CH₃CH₂COOH

Strong, broad band at 2500–3100 cm⁻¹

HOCH₂CH₂CHO

Strong broad band at 3400–3640 cm⁻¹

13.5 Only conjugated compounds absorb in the region 200 nm - 400 nm.

	Compound	Absorption at 200-400 nm?
a)	 1,3-cyclohexadiene	yes
b)	 1,4-cyclohexadiene	no
c)	H₂C=CHCOCH₃	yes
d)	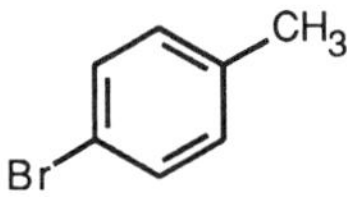 4-bromotoluene	yes
e)	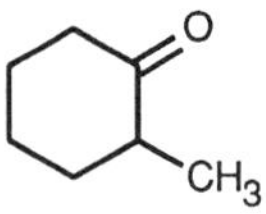 2-methylcyclohexanone	no
f)	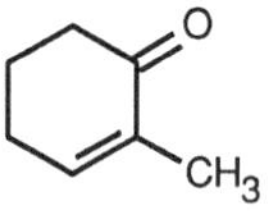 2-methyl-2-cyclohexenone	yes

13.6 Both compounds are conjugated and absorb in the ultraviolet region. 1,3,5–Hexatriene has a longer system of conjugated bonds and absorbs at a longer wavelength (lower energy) than 1,3–hexadiene.

Structure Determination 221

13.7 From Problem 13.1, we find that $\lambda = 10^{-4}$ cm is a typical value for the wavelength of infrared radiation. Use the equation $\upsilon = c / \lambda$ to find the frequency of infrared radiation.

$$\upsilon = \frac{c}{\lambda} = \frac{3 \times 10^{10} \text{ cm/sec}}{10^{-4} \text{ cm}} = 3 \times 10^{14} \text{ Hz}$$

Since υ for NMR radiation (6×10^7 Hz) is less than υ for ultraviolet radiation, the amount of energy used by NMR spectroscopy is less than that used by IR spectroscopy.

13.8

	Compound	Signals in ^{1}H NMR	Signals in ^{13}C NMR
a)	CH_4	1	1
b)	CH_3CH_3	1	1
c)	$CH_3CH_2CH_3$	2	2
d)	cyclohexane	1	1
e)	CH_3OCH_3	1	1
f)	benzene	1	1
g)	$(CH_3)_3COH$	2	2
h)	CH_3CH_2Cl	2	2
i)	$(CH_3)_2C=C(CH_3)_2$	1	2

13.9

$$\begin{array}{c} CH_3 \\ \diagdown \\ C=C \\ \diagup \diagdown \\ Cl H \end{array} \begin{array}{c} H \\ \diagup \\ \\ \end{array}$$

2–Chloropropene

The two protons on C1 are not equivalent; one proton is on the same side of the double bond as chlorine and the other proton is on the opposite side. Since 2–chloropropene has three different types of protons, it shows three signals in its ^{1}H NMR spectrum.

13.10 a) 2.1 ppm x 60 MHz = 126 Hz

b) The position of absorption in δ units is 2.1 δ for both a 60 MHz and a 100 MHz instrument. A measurement in δ units is independent of the operating frequency of the NMR spectrometer.

c) 2.1 ppm x 100 MHz = 210 Hz

13.11

$$\delta \text{ (in ppm)} = \frac{\text{observed chemical shift (in Hz)}}{(60 \text{ MHz}/10^6)}$$

a) $\delta = \dfrac{436 \text{ Hz}}{60 \text{ Hz}} = 7.27$ ppm for $CHCl_3$

b) $\delta = \dfrac{183 \text{ Hz}}{60 \text{ Hz}} = 3.05$ ppm for CH_3Cl

c) $\delta = \dfrac{208 \text{ Hz}}{60 \text{ Hz}} = 3.47$ ppm for CH_3OH

d) $\delta = \dfrac{318 \text{ Hz}}{60 \text{ Hz}} = 5.30$ ppm for CH_2Cl_2

13.12

Compound	^1H Chemical Shift
a) CH_3CH_3	0.88 δ
b) CH_3COCH_3	2.17 δ
c) C_6H_6	7.17 δ
d) $(CH_3)_3N$	2.22 δ

13.13

H₃C—⟨benzene ring⟩—CH₃   p-Xylene

There are two absorbances in the ^1H NMR spectrum of p-xylene. The four ring protons absorb at 7.0 δ, and the six methyl-group protons absorb at 2.3 δ. The peak ratio of methyl protons:ring protons is 3:2.

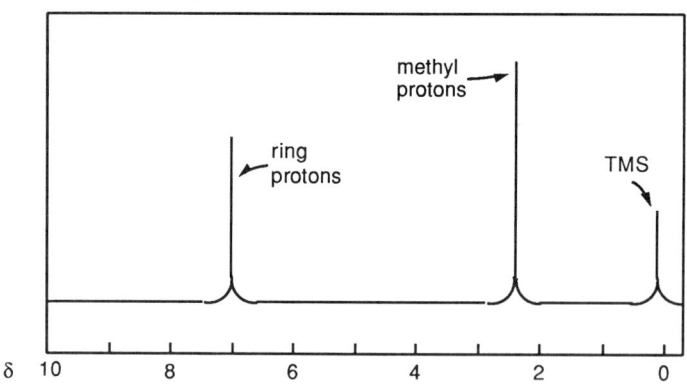

13.14

	Compound	Proton	Number of Adjacent Protons	Splitting
a)	(C$\overset{1}{H}_3$)$_3$C$\overset{2}{H}$	1	1	doublet
		2	9	multiplet
b)	C$\overset{1}{H}_3$C$\overset{2}{H}$Br$_2$	1	1	doublet
		2	3	quartet
c)	C$\overset{1}{H}_3$OC$\overset{2}{H}_2$C$\overset{3}{H}_2$Br	1	0	singlet
		2	2	triplet
		3	2	triplet
d)	C$\overset{1}{H}_3$C$\overset{2}{H}_2$$\overset{O}{\overset{||}{C}}OC\overset{3}{H}_3$	1	2	triplet
		2	3	quartet
		3	0	singlet
e)	ClC$\overset{1}{H}_2$C$\overset{2}{H}_2$C$\overset{1}{H}_2$Cl	1	2	triplet
		2	4	quintet
f)	(C$\overset{1}{H}_3$)$_2$C$\overset{2}{H}$$\overset{O}{\overset{||}{C}}OC\overset{3}{H}_3$	1	1	doublet
		2	6	septet
		3	0	singlet

13.15

a) C$_2$H$_6$O has only one kind of proton, with no neighbors.

    CH$_3$OCH$_3$

b) C$_3$H$_6$O$_2$ has two kinds of protons; neither kind has neighbors.

    CH$_3$COOCH$_3$

c) C$_3$H$_7$Cl has two kinds of protons; one kind of proton has six neighbors, and the other kind has one neighbor.

    (CH$_3$)$_2$CHCl

13.16

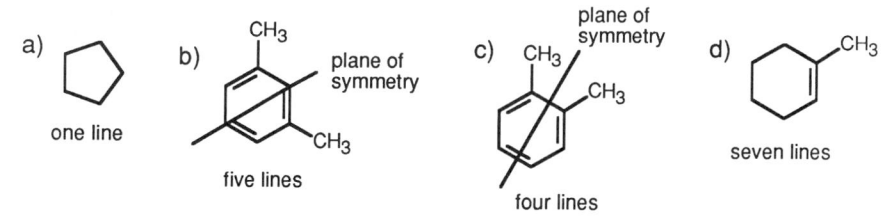

a) cyclopentane — one line

b) 1,2,4-trimethylbenzene (plane of symmetry) — five lines

c) 1,3,5-trimethylbenzene (plane of symmetry) — four lines

d) 1-methylcyclohexene — seven lines

13.17

a) CH$_3$CH$_2$CH$_2$CH$_2$CH$_2$CH=CH$_2$ seven peaks

b) (CH$_3$)$_2$CHCH$_2$CH$_2$CH$_3$ five peaks

c) (CH$_3$)$_2$CHCH$_2$Cl   three peaks

13.18

IR Absorption	Due to:
a) 1670 cm^{-1}	alkene or carbonyl group
b) 1735 cm^{-1}	ester group
c) 1540 cm^{-1}	nitro group
d) 1715 cm^{-1} 2500-3100 cm^{-1}	carboxylic acid group

13.19

	Compound	IR Absorption	Due to:
a)	C₆H₅COOH	2500-3100 cm^{-1}	O–H (carboxylic acid)
		1710 cm^{-1}	C=O (carboxylic acid)
		1600, 1500 cm^{-1}	C=C (aromatic ring)
b)	C₆H₅COOCH₃	1735 cm^{-1}	C=O (ester)
		1600, 1500 cm^{-1}	C=C (aromatic ring)
c)	HO-C₆H₄-CN	3400-3640 cm^{-1}	O–H (hydroxyl)
		2210-2260 cm^{-1}	C≡N (nitrile)
		1600, 1500 cm^{-1}	C=C (aromatic ring)
d)	cyclohexenone	1715 cm^{-1}	C=O (ketone)
		1640-1680 cm^{-1}	C=C (double bond)
e)	CH₃COCH₂CH₂COOCH₃	1735 cm^{-1}	C=O (ester)
		1715 cm^{-1}	C=O (ketone)

13.20 See Problem 13.11 for the method of solution.

a) $\delta = \dfrac{131 \text{ Hz}}{60 \text{ Hz}} = 2.18$ ppm

b) $\delta = \dfrac{287 \text{ Hz}}{60 \text{ Hz}} = 4.78$ ppm

c) $\delta = \dfrac{451 \text{ Hz}}{60 \text{ Hz}} = 7.52$ ppm

13.21 $\delta \times$ (spectrometer frequency/10^6) = observed chemical shift (in Hz)
Here, spectrometer frequency = 100 MHz

a) 2.18 x 100 Hz = 218 Hz   b) 4.78 x 100 Hz = 478 Hz
c) 7.52 x 100 Hz = 752 Hz

13.22 a) 2.10 δ x 80 Hz = 168 Hz   b) 3.45 δ x 80 Hz = 276 Hz
c) 6.30 δ x 80 Hz = 504 Hz

Structure Determination   225

13.23 Reciprocal centimeters (cm^{-1}) are the units of frequency of radiation. Since frequency is directly proportional to energy, the larger the frequency, the more energy absorbed by a bond, and the stronger the bond. A carbon-oxygen double bond absorbs energy at a higher frequency (1700 cm^{-1}) than a carbon-oxygen single bond (1000 cm^{-1}) because the C=O double bond is stronger. This answer confirms what we already know about the relative strengths of single and double bonds.

13.24 a) The *chemical shift* is the exact position at which a nucleus absorbs rf energy in an NMR spectrum.

b) If the NMR signal of nucleus A is split by the spin of adjacent nucleus B, there is reciprocal splitting of the signal of nucleus B by the spin of nucleus A. The spins of the two nuclei are said to be coupled. The distance between two individual peaks within the multiplet of A is the same as the distance between two individual peaks within the multiplet of B. This distance, measured in Hz, is called the *coupling constant*.

c) λ_{max} is the wavelength in an ultraviolet spectrum at which the percent radiation absorbed is the greatest.

d) *Spin-spin splitting* is the splitting of a single NMR resonance into multiple lines. Spin-spin splitting occurs when the effective magnetic field felt by a nucleus is influenced by the small magnetic moments of adjacent nuclei. In ^{1}H NMR the signal of a proton with *n* neighboring protons is split into *n+1* peaks. The magnitude of spin-spin splitting is given by the coupling constant *J*.

e) The *wavenumber* is the reciprocal of the wavelength in centimeters.

f) The *applied magnetic field* is the magnetic field that is externally applied to a sample by an NMR spectrometer.

13.25 a) Since the symbol "δ" indicates ppm downfield from TMS, chloroform absorbs at 7.3 ppm.

b) $$\delta = \frac{\text{observed chemical shift (\# Hz from TMS)}}{\text{spectrometer frequency (in Hz)}/10^6}$$

$$7.3 = \frac{\text{chemical shift}}{360 \text{ MHz}/10^6} = \frac{\text{chemical shift}}{360 \text{ Hz}}$$

chemical shift = 2600 Hz

c) The value of δ is still 7.3 because the chemical shift measured in δ is independent of the operating frequency of the spectrometer.

13.26

Compound	Number of absorptions in ^{13}C spectrum
a) cyclohexane ring with positions labeled 1,2,3,4,5 and C2 bearing two CH$_3$ groups (both labeled 1)	5

226   Chapter 13

Compound	Number of absorptions in ^{13}C spectrum
b) $\overset{3}{C}H_3\overset{2}{C}H_2O\overset{1}{C}H_3$	3
c) 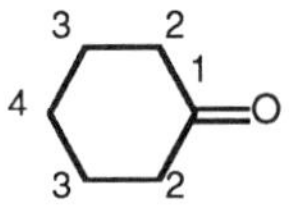 cyclohexanone with carbons labeled 1 (C=O), 2, 3, 4	4
d) 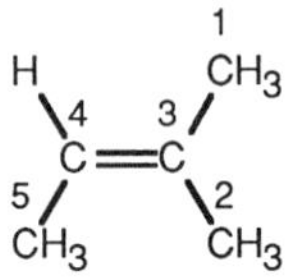 2-methyl-2-butene type structure: (CH$_3$)$_5$(H)$_4$C=C(CH$_3$)$_2$(CH$_3$)$_1$ Carbons 1 and 2 are not equivalent.	5
e) 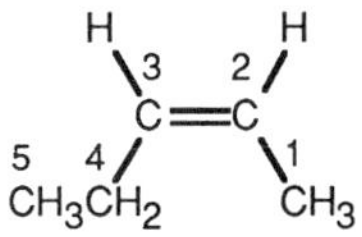 (CH$_3$CH$_2$)$_{5,4}$(H)$_3$C=C(H)$_2$(CH$_3$)$_1$	5
f) 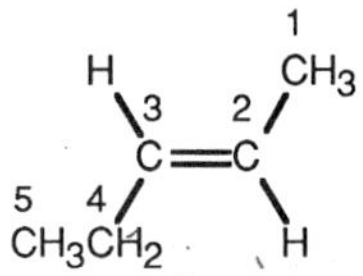 (CH$_3$CH$_2$)$_{5,4}$(H)$_3$C=C(CH$_3$)$_1$(H)$_2$	5

13.27

Compound	Types of non-equivalent protons
a) 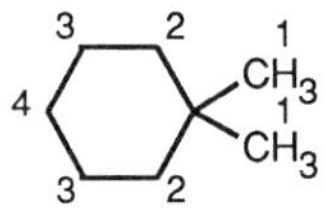 cyclohexane with gem-dimethyl: ring carbons 4, 3, 2 and C(CH$_3$)$_1$(CH$_3$)$_1$	4
b) $\overset{1}{C}H_3\overset{2}{C}H_2O\overset{3}{C}H_3$	3
c) 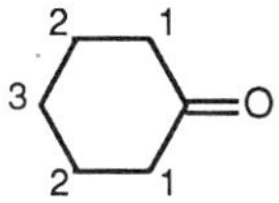 cyclohexanone with labels 1, 2, 3	3
d) 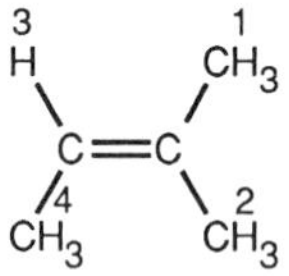 (H)$_3$(CH$_3$)$_4$C=C(CH$_3$)$_1$(CH$_3$)$_2$ Protons 1 and 2 are not equivalent.	4

Structure Determination   227

	Compound	Types of nonequivalent protons
e)	$\underset{\underset{CH_3CH_2}{5\ /4}}{\overset{\overset{H}{\overset{3}{|}}}{C}}=\underset{\underset{CH_3}{\backslash 2}}{\overset{\overset{H}{\overset{1}{|}}}{C}}$	5
f)	$\underset{\underset{CH_3CH_2}{5\ /4}}{\overset{\overset{H}{\overset{3}{|}}}{C}}=\underset{\underset{H}{\backslash 2}}{\overset{\overset{CH_3}{\overset{1}{|}}}{C}}$	5

13.28

	Compound	Protons	Chemical Shift	Rel. Peak Area	Splitting
a)	$\overset{1\ \ 2}{CH_3CHCl_2}$	1	1.0 δ	3	doublet
		2	3.9 δ	1	quartet
b)	$\overset{2\ \ \ \ \ 3\ \ 1}{CH_3COOCH_2CH_3}$	1	1.2 δ	3	triplet
		2	2.0 δ	3	singlet
		3	4.1 δ	2	quartet
c)	$\overset{1\ \ \ \ \ 3\ \ 2}{(CH_3)_3CCH_2CH_3}$	1	0.9 δ	9	singlet
		2	0.9 δ	3	triplet
		3	1.2 δ	2	quartet

The peaks from protons 1 and 2 overlap.

13.29

Lowest Chemical Shift ——————————————▶ Highest Chemical Shift

CH_4 < cyclohexane < CH_3COCH_3 < CH_2Cl_2, $H_2C=CH_2$ < benzene
0.23   1.43   2.17   5.30   5.33   7.37

13.30 a) Absorptions at 3300 cm^{-1} and 2150 cm^{-1} are due to a terminal triple bond. Possible structures:

$CH_3CH_2CH_2C≡CH$ $(CH_3)_2CHC≡CH$

b) IR absorption at 3400 cm^{-1} is due to a hydroxyl group. Since no double bond absorption is present, the compound must be a cyclic alcohol.

[cyclobutanol with OH] [cyclopropyl-CH₂OH] [cyclopropyl with OH and CH₃] [cyclopropyl-CH₃ with OH]

c) Absorption at 1715 cm^{-1} is due to a ketone. The only possible structure is $CH_3CH_2COCH_3$.

d) Absorptions at 1600 cm^{-1} and 1500 cm^{-1} are due to an aromatic ring. Possible structures:

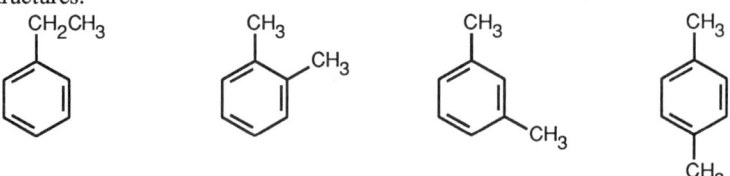

[Four aromatic structures: ethylbenzene (CH$_2$CH$_3$), 1,2-dimethylbenzene (ortho), 1,3-dimethylbenzene (meta), 1,4-dimethylbenzene (para)]

13.31 a) CH$_3$CH$_2$NHCH$_3$ (CH$_3$)$_3$N
N–H absorption at   No absorption
3300-3500 cm^{-1} at 3300-3500 cm^{-1}

b) CH$_3$COCH$_3$ H$_2$C=CHCH$_2$OH
Strong ketone absorption   Strong alcohol absorption
at 1715 cm^{-1} at 3400-3640 cm^{-1}

c) CH$_3$COCH$_3$ CH$_3$CH$_2$CHO
Strong ketone absorption   Strong aldehyde absorption
at 1715 cm^{-1} at 1725 cm^{-1}

13.32 One isomer of each pair in Problem 13.31 shows only one peak in its ^{1}H NMR spectrum. In (a), (CH$_3$)$_3$N absorbs at 2.12 δ; the other isomer has a much more complicated ^{1}H NMR spectrum. In (b) and (c), the acetone absorption occurs at 2.17 δ; the other isomers, again, show more complicated spectra.

13.33 a) The ^{13}C NMR spectrum of (CH$_3$)$_3$N shows only one peak; the spectrum of the other isomer shows 3 peaks.

b) c) The spectrum of acetone shows two peaks, one at 30 δ and one at 208 δ. The spectra of the other isomers show three peaks.

13.34

[Reaction: 1-Methylcyclohexanol with H$_2$SO$_4$ → 1-Methylcyclohexene]

1-Methylcyclohexanol   1-Methylcyclohexene

The infrared spectrum of the starting alcohol shows a broad absorption at 3400-3640 cm^{-1}, due to an O–H stretch, and another strong absorption at 1050–1100 cm^{-1}, due to a C–O stretch. The alkene product exhibits medium intensity absorbances at 1645-1670 cm^{-1} and at 3000-3100 cm^{-1}. Monitoring the *disappearance* of one of the alcohol absorptions allows one to decide when the alcohol is totally dehydrated. It is also possible to monitor the *appearance* of one of the alkene absorbances.

13.35 The absorption at 1715 cm^{-1} indicates that C$_4$H$_8$O is a ketone. The only possible structure is CH$_3$CH$_2$COCH$_3$ (2-butanone).

13.36

[1-Methylcyclohexene (A): cyclohexene ring with –CH₃ substituent on double-bond carbon]

[Methylenecyclohexane (B): cyclohexane ring with =CH₂ exocyclic group; dashed line indicating plane of symmetry]

1-Methylcyclohexene (A)   Methylenecyclohexane (B)

^{13}C: Symmetrical methylenecyclohexane (B) has only five different kinds of carbons and shows five peaks in its ^{13}C NMR spectrum. 1–Methylcyclohexene (A) has seven different kinds of carbons and shows seven peaks.

^{1}H: 1–Methylcyclohexene (A) has six different kinds of protons; methylenecyclohexane (B) has four different kinds of protons. Since several absorptions in each of the spectra overlap, it is more helpful to focus on specific absorptions of each spectrum. The ^{1}H NMR spectrum of A shows an unsplit methyl group and a vinylic proton signal of relative area 1. The vinylic absorption of B has relative area 2.

13.37

CH₃CH₂C≡CCH₂CH₃   CH₃CH=CHCH=CHCH₃

3–Hexyne (C)   2,4–Hexadiene (D)

The isomers are easily distinguished by UV spectroscopy, since only D is conjugated and absorbs in the UV region.

^{1}H NMR can also be used to identify the product. The spectrum of C consists of a quartet and a triplet. The spectrum of D is more complex, but shows four protons absorbing in the vinylic region of the spectrum; no C protons absorb in this region.

13.38

	ClCH₂CH₂CH₂Cl b a b	$\underset{a\ \ bc\ \ d}{CH_3\overset{O}{\overset{\|}{C}}CH_2CH_2Cl}$
¹³C NMR		
Number of peaks	2	4
Chemical shift	15-55 δ (a)	8-30 δ (a)
	35-80 δ (b)	15-55 δ (c)
		35-80 δ (d)
		170-210 δ (b)
¹H NMR		
Number of peaks	2	3
Chemical shift	2.2 δ (quintet) (a)	2.1 δ (singlet) (a)
	3.7 δ (triplet) (b)	2.5 δ (triplet) (c)
		3.7 δ (triplet) (d)

13.39

a) (CH₃)₄C b) (cyclopentane) c) (1,4-dioxane)

13.40

a) Possible structures for C₃H₆O

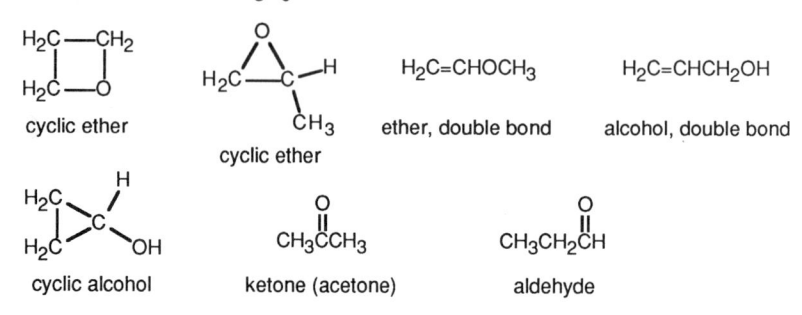

- H₂C—CH₂ / H₂C—O cyclic ether
- H₂C—C(H)—CH₃ with O (epoxide-like); cyclic ether
- H₂C=CHOCH₃ ether, double bond
- H₂C=CHCH₂OH alcohol, double bond
- H₂C—C(H)(OH)—CH₂ (cyclic) cyclic alcohol
- CH₃COCH₃ ketone (acetone)
- CH₃CH₂CHO aldehyde

b) An IR absorption at 1715 cm⁻¹ is due to a carbonyl group. Only the last two compounds show an absorption near this region.

c) The compound must be acetone, which has only one kind of proton and shows only one ¹H NMR absorption. (The aldehyde, with three different kinds of protons, would show three absorptions.)

13.41

Compound	Proton(s)	Chemical Shift
a) (CH₃)₂CHCOCH₃ (1, 2, 3)	1	0.95 δ
	2	2.43 δ
	3	2.10 δ
b) (H)(H)C=C(CH₃)(Br) (positions 2,3 H; 1 CH₃)	1	2.32 δ
	2	5.25 δ
	3	5.54 δ

13.42 Either ¹H NMR or ¹³C NMR can be used to distinguish among these isomers. In either case, it is first necessary to find the number of different kinds of protons or carbon atoms.

¹³C NMR is the preferred method for identifying these compounds; each isomer differs in the number of absorptions in its ¹³C NMR spectrum.

¹H NMR can also be used to distinguish among the isomers. The two isomers that show two ¹H NMR peaks differ in their splitting patterns.

Structure Determination 231

Compound	H₂C—CH₂ | | H₂C—CH₂	H₂C=CHCH₂CH₃	CH₃CH=CHCH₃	(CH₃)₂C=CH₂
Kinds of protons	1	5	2	2
Kinds of carbon atoms	1	4	2	3
Number of ¹H NMR peaks	1	5	2	2
Number of ¹³C NMR peaks	1	4	2	3

13.43

A   B

Distinguishing features of the ^{1}H NMR spectrum of A include one unsplit vinylic proton and a singlet methyl group. For B, distinguishing features of the ^{1}H NMR spectrum include two split vinylic protons and a singlet methyl group.

13.44 Compound A has seven different kinds of carbons and shows seven lines in its ^{13}C NMR spectrum. Compound B has five different kinds of carbons (because of symmetry) and shows five lines in its ^{13}C NMR spectrum.

13.45 Only A is conjugated and shows absorption in the UV region.

13.46

$$E = \frac{2.86 \times 10^{-3} \text{ kcal/mol}}{\lambda \text{ (in cm)}} \qquad \text{here, } \lambda = 10^{-4} \text{ cm}$$

$$E = \frac{2.86 \times 10^{-3} \text{ kcal/mol}}{10^{-4}} = 28.6 \text{ kcal/mol}$$

13.47

$$E = \frac{2.86 \times 10^{-3} \text{ kcal/mol}}{\lambda} \qquad \text{here, } \lambda = 217 \text{ nm} = 217 \times 10^{-7} \text{ cm} = 2.17 \times 10^{-5} \text{ cm}$$

$$= \frac{2.86 \times 10^{-3} \text{ kcal/mol}}{2.17 \times 10^{-5}} = 1.32 \times 10^2 \text{ kcal/mol}$$

Compare this value with the energy required for infrared excitation (Problem 13.46). More energy is required for ultraviolet excitation than for infrared excitation.

13.48

$$E = \frac{2.86 \times 10^{-3} \text{ kcal/mol}}{\lambda}$$

To find λ, use the formula

$\lambda = \frac{c}{\upsilon}$ where $c = 3 \times 10^{10}$ cm/sec

here, $\upsilon = 100$ MHz, or 10^8 Hz

so, $\lambda = 3 \times 10^2$ cm

$$E = \frac{2.86 \times 10^{-3} \text{ kcal/mol}}{3 \times 10^2}$$

$= 9.5 \times 10^{-6}$ kcal/mol

for $\upsilon = 60$ MHz, or 6×10^7 Hz

$\lambda = 5 \times 10^2$ Hz

$E = 5.7 \times 10^{-6}$ kcal/mol

Increasing the spectrophotometer frequency from 60 MHz to 100 MHz increases the amount of energy needed for resonance.

13.49

The four isomers of $C_3H_6Br_2$ are shown below, along with the number of different kinds of protons for each structure.

Structure	Kinds of protons
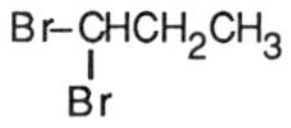 Br–CHCH$_2$CH$_3$ Br	3
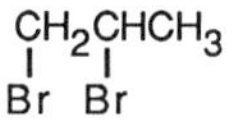 CH$_2$CHCH$_3$ Br Br	3
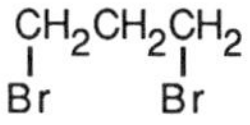 CH$_2$CH$_2$CH$_2$ Br Br	2
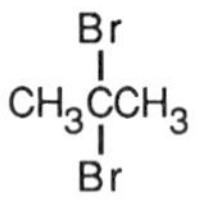 Br CH$_3$CCH$_3$ Br	1

Because the spectrum of $C_3H_6Br_2$ shows two kinds of protons, it must represent 1,3–dibromopropane. The splitting pattern shown in the spectrum (triplet, quintet) is what is expected for 1,3–dibromopropane.

13.50 The IR absorption at 1740 cm^{-1} is due to an ester group. The splitting pattern (triplet, quartet) is caused by an ethyl group. Two structures are possible at this point.

$$\underset{\text{I}}{CH_3CH_2\overset{\overset{O}{\|}}{C}OCH_2Cl} \qquad \underset{\text{II}}{CH_3CH_2O\overset{\overset{O}{\|}}{C}CH_2Cl}$$

The chemical shift of the –OCH$_2$Cl protons of I is expected to occur far downfield (5.0-6.0 δ) because of the combined effect of electronegative oxygen and chlorine. Since no absorption is observed in this region, the unknown must be II.

13.51

$$CH_3CHCH_2Br$$
$$\;\;\;\;\;|$$
$$\;\;CH_3$$

13.52

a)
$$\underset{Cl}{\overset{H_3C}{\diagdown}}C=C\underset{CH_2Cl}{\overset{H}{\diagup}} \quad Z$$

The *E* isomer is also a correct answer.

b) [benzene ring with C(CH$_3$)$_3$ substituent]

Chapter Outline

I.   Infrared radiation (Sections 13.1 - 13.2)
    A. IR spectroscopy (13.1)
    B. Identifying functional groups by IR spectroscopy (Section 13.2)

II. Ultraviolet spectroscopy (Sections 13.3 - 13.4)
    A. Nature of UV spectroscopy (Section 13.3)
    B. Interpreting UV spectroscopy (Section 13.4)
        1. Conjugated molecules
        2. Effect of conjugation on UV spectra

III. NMR spectroscopy (Sections 13.5 - 13.12)
    A. General characteristics (Sections 13.5 - 13.7)
        1. Theory of NMR (Se 13.5)
            a. Spinning nuclei align with or against applied field
            b. Energy absorption causes spin flip
        2. Nature of NMR absorptions (Section 13.6)
        3. Chemical shifts (Section 13.7)
            a. Each unique nucleus has a unique absorption
            b. Chemical shift correlates with molecular environment

234   Chapter 13

- B. ^{1}H NMR (Sections 13.8 - 13.11)
    1. Chemical shifts in ^{1}H NMR (Section 13.8)
    2. Integration in ^{1}H NMR - proton counting (Section 13.9)
    3. Spin-spin splitting (Section 13.10)
        a. Theory of spin-spin splitting
        b. Predicting spin-spin splitting - the n+1 rule
        c. Coupling constants
    4. Uses of ^{1}H NMR spectroscopy (Section 13.11)
- C. ^{13}C NMR spectroscopy (Section 13.12)
    1. Counting carbon atoms
    2. Chemical shifts
    3. Use of ^{13}C NMR to detect symmetry in molecules

Study Skills for Chapter 13

After studying this chapter, you should be able to:

1. Calculate the energy of electromagnetic radiation of various wavelengths and frequencies (Problems 13.1, 13.2, 13.7, 13.23, 13.49, 13.50, 13.51).
2. Identify the functional groups giving rise to specific IR absorptions (Problems 13.3, 13.18, 13.19, 13.30).
3. Use IR spectroscopy to identify compounds (Problems 13.4, 13.31, 13.35, 13.45).
4. Predict if compounds show UV absorption in the range 100-400 nm (Problems 13.5, 13.6).
5. Predict the number of signals appearing in the ^{1}H NMR and ^{13}C NMR spectra of compounds (Problems 13.8, 13.9, 13.16, 13.26, 13.27).
6. Calculate the relationships between delta values, chemical shifts, and spectrometer operating frequency (Problems 13.10, 13.11, 13.20, 13.21, 13.22, 13.25).
7. Predict chemical shifts (Problems 13.12, 13.29).
8. Use integration to calculate the number of protons giving rise to specific absorptions (Problem 13.13).
9. Predict the splitting patterns in NMR spectra (Problem 13.14).
10. Propose structures for compounds, given their NMR spectra (Problems 13.15, 13.17, 13.39, 13.41, 13.46, 13.47, 13.48, 13.52).
11. Describe the NMR spectra of specific compounds (Problems 13.28, 13.38).
12. Use NMR spectroscopy to distinguish between isomeric products (Problems 13.32, 13.33, 13.34, 13.36, 13.37, 13.40, 13.42, 13.43, 13.44).

Chapter 14 – Biomolecules: Carbohydrates

14.1

a)
```
   O   H
    \\ /
     C
     |
   HOCH
     |
   HCOH
     |
   CH₂OH
```
Threose
an *aldotetrose*

b)
```
   CH₂OH
     |
     C=O
     |
   HCOH
     |
   HCOH
     |
   CH₂OH
```
Ribulose
a *ketopentose*

c)
```
   CH₂OH
     |
     C=O
     |
   HOCH
     |
   HOCH
     |
   HCOH
     |
   CH₂OH
```
Tagatose
a *ketohexose*

d)
```
   O   H
    \\ /
     C
     |
    CH₂
     |
   HCOH
     |
   HCOH
     |
   CH₂OH
```
2-Deoxyribose
an *aldopentose*

14.2 As in Practice Problem 14.2, orient the molecule so that two horizontal bonds are pointing out of the page and two vertical bonds are pointing into the page. Then draw two perpendicular lines and arrange the functional groups in the same order as they are in the tetrahedral projection.

14.3 To solve this problem, you must first draw the correct tetrahedral representations of the 2–chlorobutane enantiomers. (If necessary, review Section 6.6 and Practice Problem 6.5.) Then, convert the tetrahedral representations to Fischer projections by the method used in Problem 14.2.

(*R*)–2–Chlorobutane

(*S*)–2–Chlorobutane

14.4 a) First, convert the Fischer projection to a tetrahedral representation by drawing the horizontal bonds out of the page and the vertical bonds into the page. Then use the sequence rules described in Section 6.6 to assign priorities to the four groups. Rotate the lowest priority group to the rear, and note the rotation of the arrows that go from group 1 —> 2 —> 3. If the arrows indicate clockwise rotation, the isomer is *R*; if the arrows indicate counterclockwise rotation, the isomer is *S*.

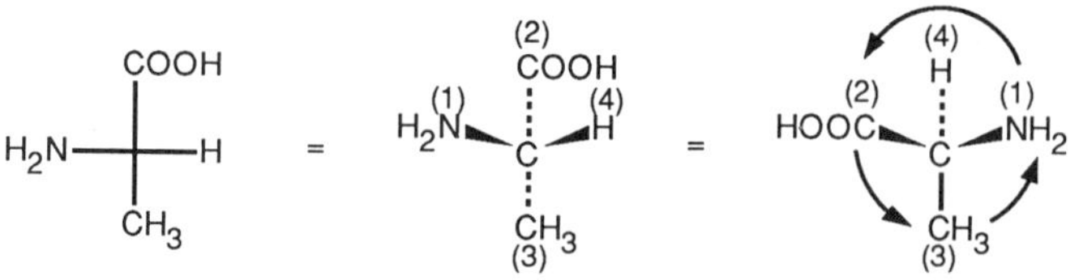

236   Chapter 14

b)

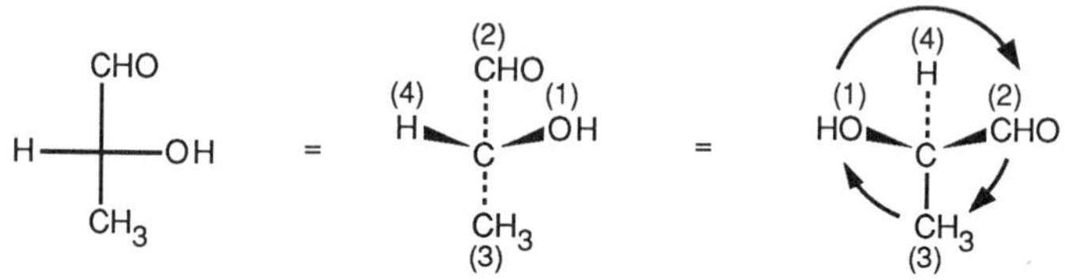

c)

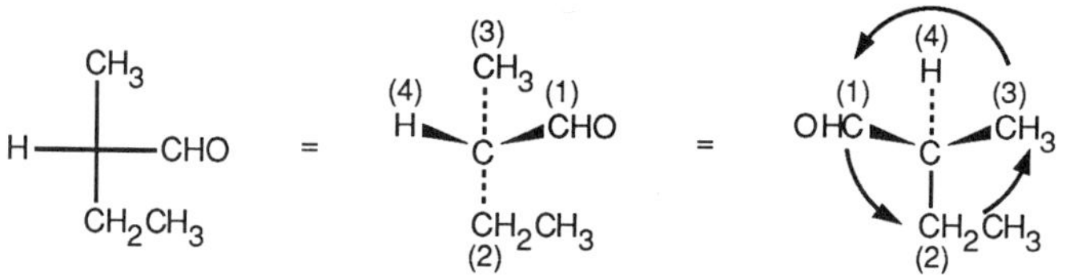

14.5 For a D sugar, the –OH group on the bottom stereogenic carbon is on the right. For an L sugar, it is on the left.

a) L-Erythrose   b) D-Xylose   c) D-Xylulose

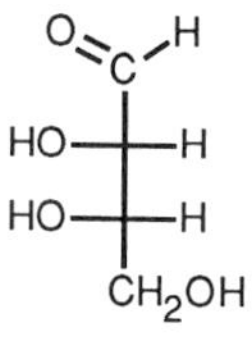

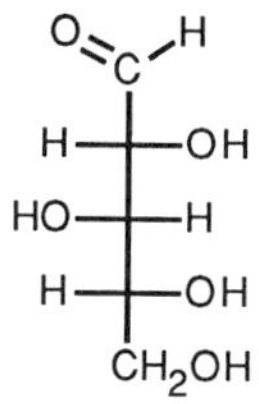

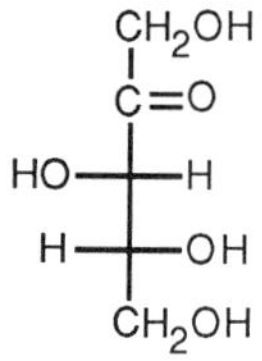

14.6 a) D-Erythrose   b) L-Xylose   c) L-Xylulose

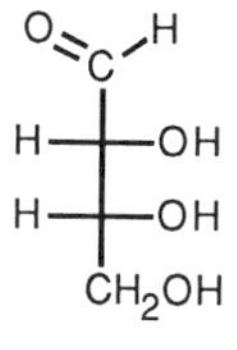

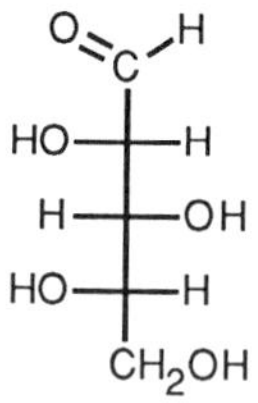

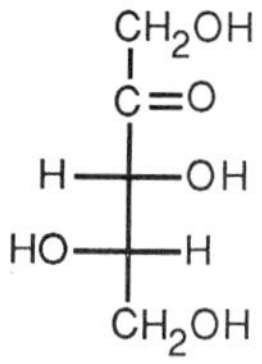

14.7

a) L-Arabinose

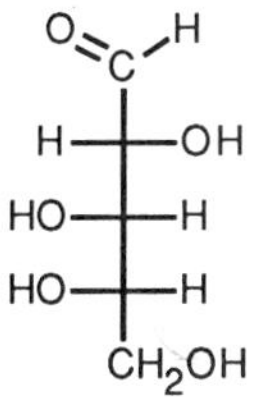

```
    O   H
     \\ //
      C
H  ─┼─ OH
HO ─┼─ H
HO ─┼─ H
     CH₂OH
```

b) L-Threose

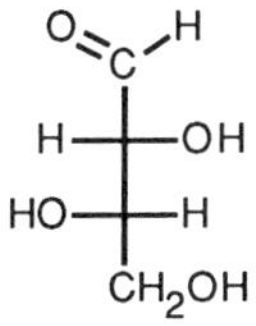

```
    O   H
     \\ //
      C
H  ─┼─ OH
HO ─┼─ H
     CH₂OH
```

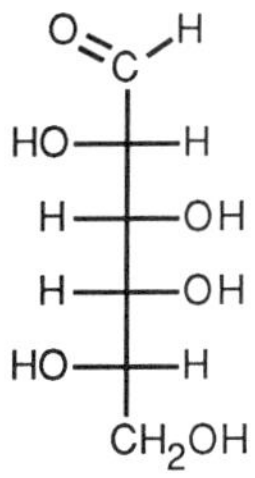

c) L-Galactose

```
    O   H
     \\ //
      C
HO ─┼─ H
H  ─┼─ OH
H  ─┼─ OH
HO ─┼─ H
     CH₂OH
```

14.8 Thirty-two aldoheptoses are possible. Sixteen are D sugars, and sixteen are L sugars.

14.9

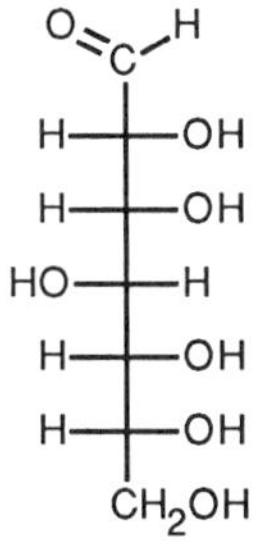

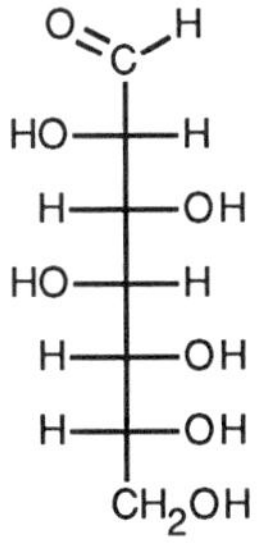

```
    O   H              O   H
     \\ //              \\ //
      C                  C
H  ─┼─ OH           HO ─┼─ H
H  ─┼─ OH           H  ─┼─ OH
HO ─┼─ H            HO ─┼─ H
H  ─┼─ OH           H  ─┼─ OH
H  ─┼─ OH           H  ─┼─ OH
     CH₂OH               CH₂OH
```

14.10

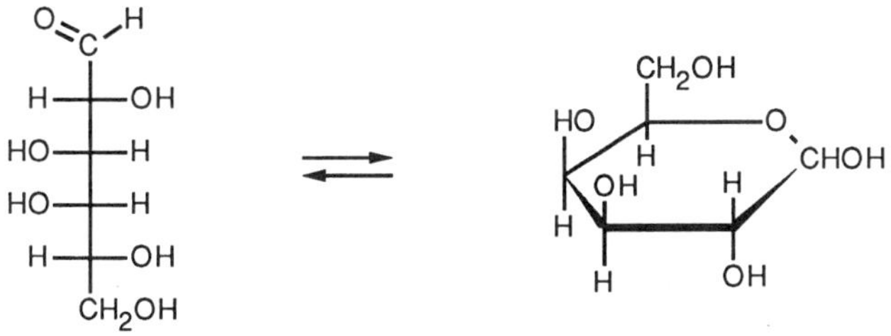

D-Galactose (open chain) ⇌ pyranose ring form

14.11

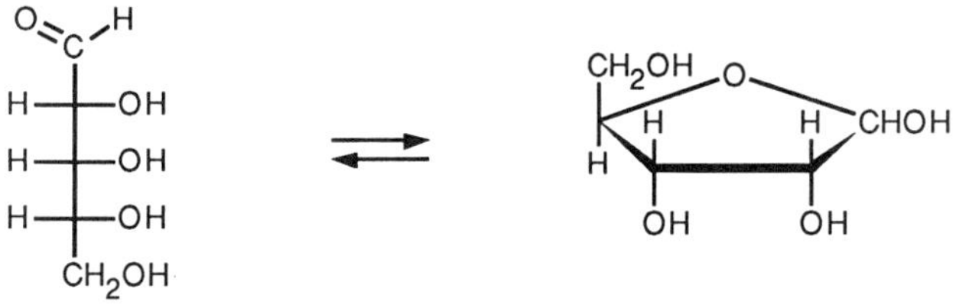

D-Ribose (open chain) ⇌ furanose ring form

238   Chapter 14

14.12

α-D-Fructofuranose   β-D-Fructofuranose

14.13 Let x be the fraction of D-glucose present as the α anomer, and y be the fraction of D-glucose present as the β anomer.

Then: $112.2°x + 18.7°y = 52.6°$ $x + y = 1; y = 1 - x$
$112.2°x + 18.7°(1-x) = 52.6°$
$93.5°x = 33.9°$
$x = .362$
$y = .638$

Thus, 36.2% of glucose is present as the α anomer and 63.8% is present as the β anomer.

14.14

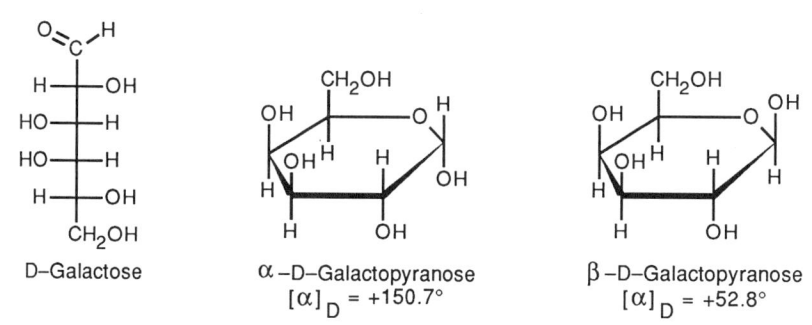

D-Galactose   α-D-Galactopyranose $[\alpha]_D = +150.7°$ β-D-Galactopyranose $[\alpha]_D = +52.8°$

Let x be the fraction of D-galactopyranose present as the α anomer and y be the fraction of D-galactopyranose present as the β anomer.

$150.7°x + 52.8°y = 80.2°$ $x + y = 1; y = 1 - x$
$150.7°x + 52.8°(1-x) = 80.2°$
$97.9°x = 27.4°$
$x = 0.280$
$y = 0.720$

28.0% of D-galactopyranose is present as the α anomer, and 72.0% is present as the β anomer.

14.15

β-D-Galactopyranose

14.16

[Structures of β-D-Galactopyranose (with "Raise" and "Lower" arrows) and β-D-Mannopyranose shown with axial/equatorial labels]

β-D-Mannopyranose

β-D-Galactopyranose and β-D-mannopyranose each have one hydroxyl group in the axial position. Galactose and mannose are of equal stability.

14.17

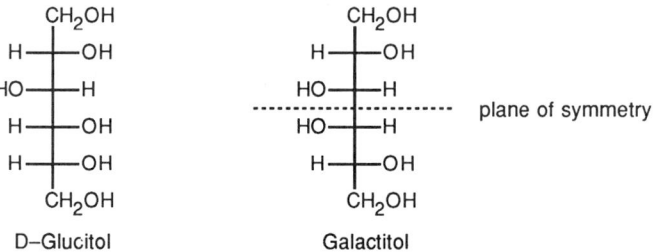

β-D-Ribofuranose

(a) CH$_3$I, Ag$_2$O → permethylated product (CH$_3$OCH$_2$, OCH$_3$, CH$_3$O, OCH$_3$)

(b) (CH$_3$CO)$_2$O, pyridine → peracetylated product (CH$_3$COCH$_2$–, –OCCH$_3$, CH$_3$CO–, –OCCH$_3$ with C=O groups)

14.18

β-D-Galactopyranose + CH$_3$CH$_2$OH $\xrightarrow{\text{H}^+ \text{ catalyst}}$ Ethyl β-D-galactopyranoside

14.19

```
   CH₂OH              CH₂OH
H──┼──OH           H──┼──OH
HO─┼──H           HO──┼──H
H──┼──OH    ········································  plane of symmetry
H──┼──OH          HO──┼──H
   CH₂OH          H──┼──OH
                     CH₂OH
 D-Glucitol          Galactitol
```

Reduction of the aldehyde group of D–galactose yields an alditol that has a plane of symmetry and is an optically inactive *meso* compound.

14.20

Reaction of an aldose with NaBH₄ produces a polyol (alditol). Because an alditol has the same functional group at both ends, the number of stereoisomers of an *n*–carbon alditol is one half the number of stereoisomers of the parent aldose, and two different aldoses can yields the same alditol. Here L–gulose and D–glucose form the same alditol (rotate the Fischer projection of L–gulitol 180° to see the identity).

14.21

Allaric acid has a plane of symmetry and is an optically inactive *meso* compound.

14.22 D–Allose and D–galactose yield *meso* aldaric acids. All other D–hexoses form optically active aldaric acids on oxidation.

14.23

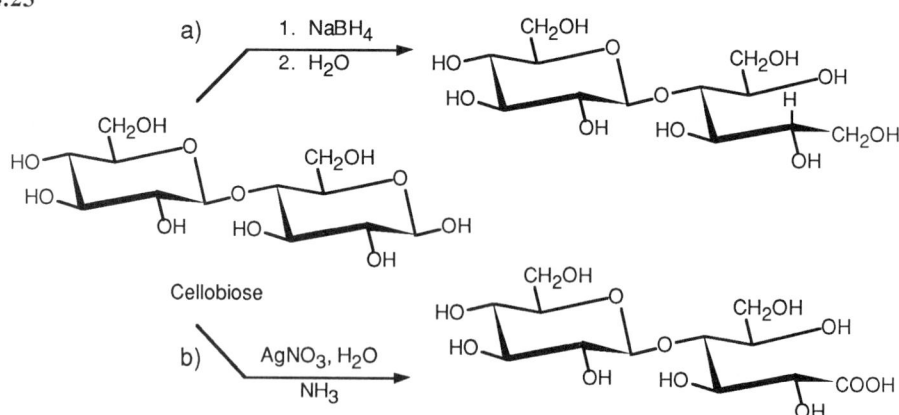

Biomolecules: Carbohydrates 241

14.24

a) 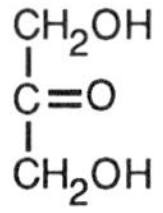
```
CH₂OH
 |
 C=O
 |
CH₂OH
```
a ketotriose

b) 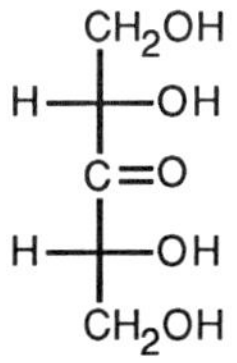
```
  CH₂OH
H─┼─OH
  C=O
H─┼─OH
  CH₂OH
```
a ketopentose

c) 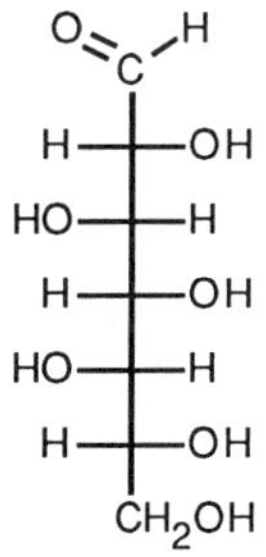
```
   O    H
    \\ //
     C
H ─┼─ OH
HO─┼─ H
H ─┼─ OH
HO─┼─ H
H ─┼─ OH
   CH₂OH
```
an aldoheptose

14.25

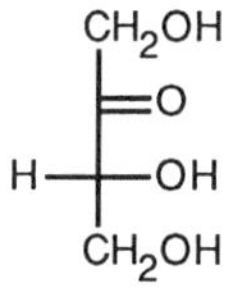

```
  CH₂OH
   ‖
   O
H─┼─OH
  CH₂OH
```
a ketotrose

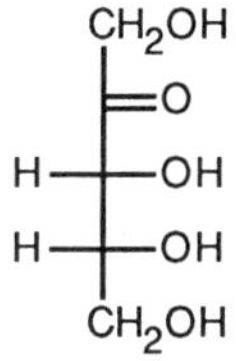

```
  CH₂OH
   ‖
   O
H─┼─OH
H─┼─OH
  CH₂OH
```
a ketopentose

14.26

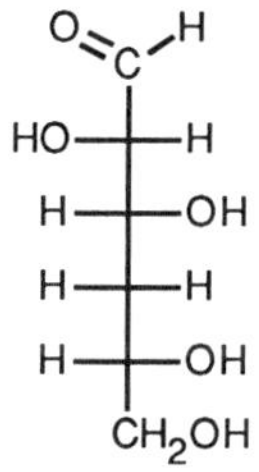

```
   O   H
    \\ //
     C
HO─┼─H
H ─┼─OH
H ─┼─H
H ─┼─OH
   CH₂OH
```
a deoxyaldohexose

14.27

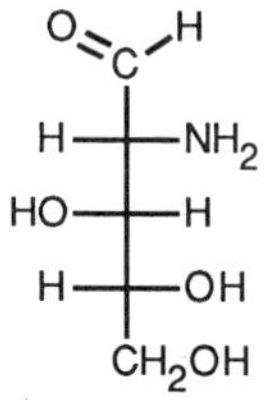

```
   O   H
    \\ //
     C
H ─┼─NH₂
HO─┼─H
H ─┼─OH
   CH₂OH
```
a five-carbon amino sugar

14.28-14.29

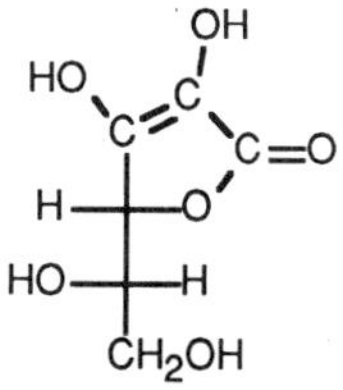

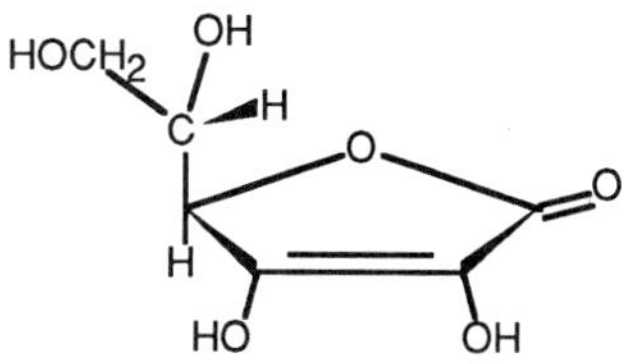

```
      OH
      |
  HO  C
   \\ // \
    C    C=O
    |    |
    H    O
H ─┼─
HO─┼─H
   CH₂OH
```

```
       OH
HOCH₂  |
     \ C  H
      \  /
       C──O
      / \
     H   \
          C=O
    HO   OH
```
L-Ascorbic acid

14.30

	Definition	Example
a)	A *monosaccharide* is a carbohydrate that cannot be hydrolyzed into smaller units.	D-Glucose (open chain Fischer projection: CHO, H–OH, HO–H, H–OH, H–OH, CH₂OH)
b)	An *anomeric* center is a stereogenic center formed when an open chain monosaccharide cyclizes to a furanose or pyranose ring.	β-D-Glucopyranose (with anomeric center labeled)
c)	A *Haworth projection* is a drawing of a pyranose or furanose in which the ring is drawn as flat. This projection allows the relationship of the ring substituents to be viewed more easily.	β-D-Glucopyranose (Haworth projection)
d)	A *Fischer projection* is a drawing of a carbohydrate in which each stereogenic center is represented as a pair of perpendicular lines. Vertical lines represent bonds going into the page, and horizontal lines represent bonds coming out of the page.	D-Erythrose (CHO, H–OH, H–OH, CH₂OH)
e)	A *glycoside* is a acetal of a carbohydrate, formed when an anomeric hydroxyl group reacts with another compound containing a hydroxyl group.	Methyl β-D-glucopyranoside

f) A *reducing sugar* is a sugar that reacts with any of several reagents to yield an oxidized sugar plus reduced reagent.

```
      O   H
       \\ /
        C
  HO──┼──H
   H──┼──OH
   H──┼──OH
       CH₂OH
    D-Arabinose
```

g) A *pyranose* is a six-membered cyclic hemiacetal ring form of a monosaccharide.

β-D-Galactopyranose

h) A *1,4' link* occurs when the anomeric hydroxyl group (at carbon 1) of a pyranose or furanose forms a glycosidic bond with the hydroxyl group at carbon 4 of a second nonosaccharide.

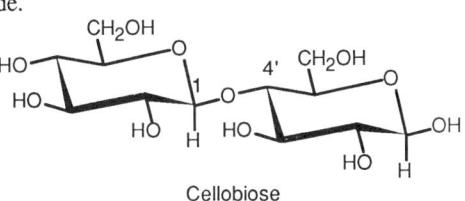

Cellobiose

i) A *D-sugar* is a sugar in which the hydroxyl group farthest from the carbonyl group points to the right in a Fischer projection.

```
       CHO
   H──┼──OH
   H──┼──OH
   H──┼──OH
       CH₂OH
    D-Ribose
```

14.31

β-D-Gulopyranose

This structure is a pyranose (6-membered ring) and is a β anomer (the C–1 hydroxyl group and the –CH₂OH groups are *cis*). It is a D-sugar because the –O– at C5 is on the right in the uncoiled form.

14.32

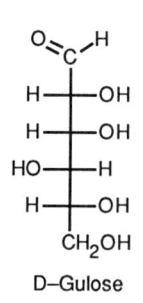

$$\begin{array}{c} \text{O}{=}\text{C}{-}\text{H} \\ \text{H}{-}\!\!-\!\!\text{OH} \\ \text{H}{-}\!\!-\!\!\text{OH} \\ \text{HO}{-}\!\!-\!\!\text{H} \\ \text{H}{-}\!\!-\!\!\text{OH} \\ \text{CH}_2\text{OH} \end{array}$$

D–Gulose

14.33

D–Ribulofuranose (β anomer)

14.34-14.35

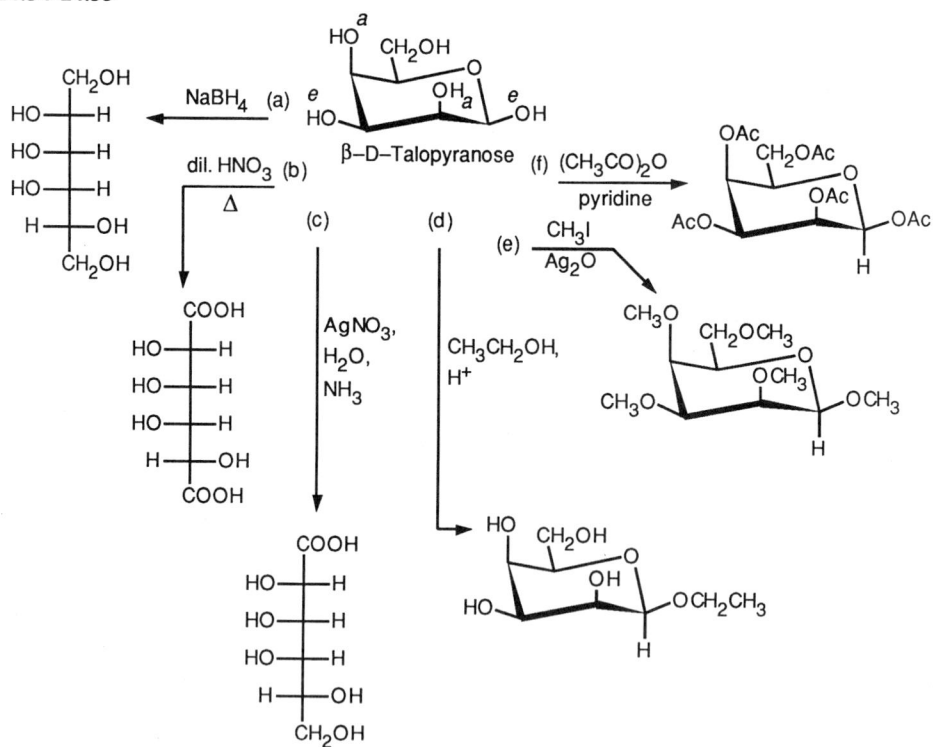

β-D-Talopyranose

(a) NaBH$_4$
(b) dil. HNO$_3$, Δ
(c) AgNO$_3$, H$_2$O, NH$_3$
(d) CH$_3$CH$_2$OH, H$^+$
(e) CH$_3$I, Ag$_2$O
(f) (CH$_3$CO)$_2$O, pyridine

14.36

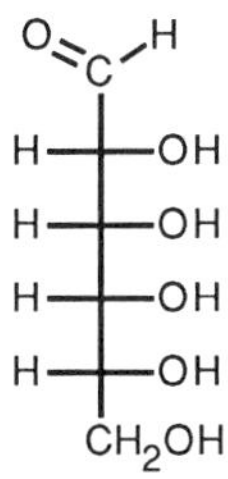

D-Allose	L-Allose
O=C-H	O=C-H
H—OH	HO—H
H—OH	HO—H
H—OH	HO—H
H—OH	HO—H
CH₂OH	CH₂OH

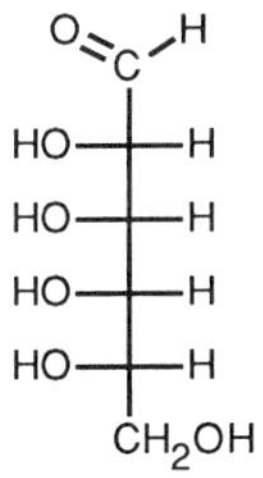

D–Allose and L–allose are enantiomers. Their physical properties, such as melting point, solubility in water, and density, are identical. Their specific rotations are equal in degree but opposite in sign.

14.37

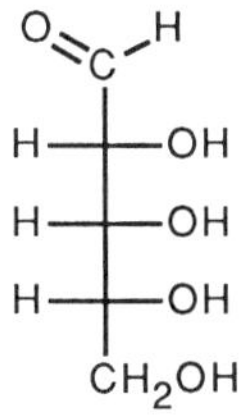

D-Ribose	L-Lyxose
O=C-H	O=C-H
H—OH	HO—H
H—OH	H—OH
H—OH	HO—H
CH₂OH	CH₂OH

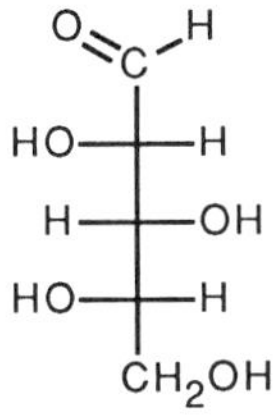

D-Ribose and L-lyxose are diastereomers and differ in all physical properties.

14.38-14.40 Four D–2–ketohexoses are possible.

```
    CH₂OH              CH₂OH              CH₂OH              CH₂OH
    C=O                C=O                C=O                C=O
  H—─OH             HO—─H              H—─OH             HO—─H
  H—─OH              H—─OH             HO—─H             HO—─H
  H—─OH              H—─OH              H—─OH              H—─OH
    CH₂OH              CH₂OH              CH₂OH              CH₂OH
   D–Psicose         D–Fructose         D–Sorbose          D–Tagatose
```

| 1. NaBH₄
| 2. H₂O

| 1. NaBH₄
| 2. H₂O

```
   CH₂OH       CH₂OH              CH₂OH       CH₂OH
  H—─OH      HO—─H               H—─OH      HO—─H
  H—─OH       H—─OH              H—─OH       H—─OH
  H—─OH  +    H—─OH             HO—─H   +  HO—─H
  H—─OH       H—─OH              H—─OH       H—─OH
   CH₂OH       CH₂OH              CH₂OH       CH₂OH
  D–Allitol   D–Altritol         D–Gulitol   D–Iditol
```

14.41

```
   CH₂OH
   C=O
  HO—─H
   H—─OH
   H—─OH
   CH₂OH
  D–Fructose
```

β–D–Fructopyranose

β–D–Fructofuranose

14.42

a)
```
        COOH
    H—─│─CH₃
        CH₂CH₃
```
(R)–2–Methylbutanoic acid

b)
```
        O═C─CH₃
    H₃C—│─H
        CH₂CH₃
```
(S)–3–Methyl–2–pentanone

14.43

a)
```
        Br                    Br
    H—─│─OCH₃   =    H₃C⋯C
        CH₃              H   OCH₃
```

b)
```
        CH₃                   CH₃
    H—─│─NH₂    =    CH₃CH₂⋯C
        CH₂CH₃              H    NH₂
```

14.44

```
     O   H
      \\ //
       C                              CH₂OH
   H ──┼── OH                     H ──┼── OH
   H ──┼── OH    1. NaBH₄         H ──┼── OH
   H ──┼── OH    ───────→         H ──┼── OH   ┈┈┈┈┈┈
   H ──┼── OH    2. H₂O           H ──┼── OH
       │                              │
      CH₂OH                          CH₂OH
    D–Allose                        Allitol
```

```
     O   H
      \\ //
       C                              CH₂OH
   H ──┼── OH                     H ──┼── OH
  HO ──┼── H     1. NaBH₄        HO ──┼── H
  HO ──┼── H     ───────→        HO ──┼── H    ┈┈┈┈┈┈
   H ──┼── OH    2. H₂O           H ──┼── OH
       │                              │
      CH₂OH                          CH₂OH
   D–Galactose                     Galactitol
```

All of the other D–hexoses yield optically active alditols on reduction with NaBH₄.

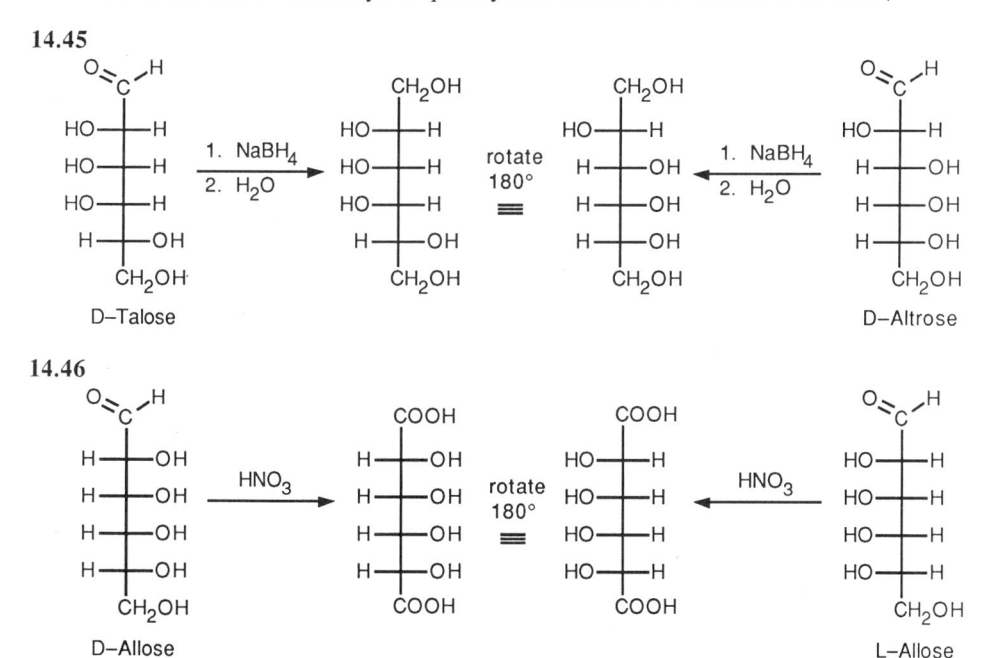

14.45

```
     O   H
      \\ //
       C                  CH₂OH              CH₂OH              O   H
                                                                 \\ //
                                                                  C
  HO ──┼── H          HO ──┼── H         HO ──┼── H          HO ──┼── H
  HO ──┼── H  1.NaBH₄ HO ──┼── H  rotate  H ──┼── OH 1.NaBH₄  H ──┼── OH
  HO ──┼── H  ──────→ HO ──┼── H  180°    H ──┼── OH ──────   H ──┼── OH
   H ──┼── OH 2. H₂O   H ──┼── OH   ≡     H ──┼── OH 2. H₂O   H ──┼── OH
       │                  │                  │                    │
      CH₂OH              CH₂OH              CH₂OH                CH₂OH
    D–Talose                                                   D–Altrose
```

14.46

```
     O   H
      \\ //
       C                  COOH               COOH               O   H
                                                                 \\ //
                                                                  C
   H ──┼── OH          H ──┼── OH        HO ──┼── H          HO ──┼── H
   H ──┼── OH   HNO₃   H ──┼── OH rotate HO ──┼── H   HNO₃  HO ──┼── H
   H ──┼── OH  ──────→ H ──┼── OH 180°   HO ──┼── H  ←─────  HO ──┼── H
   H ──┼── OH          H ──┼── OH    ≡   HO ──┼── H          HO ──┼── H
       │                  │                  │                    │
      CH₂OH              COOH               COOH                 CH₂OH
    D–Allose                                                   L–Allose
```

14.47

D-Galactose → (HNO₃) → rotate 180° ≡ ← (HNO₃) ← L-Galactose

D-Lyxose → (HNO₃) → rotate 180° ≡ ← (HNO₃) ← D-Arabinose

14.48

D-Glucose → Ruff Degradation → D-Arabinose ← Ruff Degradation ← D-Mannose

14.49 D–Galactose and D–talose must have the same configuration at C3, C4 and C5 if both yield the same aldopentose on Ruff degradation.

D-Galactose → Ruff Degradation → D-Lyxose ← Ruff Degradation ← D-Talose

14.50 Two D–aldotetroses are possible. The one with both hydroxyl groups on the right in a Fischer projection yields an optically inactive aldaric acid, and thus it must be D–erythrose. The other D–aldotetrose, D–threose, yields an optically active aldaric acid.

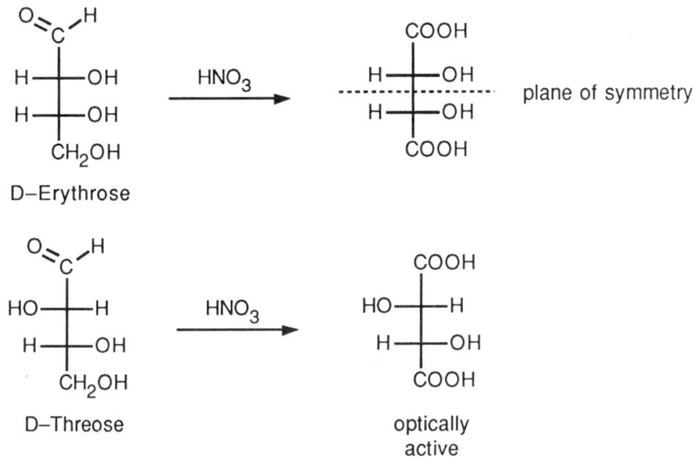

```
      O   H
       \\ //
        C
   H ——|—— OH           COOH
        |        HNO₃    |
   H ——|—— OH    ——→  H ——|—— OH   · · · · · plane of symmetry
        |              H ——|—— OH
      CH₂OH             COOH
   D–Erythrose
```

```
      O   H
       \\ //
        C
  HO ——|—— H            COOH
        |        HNO₃    |
   H ——|—— OH    ——→  HO ——|—— H
        |              H ——|—— OH
      CH₂OH             COOH
   D–Threose           optically
                        active
```

14.51

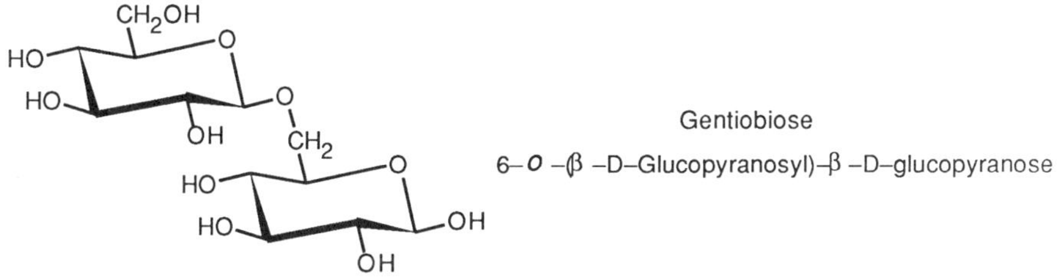

Gentiobiose
6–*O*–(β–D–Glucopyranosyl)–β–D–glucopyranose

14.52

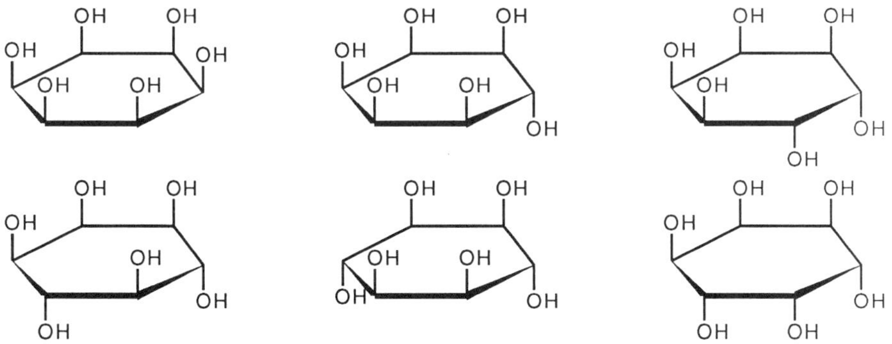

14.53 [structures of sugars shown]

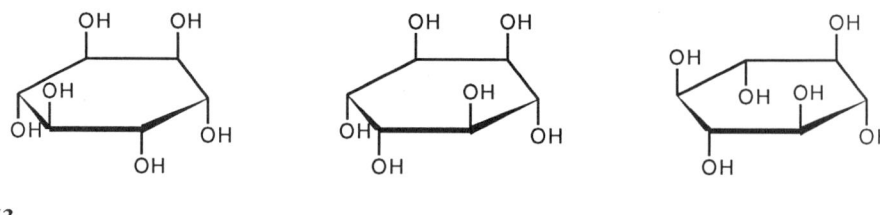

14.54 Raffinose is a nonreducing sugar because it contains no hemiacetal groups.

Chapter Outline

I. Classification of carbohydrates (Section 14.1)
II. Monosaccharides (Sections 14.2 - 14.8)
    A. Configurations of monosaccharides (Sections 14.2 - 14.4)
        1. Fischer projections (Section 14.2)
        2. D, L Sugars (Section 14.3)
        3. Configurations of the aldoses (Section 14.4)
    B. Cyclic structures of monosaccharides (Sections 14.5 - 14.7)
        1. Hemiacetal formation (Section 14.5)
            a. Mechanism of cyclization
            b. Furanose and pyranose rings
            c. Haworth projections
        2. Anomers (Section 14.6)
            a. Alpha and beta anomers
            b. Mutarotation of anomers
        3. Conformations of monosaccharides (Section 14.7)

Biomolecules: Carbohydrates 251

    C. Reactions of monosaccharides (Section 14.8)
        1. Ester and ether formation
        2. Reduction of monosaccharides to yield alditols
        3. Oxidation of monosaccharides to yield aldonic acids
        4. Oxidation of monosaccharides to yield aldaric acids
III. Disaccharides (Section 14.9)
    A. Cellobiose and maltose
    B. Sucrose
IV. Polysaccharides (Section 14.10)
    A. Cellulose - a β-linked glucose polymer
    B. Starch - an α-linked glucose polymer
V. Other important sugars (Section 14.11)
    A. Deoxy sugars
    B. Amino sugars
VI. Cell surface carbohydrates (Section 14.12)

Study Skills for Chapter 14

After studying this chapter, you should be able to:

1. Classify carbohydrates (Problems 14.1, 14.24, 14.31)
2. Identify sugars as D or L (Problems 14.5, 14.6, 14.28)
3. Draw monosaccharides in the following projections:
   Fischer projections (Problems 14.2, 14.3, 14.4, 14.6, 14.7, 14.9, 14.25, 14.26, 14.27, 14.32, 14.38, 14.42)
   Haworth projections (Problems 14.10, 14.11, 14.12, 14.29, 14.33, 14.34, 14.41, 14.52)
   Chair conformation (Problems 14.15, 14.16)
4. Calculate the equilibrium percentage of anomers from the specific rotation (Problems 14.13, 14.14)
5. Predict the products of reactions of monosaccharides (Problems 14.17, 14.18, 14.19, 14.20,. 14.21, 14.22, 14.35, 14.39, 14.40, 14.44, 14.45, 14.46, 14.47, 14.48, 14.49, 14.50)
6. Predict the products of reactions of disaccharides (Problems 14.23)
7. Deduce the structure of disaccharides (Problems 14.51, 14.53)
8. Define the terms in this chapter (Problem 14.30)

Chapter 15 – Biomolecules: Amino Acids, Peptides, and Proteins

15.1 Amino acids with aromatic rings:

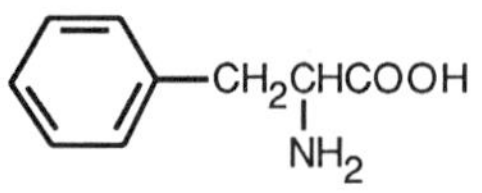

C₆H₅–CH₂CH(NH₂)COOH
Phenylalanine (Phe)

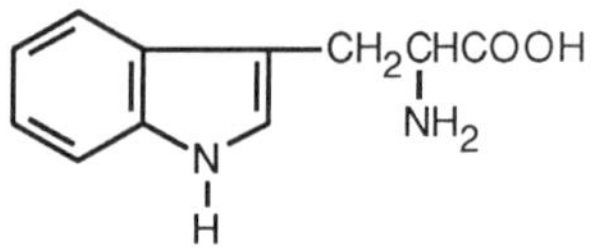

Indole-3-CH₂CH(NH₂)COOH
Tryptophan (Try)

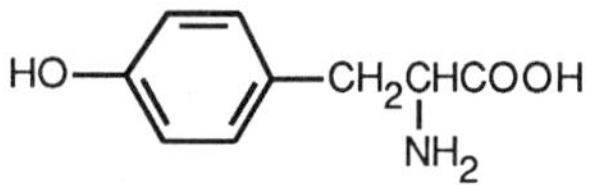

HO–C₆H₄–CH₂CH(NH₂)COOH
Tyrosine (Tyr)

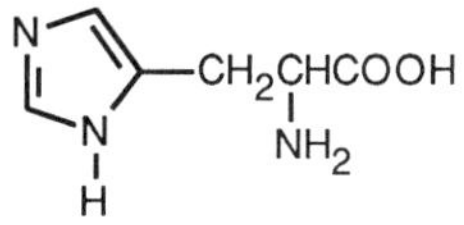

Imidazole–CH₂CH(NH₂)COOH
Histidine (His)
(an aromatic heterocycle)

Amino acids containing sulfur:

HSCH₂CH(NH₂)COOH
Cysteine (Cys)

CH₃SCH₂CH₂CH(NH₂)COOH
Methionine (Met)

Amino acids that are alcohols:

HOCH₂CH(NH₂)COOH
Serine (Ser)

CH₃CH(OH)CH(NH₂)COOH
Threonine (Thr)

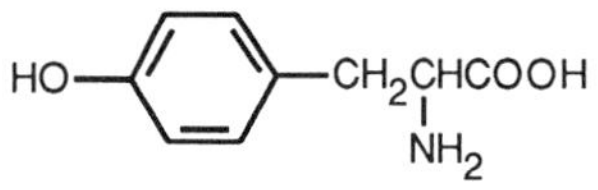

HO–C₆H₄–CH₂CH(NH₂)COOH
Tyrosine (Tyr)–a phenol

Amino acids having hydrocarbon side chains:

CH₃CH(NH₂)COOH
Alanine (Ala)

CH₃CH₂CH(CH₃)CH(NH₂)COOH
Isoleucine (Ile)

(CH₃)₂CHCH₂CH(NH₂)COOH
Leucine (Leu)

(CH₃)₂CHCH(NH₂)COOH
Valine (Val)

15.2

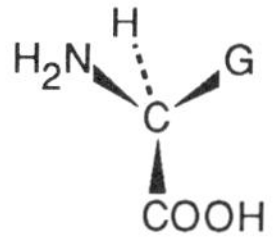

A projection of the alpha carbon of an amino acid is pictured above.

For most amino acids:

Group	*Priority*
$-NH_2$	1
$-COOH$	2
$-G$	3
$-H$	4

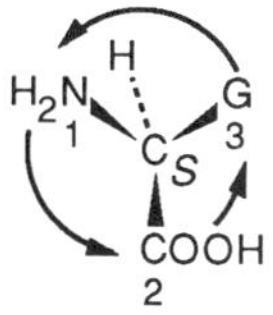

For cysteine:

Group	*Priority*
$-NH_2$	1
$-CH_2SH$	2
$-COOH$	3
$-H$	4

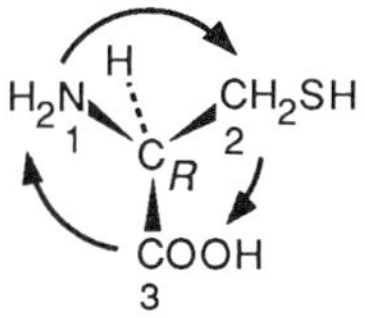

Refer to Section 6.6 if you need help.

15.3

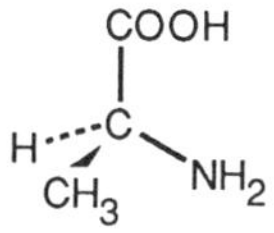

L–Alanine

15.4

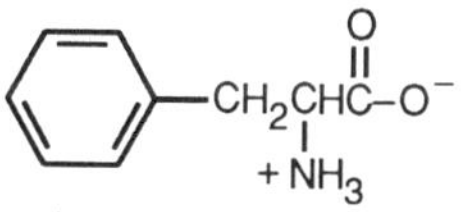

Phenylalanine

15.5

a) 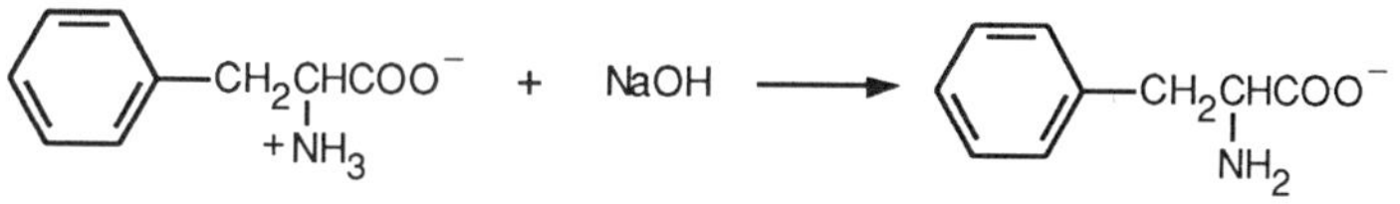

b) 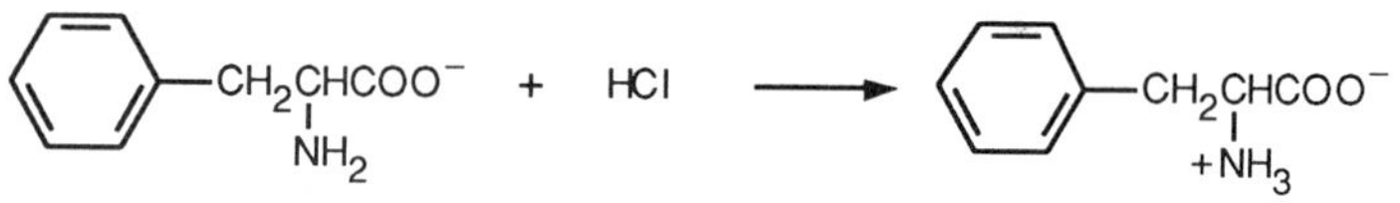

c) 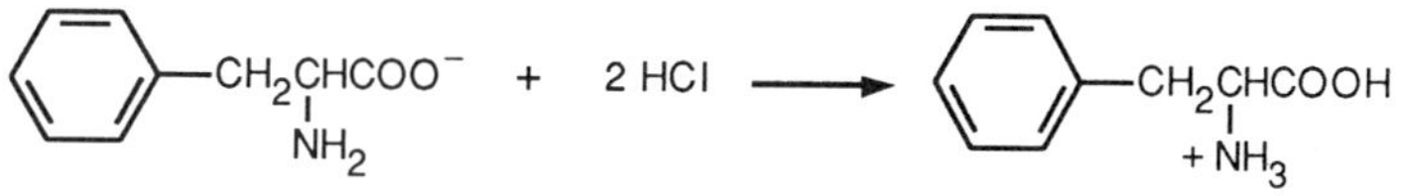

15.6

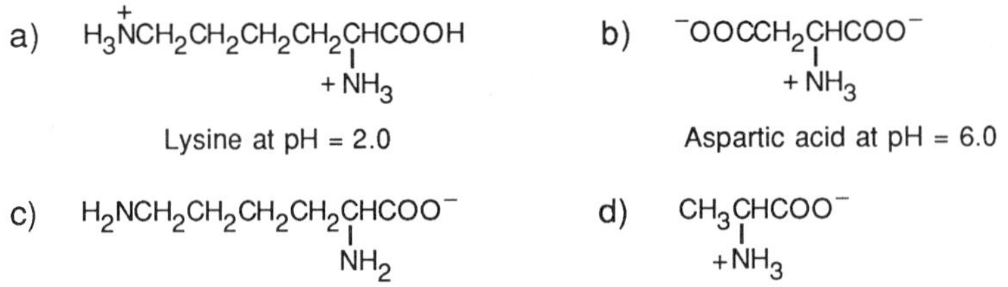

a) $H_3\overset{+}{N}CH_2CH_2CH_2CH_2\underset{\underset{+NH_3}{|}}{C}HCOOH$

Lysine at pH = 2.0

b) $^-OOCCH_2\underset{\underset{+NH_3}{|}}{C}HCOO^-$

Aspartic acid at pH = 6.0

c) $H_2NCH_2CH_2CH_2CH_2\underset{\underset{NH_2}{|}}{C}HCOO^-$

Lysine at pH = 11.0

d) $CH_3\underset{\underset{+NH_3}{|}}{C}HCOO^-$

Alanine at pH = 4.0

15.7 a) Amino acid   Isoelectric point
  Val   5.96
  Glu   3.22
  His   7.59

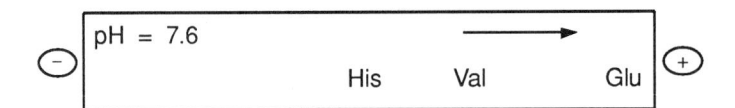

```
⊖  | pH = 7.6              →              | ⊕
   |              His    Val    Glu       |
```

b) Amino acid   Isoelectric point
  Gly   5.97
  Phe   5.48
  Ser   5.68

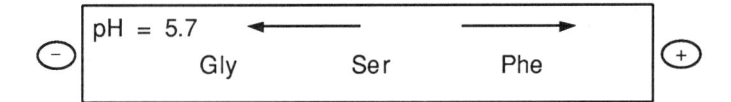

```
⊖  | pH = 5.7    ←              →         | ⊕
   |          Gly      Ser      Phe       |
```

c)
```
⊖  | pH = 6.0              →              | ⊕
   |              Gly    Ser    Phe       |
```

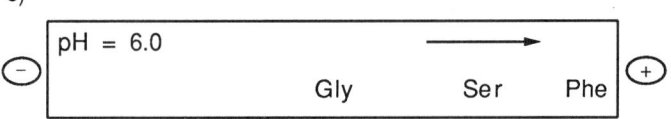

15.8

$(CH_3)_2CHCH_2$ — CH_2SH
H–NH–CH–C–NH–CH–C–OH
         ‖   ‖
         O   O
Leu——Cys

CH_2SH — $CH_2CH(CH_3)_2$
H–NH–CH–C–NH–CH–C–OH
         ‖   ‖
         O   O
Cys——Leu

15.9   Val–Tyr–Gly   Tyr–Gly–Val   Gly–Val–Tyr
           Val–Gly–Tyr   Tyr–Val–Gly   Gly–Tyr–Val

Biomolecules: Amino Acids, Peptides, and Proteins 255

15.10

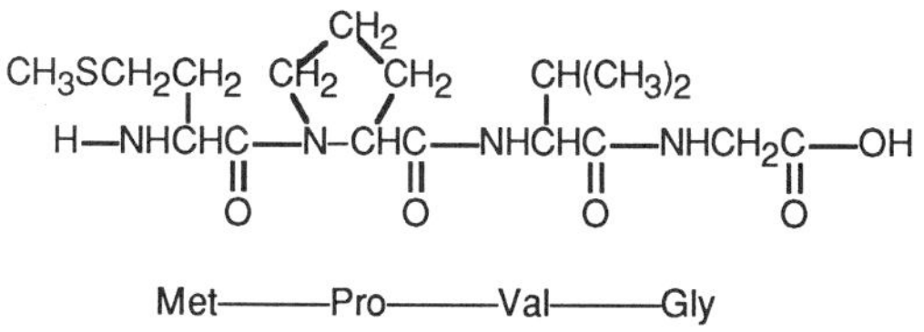

$$\text{H}-\text{NHCHC}(\text{CH}_2\text{CH}_2\text{SCH}_3)-\text{N}(\text{CH}_2\text{CH}_2\text{CH}_2)-\text{CHC}-\text{NHCHC}(\text{CH}(\text{CH}_3)_2)-\text{NHCH}_2\text{C}-\text{OH}$$

with carbonyl (C=O) groups at each peptide linkage.

Met——Pro——Val——Gly

15.11

2 Ninhydrin (with OH, OH groups) + $(CH_3)_2CHCHCOOH$ (with NH_2) — Valine — $\xrightarrow{^-OH}$ $(CH_3)_2CHCHO$ + CO_2 + $3 H_2O$ + (diketohydrindylidene–diketohydrindamine product)

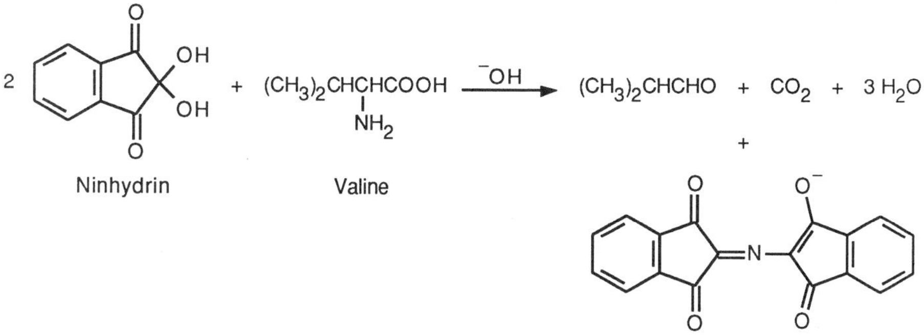

15.12

Asp–Arg + Val–Tyr–Ile–His–Pro–Phe
↑
Trypsin

Asp–Arg–⦃–Val–Tyr–⦃–Ile–His–Pro–Phe

↓ Chymotrypsin

Asp–Arg–Val–Tyr + Ile–His–Pro–Phe

Trypsin cleaves peptide bonds at the carboxyl side of *lysine* and *arginine*. Chymotrypsin cleaves peptide bonds at the carboxyl side of *phenylalanine, tyrosine* and *tryptophan*.

15.13 Arg–Pro
  Pro–Leu–Gly
    Gly–Ile–Val

The complete sequence:

Arg–Pro–Leu–Gly–Ile–Val

15.14

```
    Ala              Ala
   /   \            /   \
 Leu — Phe       Phe — Leu
```

The tripeptide is cyclic.

15.15

$$\text{Leu} = \begin{array}{c} \text{CH(CH}_3)_2 \\ | \\ \text{CH}_2 \\ | \\ \text{H}_2\text{N-CH-COOH} \end{array} \qquad R = \begin{array}{c} \text{CH(CH}_3)_2 \\ | \\ \text{CH}_2 \\ | \end{array}$$

1. Protect the amino group of leucine.

$$(\text{CH}_3)_3\text{COCOCOC(CH}_3)_3 + \underset{\text{Leu}}{\text{H}_2\overset{R}{\text{N}}\text{CHCOOH}} \xrightarrow{(\text{C}_2\text{H}_5)_3\text{N}} (\text{CH}_3)_3\text{COC-NH-}\overset{R}{\text{CHCOOH}} + \text{CO}_2 + (\text{CH}_3)_3\text{COH}$$

2. Protect the carboxylic acid group of alanine.

$$\underset{\text{Ala}}{\text{H}_2\overset{\text{CH}_3}{\text{N}}\text{CHCOOH}} + \text{CH}_3\text{OH} \xrightarrow{\text{H}^+} \text{H}_2\overset{\text{CH}_3}{\text{N}}\text{CHCOOCH}_3 + \text{H}_2\text{O}$$

3. Couple the protected amino acids with DCC.

$$(\text{CH}_3)_3\text{COCNHCHCOOH} + \text{H}_2\text{N-CHCOOCH}_3 + \text{(DCC)} \longrightarrow$$

$$(\text{CH}_3)_3\text{COCNHCHC-NHCHCOOCH}_3 + \text{NH-C-NH (dicyclohexylurea)}$$

4. Remove the leucine protecting group.

$$(\text{CH}_3)_3\text{COCNHCHC-NHCHCOOCH}_3 \xrightarrow{\text{CF}_3\text{COOH}} \overset{+}{\text{H}}_3\text{NCHC-NHCHCOOCH}_3 + (\text{CH}_3)_2\text{C=CH}_2 + \text{CO}_2$$

5. Remove the alanine protecting group.

$$\overset{+}{\text{H}}_3\text{NCHC-NHCHCOOCH}_3 \xrightarrow{\text{NaOH, H}_2\text{O}} \underset{\text{Leu—Ala}}{\text{H}_2\text{NCHC-NHCHCOO}^-} + \text{CH}_3\text{OH}$$

15.16

For simplicity, call $(CH_3)_3COCOCOC(CH_3)_3$
$\overset{O}{\underset{\|}{C}}\overset{O}{\underset{\|}{C}}$ "TBDC" in this problem.

1. Phe + TBDC $\xrightarrow{N(CH_2CH_3)_3}$ BOC–Phe

2. Gly + CH₃OH $\xrightarrow{H^+}$ Gly–OCH₃

3. BOC–Phe + Gly–OCH₃ $\xrightarrow{DCC}$ BOC–Phe–Gly–OCH₃

4. BOC–Phe–Gly–OCH₃ $\xrightarrow{CF_3COOH}$ Phe–Gly–OCH₃

5. Val + TBDC $\xrightarrow{N(CH_2CH_3)_3}$ BOC–Val

6. BOC–Val + Phe–Gly–OCH₃ $\xrightarrow{DCC}$ BOC–Val–Phe–Gly–OCH₃

7. BOC–Val–Phe–Gly–OCH₃ $\xrightarrow{CF_3COOH}$ Val–Phe–Gly–OCH₃

8. Val–Phe–Gly–OCH₃ $\xrightarrow{NaOH, H_2O}$ Val–Phe–Gly

15.17 a) Pyruvate decarboxylase is a lyase.

b) Chymotrypsin is a hydrolase.

c) Alcohol dehydrogenase is an oxidoreductase.

15.18 When referring to an amino acid, the prefix "α" indicates that the amino group is bonded to the carbon atom next to (alpha to) the carboxylic acid group.

15.19 a) Ser = serine   b) Thr = threonine   c) Pro = proline
d) Phe = phenylalanine   e) Glu = glutamic acid

15.20 a) Nucleoproteins contain RNA and protein.
b) Glycoproteins contain carbohydrate and protein.
c) Lipoproteins contain lipids and protein.

15.21 The disulfide bridges that cysteine forms help to stabilize a protein's tertiary structure.

15.22

$$H_2N-CH(CH_2-C_6H_4-OH)-CO-NH-CH_2-CO-NH-CH_2-CO-NH-CH(CH_2-C_6H_5)-CO-NH-CH(CH_2CH_2SCH_3)-COOH$$

Tyr—Gly—Gly—Phe—Met

15.23 a) *Hydrolases* catalyze the hydrolysis of substrates.
b) *Lyases* catalyze the addition of a small molecule to a substrate, or the reverse reaction.
c) *Transferases* catalyze the transfer of a group from one substrate to another.

15.24 a) A *protease* catalyzes the hydrolysis of an amide group.
b) A *kinase* catalyzes the transfer of a phosphate group.
c) A *carboxylase* catalyzes the addition of CO_2 to a substrate.

15.25

(R)-Serine: COOH, H—C—NH₂, CH₂OH

(R)-Alanine: COOH, H—C—NH₂, CH₃

15.26

(S)-Proline

15.27 a) *Amphoteric* compounds can react either as acids or as bases, depending on the circumstances.
b) The *isoelectric point* is the pH at which a solution of an amino acid or protein is electronically neutral.
c) A *peptide* is an amide polymer composed of from two to fifty amino acid residues.
d) The *N-terminus* of a peptide or a protein is the end amino acid residue that forms no peptide bond with its amine group.
e) The *C-terminus* of a peptide or protein is the end amino acid residue that forms no peptide bond with its carboxyl group.
f) A *zwitterion* is a compound that contains both positively charged and negatively charged portions.

Biomolecules: Amino Acids, Peptides, and Proteins   259

15.28 a) Val–Leu–Ser   Ser–Val–Leu
       Val–Ser–Leu   Leu–Val–Ser
       Ser–Leu–Val   Leu–Ser–Val

   b) Ser–Leu–Leu–Pro   Leu–Leu–Ser–Pro
      Ser–Leu–Pro–Leu   Leu–Leu–Pro–Ser
      Ser–Pro–Leu–Leu   Leu–Ser–Leu–Pro
      Pro–Leu–Leu–Ser   Leu–Ser–Pro–Leu
      Pro–Leu–Ser–Leu   Leu–Pro–Leu–Ser
      Pro–Ser–Leu–Leu   Leu–Pro–Ser–Leu

15.29

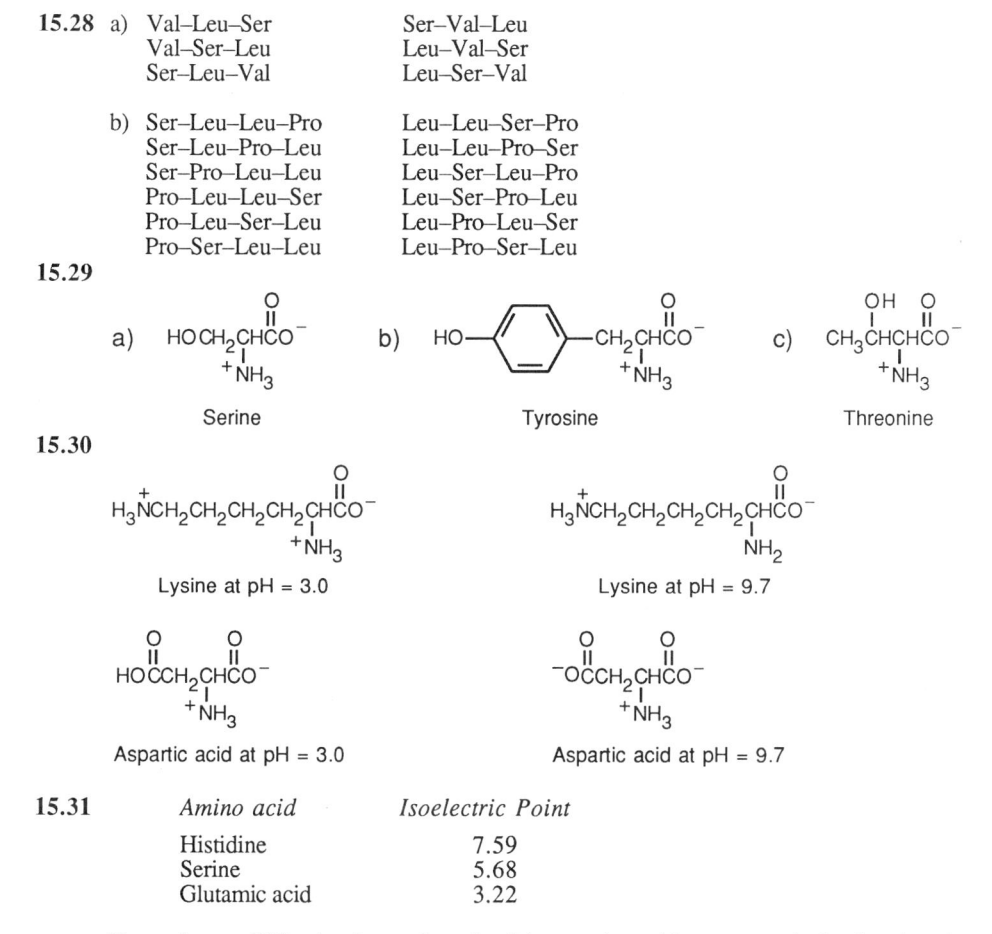

a) HOCH$_2$CH(+NH$_3$)CO$_2^-$ Serine

b) HO–C$_6$H$_4$–CH$_2$CH(+NH$_3$)CO$_2^-$ Tyrosine

c) CH$_3$CH(OH)CH(+NH$_3$)CO$_2^-$ Threonine

15.30

H$_3$+NCH$_2$CH$_2$CH$_2$CH$_2$CH(+NH$_3$)CO$_2^-$
Lysine at pH = 3.0

H$_3$+NCH$_2$CH$_2$CH$_2$CH$_2$CH(NH$_2$)CO$_2^-$
Lysine at pH = 9.7

HOOCCH$_2$CH(+NH$_3$)CO$_2^-$
Aspartic acid at pH = 3.0

$^-$OOCCH$_2$CH(+NH$_3$)CO$_2^-$
Aspartic acid at pH = 9.7

15.31

Amino acid	Isoelectric Point
Histidine	7.59
Serine	5.68
Glutamic acid	3.22

The optimum pH for the electrophoresis of three amino acids occurs at the isoelectric point of the amino acid intermediate in acidity. At this pH, the least acidic amino acid migrates toward the negative electrode, the most acidic amino acid migrates toward the positive electrode, and the amino acid intermediate in acidity does not migrate. In this example, electrophoresis at pH = 5.7 allows the maximum separation of the three amino acids.

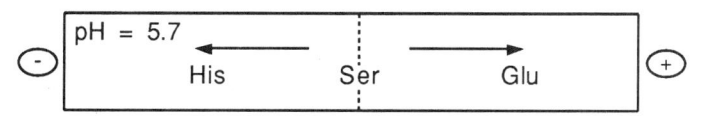

⊖ | pH = 5.7   ←His   Ser   Glu→ | ⊕

15.32 Amino acids with hydrocarbon side chains (valine and isoleucine) are more likely to be found on the inside of a globular protein, whereas amino acids with charged side chains (aspartic acid and lysine) are more likely to be found on the outside of a globular protein.

260 Chapter 15

15.33

a) $(CH_3)_2CHCH-COO^-$ with $+NH_3$ → CH_3CH_2OH, H^+ → $(CH_3)_2CHCHCOOCH_2CH_3$ with $+NH_3$

L-Valine

b) $(CH_3)_2CHCH-COO^-$ with $+NH_3$ → $NaOH, H_2O$ → $(CH_3)_2CHCHCOO^-$ with NH_2

c) $(CH_3)_2CHCH-COO^-$ with $+NH_3$ → $(CH_3)_3COCOCOC(CH_3)_3$ / $(CH_3CH_2)_3N$ → $(CH_3)_2CHCHCO^-$ with $NHCOC(CH_3)_3$

15.34

a)
```
        CH3      C6H5     SH
         |        |        |
        CHCH3    CH2      CH2
         |        |        |
H—NHCHC—NHCHC—NHCHC—OH
         ||       ||       ||
         O        O        O
```
Val——Phe——Cys

b)
```
    HOOCCH2            CH2CH3  CH(CH3)2
      |        CH2       |        |
     CH2  CH2   CH2     CHCH3    CH2
      |    \  /   |       |        |
H—NHCHC—N——CHC—NHCHC—NHCHC—OH
      ||        ||       ||       ||
      O         O        O        O
```
Glu——Pro——Ile——Leu

15.35-15.36

```
      CHO                COOH              COOH              COOH
       |S                 |S                |S                |R
 HO—  —H           H2N—  —H          H2N—  —H           H—  —NH2
       |                  |                 |                 |
  H—  —OH            H—  —OH          HO—  —H            H—  —OH
       |R                 |R                |S                |R
      CH2OH              CH3               CH3               CH3

   D-Threose          L-Threonine

                                        Diastereomers of L-Threonine
```

15.37 Only primary amines can form the extensively conjugated purple ninhydrin product. A secondary amine such as proline yields a product containing a shorter system of conjugated bonds, which absorbs at a shorter wavelength (440 nm vs. 570 nm).

15.38 100 g of Cytochrome c contains 0.43 g iron.

100 g of Cytochrome c contains:

$$\frac{0.43 \text{ g Fe}}{55.8 \text{ g/mol Fe}} = 0.0077 \text{ moles Fe}$$

Biomolecules: Amino Acids, Peptides, and Proteins   261

$$\frac{100 \text{ g Cytochrome } c}{0.0077 \text{ moles Fe}} = \frac{X \text{ g Cytochrome } c}{1 \text{ mole Fe}}$$

13,000 g/mol Fe = X

Cytochrome c has a minimum molecular weight of 13,000.

15.39

a)

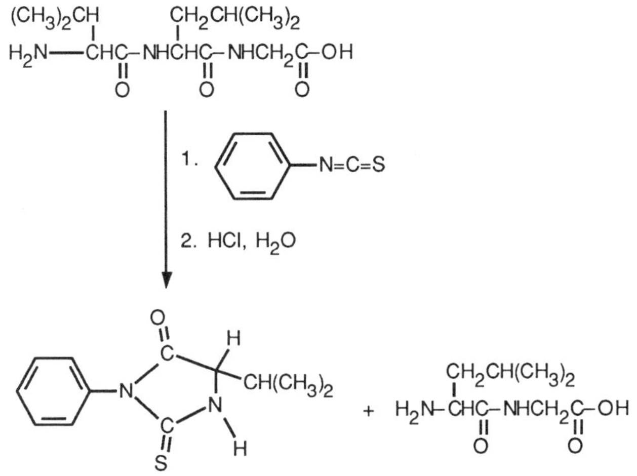

$(CH_3)_2CH \quad CH_2CH(CH_3)_2$
$H_2N-CHC-NHCHC-NHCH_2C-OH$
       ‖   ‖   ‖
       O   O   O

1. ⟨C₆H₅⟩—N=C=S
2. HCl, H₂O

↓

Phenylthiohydantoin of Val (ring with N-phenyl, C=O, C=S, NH, CH(CH₃)₂)

+ $H_2N-CHC-NHCH_2C-OH$ with side chain $CH_2CH(CH_3)_2$
       ‖   ‖
       O   O

b)

Proline-Phe-... structure:
$H_2N-CHC—N—CHC—NHCHC-OH$
with CH₃ on first carbon; pyrrolidine ring (H₃C, H₂C, CH₂, CH₂) on N; CH₂C₆H₅ side chain
       ‖   ‖   ‖
       O   O   O

1. ⟨C₆H₅⟩—N=C=S
2. HCl, H₂O

↓

PTH of Ala (ring with N-phenyl, C=O, CH-CH₃, NH, C=S)

+ proline-phenylalanine dipeptide:
$HN—CHC—NHCHCOH$ with pyrrolidine ring (H₂C, CH₂, CH₂) and CH₂C₆H₅ side chain
     ‖   ‖
     O   O

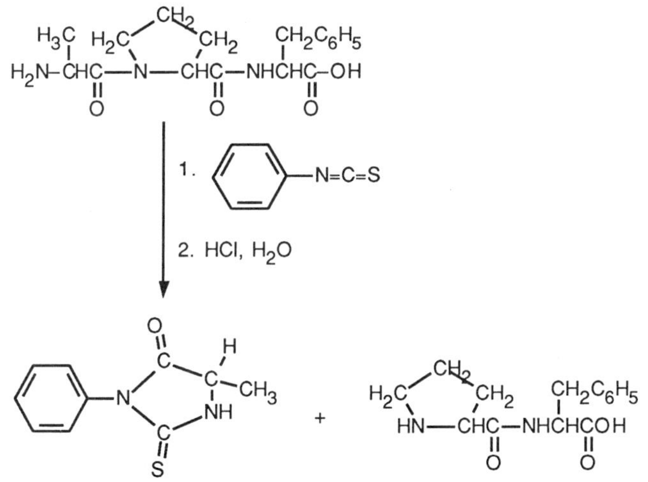

15.40

$$H_2N-\overset{\overset{\ddot{N}H}{\|}}{C}-\ddot{N}HR \xrightarrow{H^+} \left[H_2N-\overset{\overset{+NH_2}{\|}}{C}-\ddot{N}HR \longleftrightarrow H_2\overset{+}{N}=\overset{\overset{\ddot{N}H_2}{|}}{C}-\ddot{N}HR \longleftrightarrow H_2\ddot{N}-\overset{\overset{\ddot{N}H_2}{|}}{\underset{+}{C}}=NHR \longleftrightarrow H_2\ddot{N}-\overset{\overset{\ddot{N}H_2}{|}}{\underset{+}{C}}-\ddot{N}HR \right]$$

The protonated guanidino group can be stabilized by resonance.

15.41 As in Problem 15.16, use the abbreviation "TBDC" for di–*tert*-butyl dicarbonate.

1. Ala + TBDC $\xrightarrow{N(CH_2CH_3)_3}$ BOC–Ala

2. Val + CH₃OH $\xrightarrow{H^+}$ Val–OCH₃

3. BOC–Ala + Val–OCH₃ $\xrightarrow{DCC}$ BOC–Ala–Val–OCH₃

4. BOC–Ala–Val–OCH₃ $\xrightarrow{CF_3COOH}$ Ala–Val–OCH₃

5. Phe + TBDC $\xrightarrow{N(CH_2CH_3)_3}$ BOC–Phe

6. BOC–Phe + Ala–Val–OCH₃ $\xrightarrow{DCC}$ BOC–Phe–Ala–Val–OCH₃

7. BOC–Phe–Ala–Val–OCH₃ $\xrightarrow{CF_3COOH}$ Phe–Ala–Val–OCH₃

8. Phe–Ala–Val–OCH₃ $\xrightarrow{NaOH, H_2O}$ Phe–Ala–Val

15.42

a) $H_2N\overset{\overset{R}{|}}{C}HCOOH + H_2N\overset{\overset{R}{|}}{C}HCOOH \longrightarrow H_2N\overset{\overset{R}{|}}{C}H\overset{\overset{O}{\|}}{C}-NH\overset{\overset{R}{|}}{C}H\overset{\overset{O}{\|}}{C}-OH$

+

cyclohexyl–N=C=N–cyclohexyl (DCC)

+

cyclohexyl–NH–C(=O)–NH–cyclohexyl

In this step, a dipeptide is formed from two amino acid residues.

Biomolecules: Amino Acids, Peptides, and Proteins 263

b)

$$\text{Cy-N=C=N-Cy} + \underset{\text{HN}}{\overset{\text{RCH—C-OH}}{|}}\underset{\overset{\text{C}}{||}O}{\overset{\text{NH}_2}{|}}\text{CHR} \longrightarrow \text{Cy-NH-C(=O)-NH-Cy} + \text{a 2,5-diketopiperazine}$$

DCC couples the carboxylic acid end of the dipeptide to the amino end to yield the 2,5-diketopiperazine.

15.43

Phe–⦃–Leu–Met–Lys–⫶–Tyr–⦃Asp–Gly–Gly–Arg–⫶Val–Ile–Pro–Tyr

        cleaved by trypsin = - - - - -
        cleaved by chymotrypsin = ∿∿∿

15.44

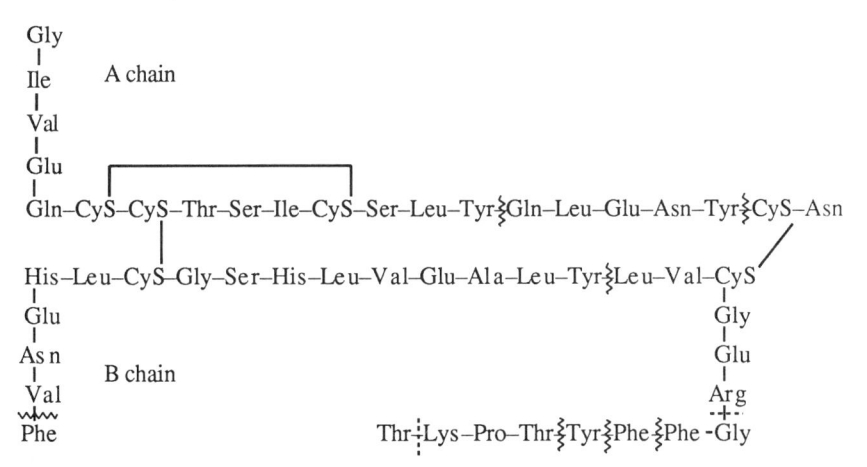

```
Gly
 |
Ile       A chain
 |
Val
 |
Glu
 |
Gln–CyS–CyS–Thr–Ser–Ile–CyS–Ser–Leu–Tyr⦃Gln–Leu–Glu–Asn–Tyr⦃CyS–Asn
         |
His–Leu–CyS–Gly–Ser–His–Leu–Val–Glu–Ala–Leu–Tyr⦃Leu–Val–CyS
 |                                                          |
Glu                                                        Gly
 |                                                          |
Asn       B chain                                          Glu
 |                                                          |
Val                                                        Arg
∿∿∿                                                        -⫶-
Phe                          Thr⫶Lys–Pro–Thr⦃Tyr⦃Phe⦃Phe -Gly
```

        cleaved by trypsin = - - - - -
        cleaved by chymotrypsin = ∿∿∿

15.45 Gly–Asp–Phe–Pro
　　　　　　Phe–Pro–Val
　　　　　　　　　Val–Pro–Leu

The complete sequence:

Gly–Asp–Phe–Pro–Val–Pro–Leu

15.46 a) Arg–Pro
　　　　　　Pro–Leu–Gly
　　　　　　　　　Gly–Ile–Val

The complete sequence:

Arg–Pro–Leu–Gly–Ile–Val

b) Val–Met–Trp
　　　Trp–Asp–Val
　　　　　Val–Leu

The complete sequence:

Val–Met–Trp–Asp–Val–Leu

15.47 A proline residue in a polypeptide chain interrupts α–helix formation. The amide nitrogen of proline has no hydrogen that can contribute to the hydrogen-bonded structure of an α–helix. In addition, the pyrrolidine ring of proline restricts rotation about the C–N bond and reduces flexibility in the polypeptide chain.

15.48

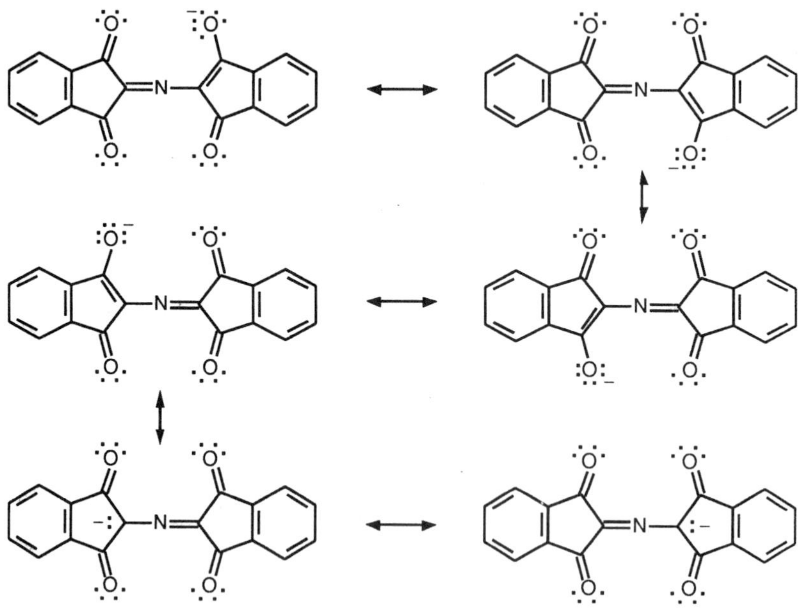

Not included are the resonance forms that involve the aromatic rings.

15.49 Ser–Ile–Arg–Val–Val–Pro–Tyr–Leu–Arg

15.50 Cys–Tyr
 Tyr–Ile–Glu
  Ile–Glu
   Glu–Asp–Cys
    Asp–Cys
     Cys–Pro–Leu
      Leu–Gly

The complete structure of reduced oxytocin:

Cys–Tyr–Ile–Gln–Asn–Cys–Pro–Leu–Gly(–NH$_2$)

Oxidized oxytocin:

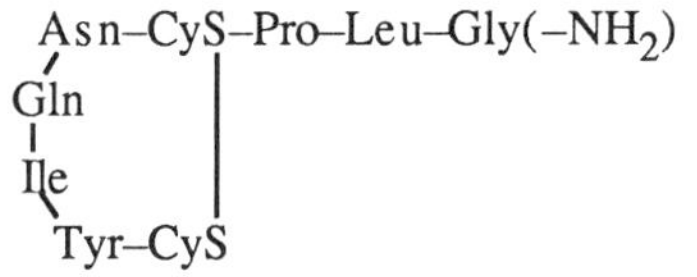

$$\begin{array}{c} \text{Asn–CyS–Pro–Leu–Gly(–NH}_2\text{)} \\ \diagup \qquad \qquad | \\ \text{Gln} \qquad \qquad | \\ | \qquad \qquad \quad | \\ \text{Ile} \qquad \qquad \quad | \\ \diagdown \qquad \qquad | \\ \text{Tyr–CyS} \end{array}$$

The C–terminal end of oxytocin is an amide, but this can't be determined from the information given.

15.51

a) 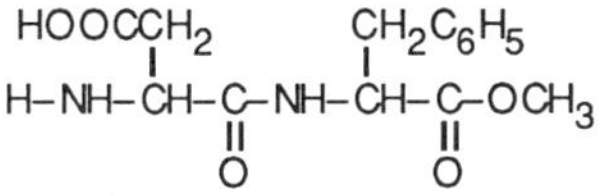
$$\text{H–NH–CH–C–NH–CH–C–OCH}_3$$
with HOOCCH$_2$ and CH$_2$C$_6$H$_5$ side chains, C=O groups — Aspartame

b) 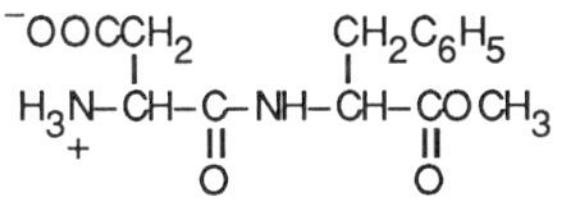
$$\text{H}_3\overset{+}{\text{N}}\text{–CH–C–NH–CH–COCH}_3$$
with $^-$OOCCH$_2$ and CH$_2$C$_6$H$_5$ side chains — At pH = 5.9

c) 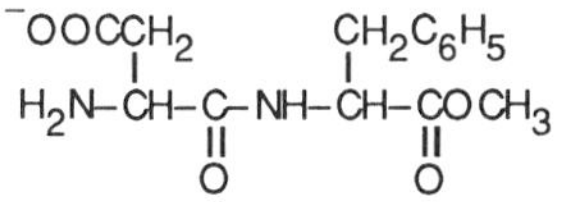
$$\text{H}_2\text{N–CH–C–NH–CH–COCH}_3$$
with $^-$OOCCH$_2$ and CH$_2$C$_6$H$_5$ side chains — At pH = 7.6

d)  Aspartame $\xrightarrow{\text{H}_3\text{O}^+}$ H$_2$N–CH(CH$_2$COOH)–COOH + H$_2$N–CH(CH$_2$C$_6$H$_5$)–COOH + CH$_3$OH

Chapter Outline

I. Amino acids (Secs. 15.1 - 15.3)
    A. Structure of amino acids (Section 15.1 - 15.2)
        1. Stereochemistry of α–amino acids (Section 15.1)
        2. Side chains of α–amino acids
        3. Dipolar structure of α–amino acids (Section 15.2)
    B. Isoelectric points (Section 15.3)
        1. Correlation of isoelectric point and side-chain structure
        2. Electrophoresis

II. Peptides (Secs. 15.4 - 15.8)
    A. Introduction to peptides (Section 15.4)
        1. Writing peptide structures
        2. N terminus and C terminus
    B. Covalent bonding in peptides (Section 15.5)
        1. Amide bonds
        2. Disulfide bonds
    C. Peptide structure determination (Secs. 15.6 - 15.7)
        1. Amino-acid analysis (Section 15.6)
        2. Peptide sequencing (Section 15.7)
            a. Edman degradation for N terminus
            b. Chemical and enzymatic hydrolysis
    D. Peptide synthesis (Section 15.8)
        1. Protection of amino acids
            a. Protection of amino group as BOC derivative
            b. Protection of carboxyl group as methyl ester
        2. Coupling of protected amino acids by DCC
        3. Cleavage of protecting groups

III. Proteins (Secs. 15.9 - 15.10)
    A. Classification of proteins (Section 15.9)
        1. Simple and conjugated proteins
        2. Fibrous and globular proteins
    B. Protein structure (Section 15.10)

1. Primary structure - the amino acid sequence
2. Secondary structure - coiling of the peptide backbone
   a. α–Helix
   b. β–Pleated sheet
3. Tertiary structure - overall three-dimensional shape
   a. Hydrophobic interactions
   b. Salt bridges
4. Quaternary structure - protein aggregates

IV. Enzymes (Sections 15.11 - 15.12)
   A. Introduction to enzymes (Section 15.11)
   B. Structure and classification of enzymes (Section 15.12)
      1. Holoenzymes: apoenzymes + cofactors
      2. Enzyme classes

Study Skills for Chapter 15

After studying this chapter, you should be able to:

1. Identify the common amino acids (Problem 15.1, 15.19).
2. Draw α–amino acids with the correct stereochemistry and in dipolar form (Problems 15.2, 15.3, 15.4, 15.6, 15.15.25, 15.26, 15.29, 15.30, 15.35, 15.36).
3. Describe the separation of a mixture of amino acids by electrophoresis (Problems 15.7, 15.31).
4. Draw the structures of simple peptides (Problems 15.8, 15.9, 15.10, 15.22, 15.28, 15.34, 15.51)
5. Deduce the structure of peptides (Problems 15.12, 15.13, 15.14, 15.43, 15.44, 15.45, 15.46, 15.49, 15.50)
6. Outline the scheme of peptide synthesis (Problems 15.15, 15.16, 15.41).
7. Define the terms in this chapter (Problem 15.18, 15.27).
8. Draw structures of reaction products of amino acids and peptides (Problems 15.5, 15.11, 15.33, 15.39, 15.42).
9. Classify enzymes by structure and function (Problems 15.17, 15.20, 15.23, 15.42)

Chapter 16 – Biomolecules: Lipids and Nucleic Acids

16.1

$$CH_3(CH_2)_{20}\overset{O}{\underset{\|}{C}}O(CH_2)_{27}CH_3 \xrightarrow[\text{2. } H_3O^+]{\text{1. NaOH, } H_2O} CH_3(CH_2)_{20}\overset{O}{\underset{\|}{C}}OH + HO(CH_2)_{27}CH_3$$

carboxylic acid   alcohol

16.2

a)
$$CH_2O\overset{O}{\underset{\|}{C}}CH_2CH_2CH_2CH_2CH_2CH_2CH_2\overset{H}{\underset{}{C}}{=}\overset{H}{\underset{}{C}}CH_2CH_2CH_2CH_2CH_2CH_2CH_3$$
$$|$$
$$CHO\overset{O}{\underset{\|}{C}}CH_2CH_2CH_2CH_2CH_2CH_2CH_2\overset{H}{\underset{}{C}}{=}\overset{H}{\underset{}{C}}CH_2CH_2CH_2CH_2CH_2CH_2CH_3$$
$$|$$
$$CH_2O\overset{O}{\underset{\|}{C}}CH_2CH_2CH_2CH_2CH_2CH_2CH_2\overset{H}{\underset{}{C}}{=}\overset{H}{\underset{}{C}}CH_2CH_2CH_2CH_2CH_2CH_2CH_3$$

Glyceryl trioleate

b)
$$CH_2O\overset{O}{\underset{\|}{C}}CH_2CH_2CH_2CH_2CH_2CH_2\overset{H}{\underset{}{C}}{=}\overset{H}{\underset{}{C}}CH_2CH_2CH_2CH_2CH_2CH_2CH_2CH_3$$
$$|$$
$$CHO\overset{O}{\underset{\|}{C}}CH_2CH_2CH_2CH_2CH_2CH_2CH_2CH_2CH_2CH_2CH_2CH_2CH_2CH_2CH_2CH_3$$
$$|$$
$$CH_2O\overset{O}{\underset{\|}{C}}CH_2CH_2CH_2CH_2CH_2CH_2CH_2CH_2CH_2CH_2CH_2CH_2CH_2CH_2CH_2CH_3$$

Glyceryl monooleate distearate (one isomer)

Glyceryl monooleate distearate is higher melting because it contains only one unsaturated fatty acid; glyceryl trioleate contains three.

16.3

$$\begin{array}{l}CH_2O\overset{O}{\overset{\|}{C}}(CH_2)_{14}CH_3 \\ | \\ CHO\overset{O}{\overset{\|}{C}}(CH_2)_{16}CH_3 \\ | \\ CH_2O\overset{O}{\overset{\|}{C}}(CH_2)_{14}CH_3 \end{array}$$
optically inactive

$$\begin{array}{l}CH_2O\overset{O}{\overset{\|}{C}}(CH_2)_{14}CH_3 \\ | \\ *\,CHO\overset{O}{\overset{\|}{C}}(CH_2)_{14}CH_3 \\ | \\ CH_2O\overset{O}{\overset{\|}{C}}(CH_2)_{16}CH_3 \end{array}$$
optically active

1. $^-OH, H_2O$
2. H_3O^+

$$\begin{array}{l}CH_2OH \\ | \\ CHOH \\ | \\ CH_2OH\end{array} \quad + \quad 2\,CH_3(CH_2)_{14}COOH \quad + \quad CH_3(CH_2)_{16}COOH$$

Palmitic acid Stearic acid

Four different groups are bonded to the central glycerol carbon atom in the optically active fat.

16.4

$$CH_3(CH_2)_7CH=CH(CH_2)_7\overset{O}{\overset{\|}{C}}O^-\ Mg^{2+}\ ^-O\overset{O}{\overset{\|}{C}}(CH_2)_7CH=CH(CH_2)_7CH_3$$
Magnesium oleate

The double bonds are *cis*.

16.5

$$\begin{array}{l}CH_2O\overset{O}{\overset{\|}{C}}(CH_2)_{14}CH_3 \\ | \\ CHO\overset{O}{\overset{\|}{C}}(CH_2)_7CH=CH(CH_2)_7CH_3\ cis \\ | \\ CH_2O\overset{O}{\overset{\|}{C}}(CH_2)_7CH=CH(CH_2)_7CH_3\ cis \end{array}$$

Glyceryl monopalmitate dioleate

$\xrightarrow{NaOH,\ H_2O}$

$$\begin{array}{l}CH_2OH \\ | \\ CHOH \\ | \\ CH_2OH\end{array} \quad +$$

Glycerol

$$Na^+\ ^-O\overset{O}{\overset{\|}{C}}(CH_2)_{14}CH_3$$
Sodium palmitate

$$2\,Na^+\ ^-O\overset{O}{\overset{\|}{C}}(CH_2)_7CH=CH(CH_2)_7CH_3$$
Sodium oleate

270   Chapter 16

16.6 The hydroxyl group in cholesterol is equatorial.

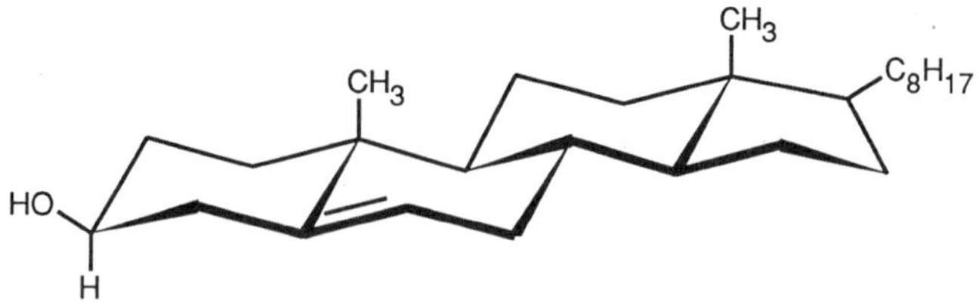

16.7

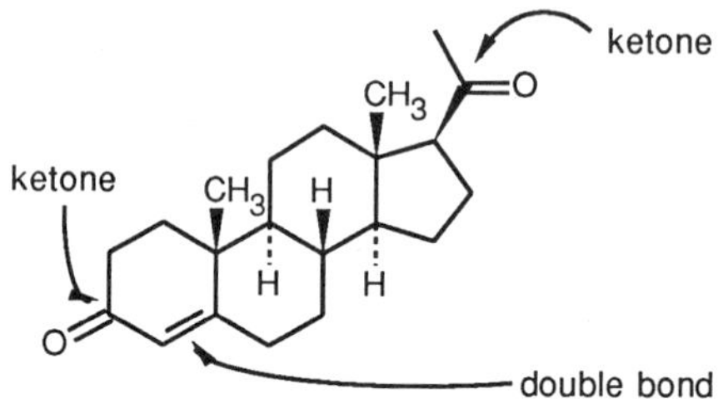

Progesterone (with labels: ketone, ketone, double bond)

16.8

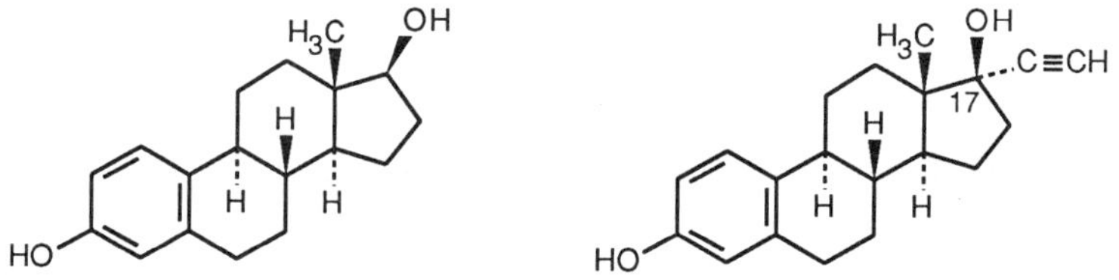

Estradiol

Ethynylestradiol

There is only one difference between estradiol and ethynylestradiol; ethynylestradiol has a –C≡C–H group at C17 that is not present in estradiol. Both compounds are estrogens because they both have a tetracyclic steroid skeleton with an aromatic A ring.

16.9

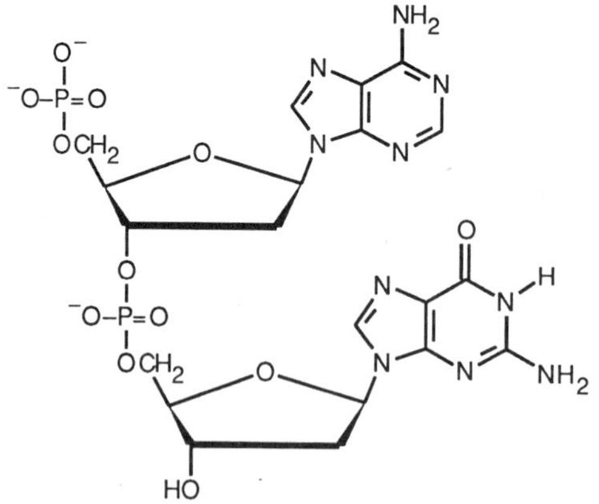

2'–Deoxyadenosine (A)

2'–Deoxyguanosine (G)

16.10

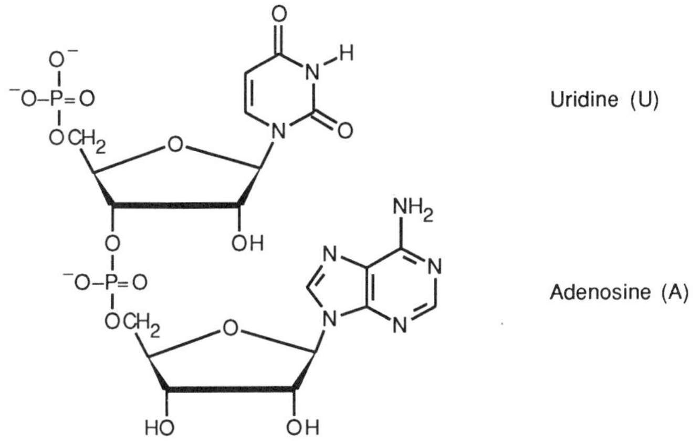

[Structure showing Uridine (U) and Adenosine (A) linked via phosphate groups]

Uridine (U)

Adenosine (A)

16.11

Original DNA: G–G–C–T–A–A–T–C–C–G–T
Complement: C–C–G–A–T–T–A–G–G–C–A

16.12

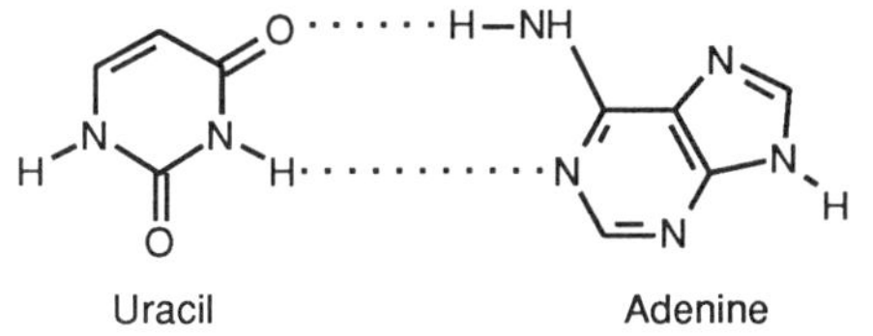

[Structure showing base pairing between Uracil and Adenine with hydrogen bonds]

Uracil   Adenine

16.13

DNA: G–A–T–T–A–C–C–G–T–A is complementary to:
RNA: C–U–A–A–U–G–G–C–A–U

16.14

RNA: U–U–C–G–C–A–G–A–G–U
DNA: A–A–G–C–G–T–C–T–C–A

16.15 Several different codons can code for the same amino acid.

Amino acid:	Ala	Phe	Leu	Tyr
Codon sequence:	GCU	UUU	UUA	UAU
	GCC	UUC	UUG	UAC
	GCA		CUU	
	GCG		CUC	
			CUA	
			CUG	

16.16-16.18

The mRNA base sequence:	CUU– AUG– GCU– UGG– CCC– UAA
The amino acid sequence:	Leu – Met – Ala – Trp – Pro – (stop)
The tRNA sequences:	GAA UAC CGA ACC GGG AUU
The DNA sequence:	GAA– TAC– CGA– ACC– GGG – ATT

16.19 Remember:

1. Only a few of the many possible splittings occur in each reaction.
2. Cleavage occurs at both sides of the reacting nucleotide.

^{32}P–A–A–C–A–T–G–G–C–G–C–T–T–A–T–G–A–C–G–A

reaction   *fragments*

a)   A ^{32}P–A
^{32}P–A–A–C
^{32}P–A–A–C–A–T–G–G–C–G–C–T–T
^{32}P–A–A–C–A–T–G–G–C–G–C–T–T–A–T–G
^{32}P–A–A–C–A–T–G–G–C–G–C–T–T–A–T–G–A–C–G

b)   G ^{32}P–A–A–C–A–T
^{32}P–A–A–C–A–T–G
^{32}P–A–A–C–A–T–G–G–C
^{32}P–A–A–C–A–T–G–G–C–G–C–T–T–A–T
^{32}P–A–A–C–A–T–G–G–C–G–C–T–T–A–T–G–A–C

c)   C ^{32}P–A–A
^{32}P–A–A–C–A–T–G–G
^{32}P–A–A–C–A–T–G–G–C–G
^{32}P–A–A–C–A–T–G–G–C–G–C–T–T–A–T–G–A

d)   C + T ^{32}P–A–A
^{32}P–A–A–C–A
^{32}P–A–A–C–A–T–G–G
^{32}P–A–A–C–A–T–G–G–C–G
^{32}P–A–A–C–A–T–G–G–C–G–C
^{32}P–A–A–C–A–T–G–G–C–G–C–T
^{32}P–A–A–C–A–T–G–G–C–G–C–T–T–A
^{32}P–A–A–C–A–T–G–G–C–G–C–T–T–A–T–G–A

Biomolecules: Lipids and Nucleic Acids 273

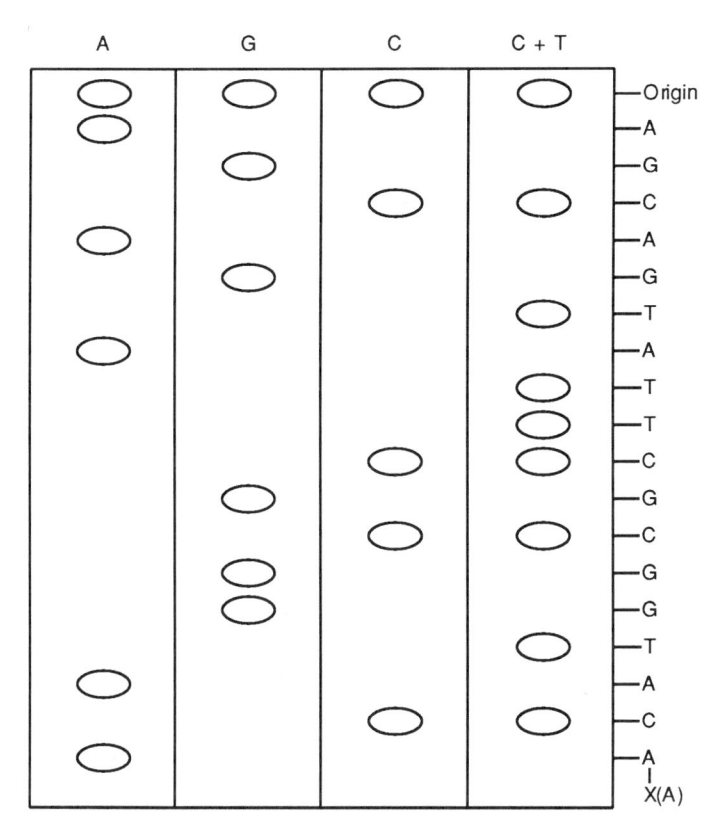

16.20

	A	G	C	C + T	
	○	○	○	○	—Origin
	○				—A
		○			—G
			○	○	—C
	○				—A
		○			—G
				○	—T
	○				—A
				○	—T
				○	—T
			○	○	—C
		○			—G
			○	○	—C
		○			—G
		○			—G
				○	—T
	○				—A
			○	○	—C
	○				—A
					|
					X(A)

16.21 The complete sequence:

X–T–C–A–G–C–G–A–T–T–C–G–G–T–A–C

16.22

a)
$$\begin{array}{l} \text{CH}_2\text{OC(CH}_2)_{14}\text{CH}_3 \\ | \\ \text{CHOC(CH}_2)_7\text{CH=CH(CH}_2)_7\text{CH}_3 \text{ (cis)} \\ | \\ \text{CH}_2\text{OC(CH}_2)_{16}\text{CH}_3 \end{array}$$
(each OC with C=O)

a fat

b)
$$\begin{array}{l} \text{CH}_2\text{OC(CH}_2)_7\text{CH=CH(CH}_2)_7\text{CH}_3 \\ | \\ \text{CHOC(CH}_2)_7\text{CH=CH(CH}_2)_7\text{CH}_3 \\ | \\ \text{CH}_2\text{OC(CH}_2)_7\text{CH=CH(CH}_2)_7\text{CH}_3 \end{array}$$

a vegetable oil
(all double bonds *cis*)

274 Chapter 16

c) A steroid (Estradiol)

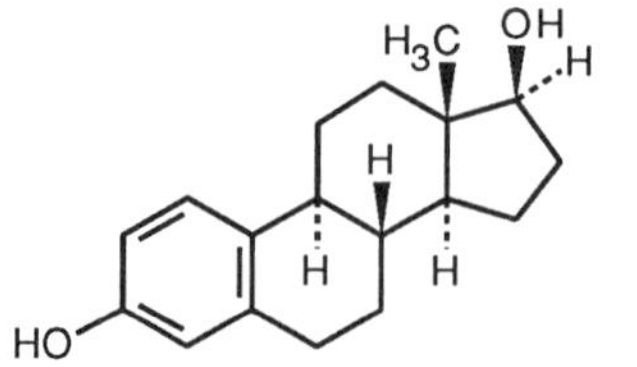

16.23

a) $CH_3(CH_2)_{16}\overset{O}{\underset{\|}{C}}-O^-Na^+$
Sodium stearate

b) $CH_3(CH_2)_4CH=CHCH_2CH=CH(CH_2)_7\overset{O}{\underset{\|}{C}}OCH_2CH_3$
Ethyl linoleate

c)
$$CH_2O\overset{O}{\underset{\|}{C}}(CH_2)_{14}CH_3$$
$$CHO\overset{O}{\underset{\|}{C}}(CH_2)_7CH=CH(CH_2)_7CH_3$$
$$CH_2O\overset{O}{\underset{\|}{C}}(CH_2)_7CH=CH(CH_2)_7CH_3$$

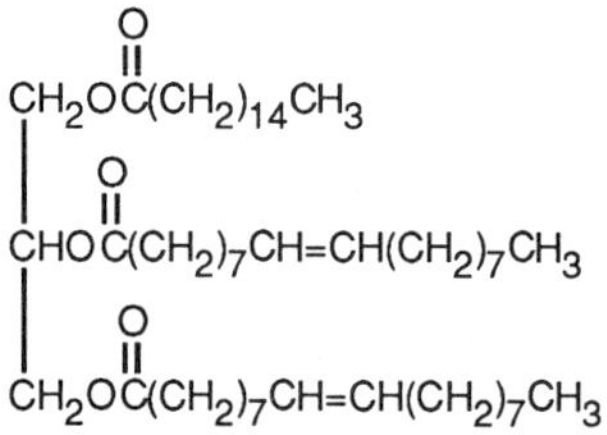

or

$$CH_2O\overset{O}{\underset{\|}{C}}(CH_2)_7CH=CH(CH_2)_7CH_3$$
$$CHO\overset{O}{\underset{\|}{C}}(CH_2)_{14}CH_3$$
$$CH_2O\overset{O}{\underset{\|}{C}}(CH_2)_7CH=CH(CH_2)_7CH_3$$

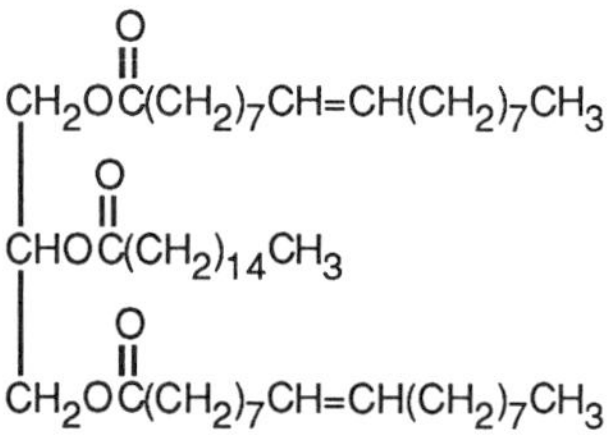

Glyceryl palmitodioleate

16.24

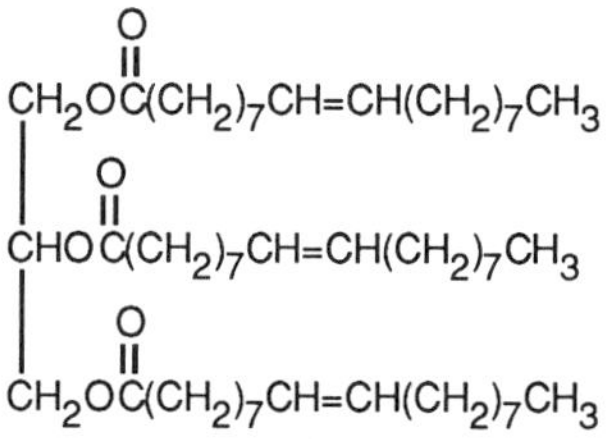

$$CH_2O\overset{O}{\underset{\|}{C}}(CH_2)_7CH=CH(CH_2)_7CH_3$$
$$CHO\overset{O}{\underset{\|}{C}}(CH_2)_7CH=CH(CH_2)_7CH_3$$
$$CH_2O\overset{O}{\underset{\|}{C}}(CH_2)_7CH=CH(CH_2)_7CH_3$$

Glyceryl trioleate

a) Glyceryl trioleate $\xrightarrow[CCl_4]{Br_2}$

$$CH_2O\overset{O}{\underset{\|}{C}}(CH_2)_7CHBrCHBr(CH_2)_7CH_3$$
$$CHO\overset{O}{\underset{\|}{C}}(CH_2)_7CHBrCHBr(CH_2)_7CH_3$$
$$CH_2O\overset{O}{\underset{\|}{C}}(CH_2)_7CHBrCHBr(CH_2)_7CH_3$$

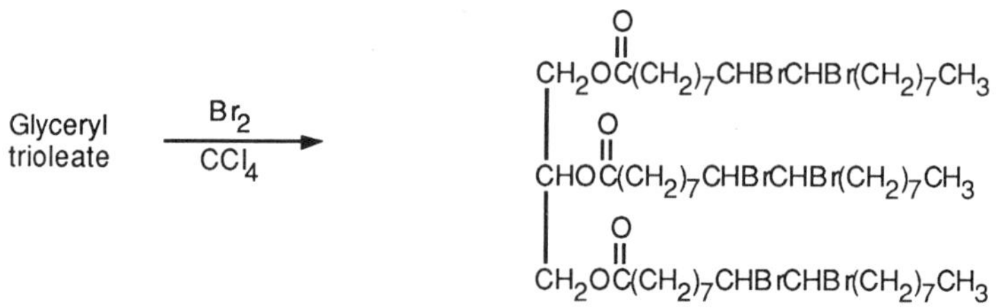

Biomolecules: Lipids and Nucleic Acids 275

b) Glyceryl trioleate $\xrightarrow{\text{H}_2, \text{Pd catalyst}}$

$$\begin{array}{l} \text{CH}_2\text{OC(=O)(CH}_2)_{16}\text{CH}_3 \\ |\\ \text{CHOC(=O)(CH}_2)_{16}\text{CH}_3 \\ |\\ \text{CH}_2\text{OC(=O)(CH}_2)_{16}\text{CH}_3 \end{array}$$

c) Glyceryl trioleate $\xrightarrow{\text{NaOH, H}_2\text{O}}$

$$\begin{array}{l} \text{CH}_2\text{OH} \\ |\\ \text{CHOH} \\ |\\ \text{CH}_2\text{OH} \end{array} + 3\ \text{CH}_3(\text{CH}_2)_7\text{CH=CH(CH}_2)_7\text{COO}^-\text{Na}^+$$

d) Glyceryl trioleate $\xrightarrow{\text{1. O}_3;\ \text{2. Zn, CH}_3\text{COOH}}$

$$\begin{array}{l} \text{CH}_2\text{OC(=O)(CH}_2)_7\text{CHO} \\ |\\ \text{CHOC(=O)(CH}_2)_7\text{CHO} \\ |\\ \text{CH}_2\text{OC(=O)(CH}_2)_7\text{CHO} \end{array} + 3\ \text{CH}_3(\text{CH}_2)_7\text{CHO}$$

e) Glyceryl trioleate $\xrightarrow{\text{1. LiAlH}_4;\ \text{2. H}_3\text{O}^+}$

$$\begin{array}{l} \text{CH}_2\text{OH} \\ |\\ \text{CHOH} \\ |\\ \text{CH}_2\text{OH} \end{array} + 3\ \text{CH}_3(\text{CH}_2)_7\text{CH=CH(CH}_2)_7\text{CH}_2\text{OH}$$

16.25

$$\textit{cis-}\text{CH}_3(\text{CH}_2)_7\text{CH=CH(CH}_2)_7\text{COOH}$$
Oleic acid

a) Oleic acid $\xrightarrow{\text{CH}_3\text{OH, H}^+}$ CH$_3$(CH$_2$)$_7$CH=CH(CH$_2$)$_7$COOCH$_3$
Methyl oleate

b) Methyl oleate (part a) $\xrightarrow{\text{H}_2, \text{Pd catalyst}}$ CH$_3$(CH$_2$)$_{16}$COOCH$_3$
Methyl stearate

c) Oleic acid $\xrightarrow{\text{1. O}_3;\ \text{2. Zn, CH}_3\text{COOH}}$ CH$_3$(CH$_2$)$_7$CHO + OCH(CH$_2$)$_7$COOH
Nonanal   9-Oxononanoic acid

d) OHC(CH$_2$)$_7$COOH (part c) $\xrightarrow{\text{Tollens' reagent}}$ HOOC(CH$_2$)$_7$COOH
Nonanedioic acid

276   Chapter 16

16.26

CH₃CH₂CH₂CH₂\C=C/H \C=C/H \C=C/H
with substituents leading to CH₂CH₂CH₂CH₂CH₂CH₂CH₂COOH

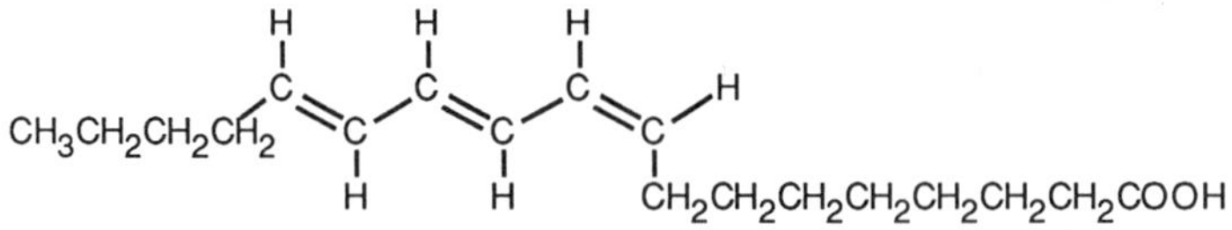

(9Z,11E,13E)–Octadecatrienoic acid
(Eleostearic acid)

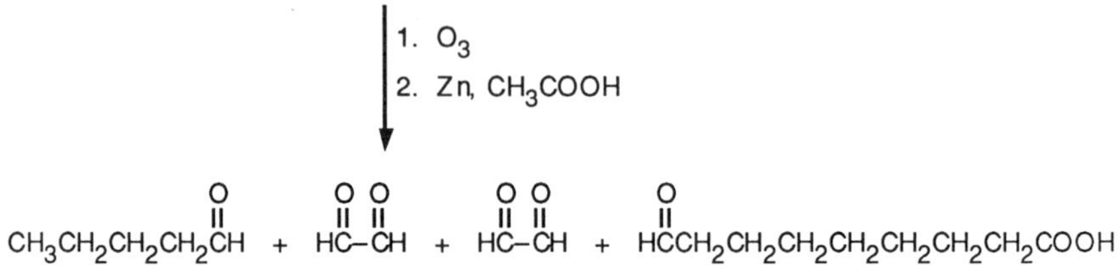

$$\xrightarrow[\text{2. Zn, CH}_3\text{COOH}]{\text{1. O}_3}$$

CH₃CH₂CH₂CH₂CHO + HC(O)–CH(O) + HC(O)–CH(O) + H(O)CCH₂CH₂CH₂CH₂CH₂CH₂COOH

The stereochemistry of the double bonds can't be determined from the information given.

16.27

CH₃(CH₂)₇C≡C(CH₂)₇COOH $\xrightarrow[\text{Lindlar catalyst}]{\text{H}_2}$ CH₃(CH₂)₇CH=CH(CH₂)₇COOH (*cis*)

Stearolic acid   Oleic acid

16.28

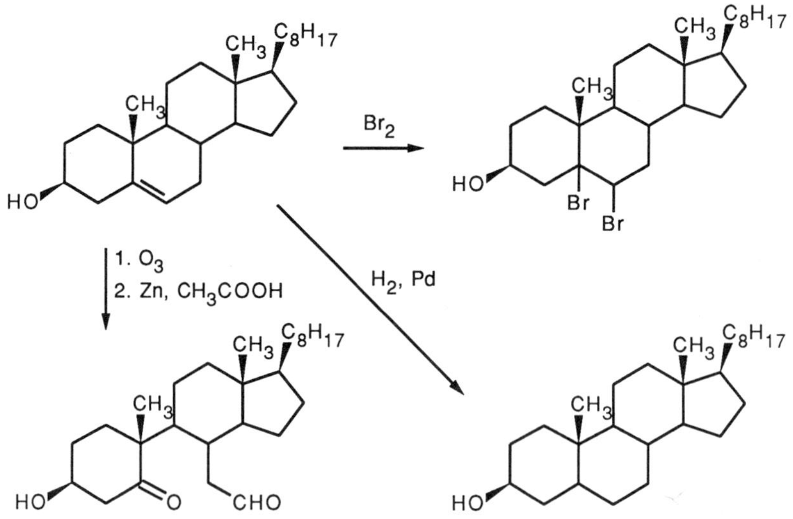

[Steroid structures showing reactions of a cholesterol-type molecule with Br₂ (giving dibromide), H₂/Pd (giving saturated steroid), and 1. O₃, 2. Zn, CH₃COOH (giving ring-opened keto-aldehyde)]

16.29

$$\begin{array}{c} CH_2OCR \\ | \\ CHOCR' \\ | \\ CH_2OCR'' \end{array} \;\; + \;\; 3\,NaOH \;\; \xrightarrow{H_2O} \;\; \begin{array}{c} CH_2OH \\ | \\ CHOH \\ | \\ CH_2OH \end{array} \;\; + \;\; \begin{array}{c} RCO^-Na^+ \\ R'CO^-Na^+ \\ R''CO^-Na^+ \end{array}$$

(each carboxylate with C=O)

Molar mass 1500 g   Molar mass 40.0 g

Since three moles of NaOH are needed to saponify one mole of soybean oil, 3 × 40.0 g = 120 g NaOH are needed to saponify 1500 g of oil.

$$\text{Grams NaOH} = \frac{5.00 \text{ g oil}}{1500 \frac{\text{g oil}}{\text{mol oil}}} \times 120 \frac{\text{g NaOH}}{\text{mol oil}} = 0.400 \text{ g NaOH}$$

Thus, 0.400 g of NaOH is needed to saponify 5.00 g of soy oil.

16.30
a) A *steroid* is an organic molecule whose structure is based on a specific tetracyclic skeleton.

b) *DNA* is a biological polymer whose monomer units are nucleotides. A nucleotide is composed of a heterocyclic amine base, the sugar deoxyribose, and a phosphate group. DNA is the transmitter of the genetic code of all complex living organisms.

c) A *base pair* is a specific pair of heterocyclic amine bases that hydrogen-bond to each other in a DNA double helix and during protein synthesis.

d) A *codon* is a sequence of three mRNA nucleotides that specifies a particular amino acid to be used in protein synthesis.

e) A *lipid* is a naturally occurring organic molecule that is insoluble in water but soluble in organic solvents.

f) *Transcription* is the process by which the genetic message contained in DNA is read by RNA and carried from the nucleus to the ribosomes.

16.31 The percent of A always equals the percent of T, since A and T are complementary. A similar relationship is also true for G and C. Thus, sea urchin DNA contains about 32% each of A and T, and 18% each of G and C.

16.32

Original DNA: G–A–A–G–T–T–C–A–T–G–C
Complement: C–T–T–C–A–A–G–T–A–C–G

16.33

Amino Acid:	Ile	Asp	Thr
Codon Sequence:	AUU	GAU	ACU
	AUC	GAC	ACC
	AUA		ACA
			ACG

16.34-16.35

UAC is the codon for tyrosine. It was transcribed from ATG of the DNA chain.

[Structural diagrams: mRNA codon U-A-C (with uracil, adenine, cytosine on ribose-phosphate backbone with OH groups) and DNA A-T-G (with adenine, thymine, guanine on deoxyribose-phosphate backbone)]

16.36-16.38

mRNA codon:	a) AAU	b) GAG	c) UCC	d) CAU	e) ACC
Amino acid	Asn	Glu	Ser	His	Thr
DNA sequence:	TTA	CTC	AGG	GTA	TGG
tRNA anticodon:	UUA	CUC	AGG	GUA	UGG

Biomolecules: Lipids and Nucleic Acids   279

16.39

	Normal	Mutated
DNA:	T-A-A-C-C-G-G-A-T	T-G-A-C-C-G-G-T-A
mRNA:	A-U-U-G-G-C-C-U-A	A-C-U-G-G-C-C-U-A
Amino Acids:	Ile — Gly — Leu	Thr — Gly — Leu

In the mutated protein, a threonine residue replaces an isoleucine residue.

16.40 Metenkephalin: Tyr — Gly — Gly — Phe — Met   is coded by:

mRNA: UAU GGU GGU UUU AUG UAA (stop)

UAC GGC GGC UUC   UAG

GGA GGA

GGG GGG

16.41 Using the first set of base pairs to solve this problem:

DNA: ATA – CCA – CCA – AAA – TAC – ATT

16.42 The chain containing 21 amino acids needs 21 x 3 = 63 bases to code for it. In addition, a three base codon is needed to terminate the chain. Thus, 66 bases are needed to code for the first chain. The chain containing 30 amino acids needs 30 x 3 = 90 bases, plus a three base "stop" codon, for a total of 96 bases.

16.43 Position 9:   Horse amino acid = Gly   Human amino acid = Ser

mRNA
codons: GGU GGC GGA GGG   UCU UCC UCA UCG AGU AGC

DNA
bases: <u>CCA</u> <u>CCG</u> CCT CCC   AGA AGG ACT ACC <u>TCA</u> <u>TCG</u>

The underlined horse DNA base triplets differ from their human counterparts by only one base.

Position 30:   Horse amino acid = Ala   Human amino acid = Thr

mRNA codons   GCU GCC GCA GCG   ACU ACC ACA ACG

DNA bases: CGA CGG CGT CGC   TGA TGG TGT TGC

Each group of three DNA bases from horse insulin has a counterpart in human insulin that differs from it by only one base. It is possible that horse insulin DNA differs from human insulin DNA by only two bases out of 165!

16.44-16.45

mRNA codon:	CUA– GAC – CGU –UCC – AAG – UGA
Amino acid:	Leu – Asp – Arg – Ser – Lys – (Stop)
tRNA anticodons:	GAU CUG GCA AGG UUC ACU
DNA sequence:	GAT– CTG– GCA– AGG – TTC – ACT
DNA complement:	CTA– GAC– CGT– TCC – AGG – TGA

16.46

Angiotensin II: Asp – Arg – Val – Tyr – Ile – His – Pro – Phe – (Stop)

mRNA codon:
```
GAU  CGU  GUU  UAU  AUU  CAU  CCU  UUU  UAA
GAC  CGC  GUC  UAC  AUC  CAC  CCC  UUC  UAG
     CGA  GUA       AUA       CCA       UGA
     CGG  GUG                 CCG
     AGA
     AGG
```

16.47-16.48

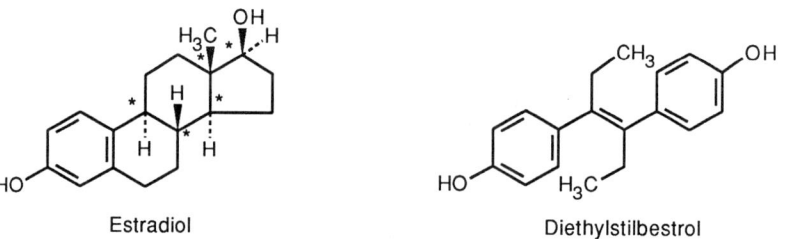

Estradiol

Diethylstilbestrol

Estradiol has five stereogenic centers.

16.49

a) Estradiol $\xrightarrow{\text{1. NaOH} \atop \text{2. CH}_3\text{I}}$

b) Estradiol $\xrightarrow{\text{CH}_3\text{COCl} \atop \text{pyridine}}$

c)

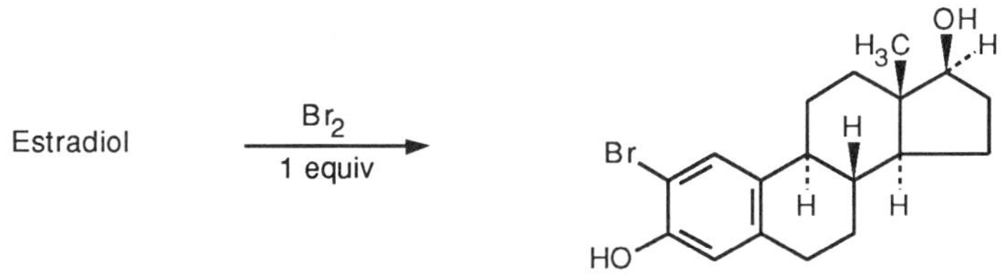

Estradiol $\xrightarrow[\text{1 equiv}]{\text{Br}_2}$ [brominated estradiol structure with Br at position ortho to HO]

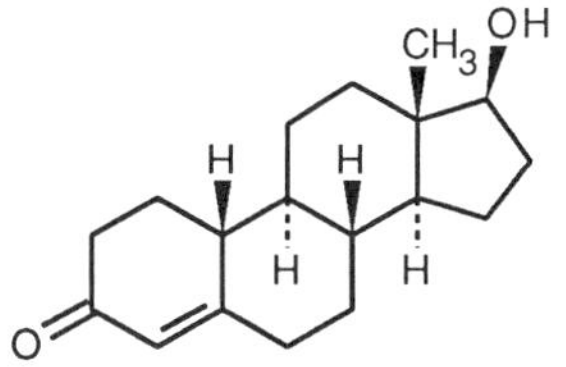

16.50

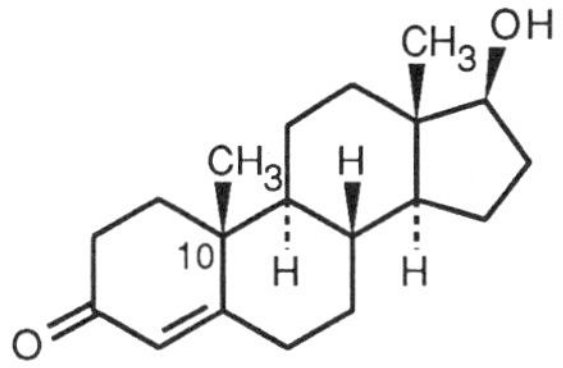

Testosterone and Nandrolone structures shown.

Testosterone and nandrolone are identical, except for the methyl group at C10 of testosterone, which is replaced by a –H in nandrolone. Both steroids have the same carbon skeleton, both have the same enone group in the A ring, and both have a hydroxyl group at C17 of the D ring.

Chapter Outline

I. Lipids (Sections 16.1 - 16.5)
   A. Fats and vegetable oils - triacylglycerols (Section 16.2)
   B. Soaps (Section 16.3)
   C. Phospholipids (Section 16.4)
      1. Phosphoglycerides
         a. Lecithins
         b. Cephalins
      2. Sphingolipids
   D. Steroids (Section 16.5)
      1. Steroid structure
      2. Steroid variety
         a. Sex hormones
         b. Adrenocortical hormones
         c. Synthetic steroids

II. Nucleic acids (Sections 16.6 - 16.13)
   A. Introduction (Section 16.6)
      1. Structure of nucleic acids
      2. Structure of nucleic acid components

282   Chapter 16

    B. Structure of DNA (Sections 16.7 - 16.8)
        1. Formation of DNA chain (Section 16.7)
        2. Double-helix structure (Section 16.8)
        3. Complementary base-pairing
    C. Nucleic acids and heredity (Sections 16.9 - 16.12)
        1. Replication of DNA (Section 16.10)
        2. Structure and biosynthesis of RNA - transcription (Section 16.11)
        3. Translation of RNA - protein biosynthesis (Section 16.12)
            a. The genetic code
            b. Structure and function of transfer RNA
    D. Sequencing of DNA (Section 16.13)
        1. Cleavage of DNA with restriction enzymes
        2. Labeling of DNA with radioactive phosphate
        3. Selective cleavages of the DNA chain
        4. Reading the DNA sequence

Study Skills for Chapter 16

After studying this chapter, you should be able to:

1. Draw the structures of fats, oils, soaps, and steroids (Problems 16.2, 16.3, 16.4, 16.22, 16.23).
2. Formulate reactions of lipids (Problems 16.1, 16.5, 16.24, 16.25, 16.26, 16.27, 16.28, 16.29, 16.49).
3. Understand the structure and stereochemistry of steroids (Problems 16.6, 16.7, 16.8, 16.47, 16.48, 16.50).
4. Given a DNA or RNA strand, draw its complementary strand (Problems 16.11, 16.13, 16.14, 16.18, 16.32, 16.35, 16.37, 16.41).
5. Draw the structure of a given DNA or RNA fragment (Problems 16.9, 16.10, 16.34).
6. List the codon sequence for a given amino acid or peptide (Problems 16.15, 16.33, 16.40, 16.46).
7. Deduce an amino acid sequence from a given mRNA base sequence (Problems 16.16, 16.36, 16.39, 16.44).
8. Draw the anticodon sequence of tRNA, given the mRNA sequence (Problems 16.17, 16.38, 16.45).
9. Deduce a DNA sequence from an electrophoresis pattern (Problems 16.20, 16.21).
10. Define the important terms in this chapter (Problem 16.30).

Chapter 17 – The Organic Chemistry of Metabolic Pathways

17.1

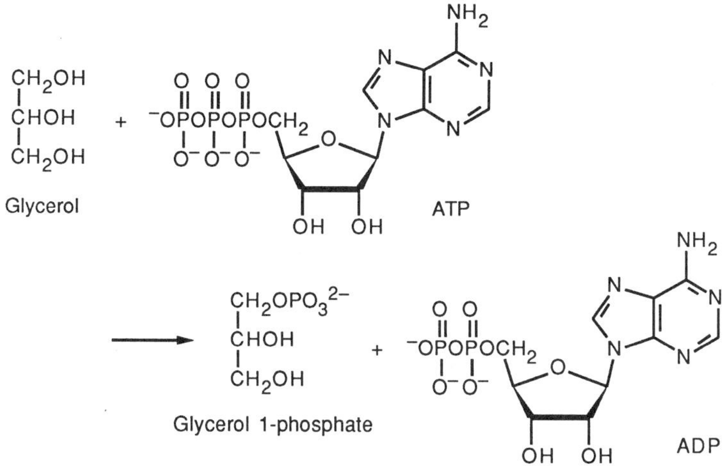

Glycerol + ATP ⟶ Glycerol 1-phosphate + ADP

17.2

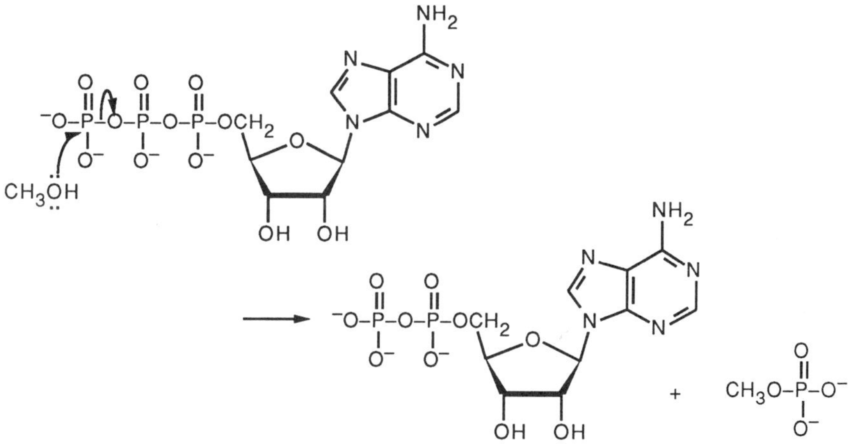

17.3

$$CH_3CH_2-CH_2CH_2-CH_2CH_2-CH_2\overset{O}{\underset{\|}{C}}SCoA$$

Caprylyl CoA

↓ (turn 4)

$$CH_3CH_2-CH_2CH_2-CH_2\overset{O}{\underset{\|}{C}}SCoA \;+\; CH_3\overset{O}{\underset{\|}{C}}SCoA$$

Hexanoyl CoA

↓ (turn 5)

$$CH_3CH_2-CH_2\overset{O}{\underset{\|}{C}}SCoA \;+\; CH_3\overset{O}{\underset{\|}{C}}SCoA$$

Butyryl CoA

↓ (turn 6)

$$CH_3\overset{O}{\underset{\|}{C}}SCoA \;+\; CH_3\overset{O}{\underset{\|}{C}}SCoA$$

17.4

a) $CH_3CH_2\!\downarrow\!CH_2CH_2\!\downarrow\!CH_2CH_2\!\downarrow\!CH_2CH_2\!\downarrow\!CH_2CH_2\!\downarrow\!CH_2CH_2\!\downarrow\!CH_2CH_2\!\downarrow\!CH_2\overset{O}{\underset{\|}{C}}OH$

↓

$8\;CH_3\overset{O}{\underset{\|}{C}}SCoA$

Seven turns of the spiral are needed.

b) $CH_3CH_2-(CH_2CH_2)_8-CH_2\overset{O}{\underset{\|}{C}}OH \longrightarrow 10\;CH_3\overset{O}{\underset{\|}{C}}SCoA$

Nine turns of the spiral are needed.

17.5 ATP is produced in step 7 (1,3-diphosphoglycerate —> 3-phosphoglycerate) and in step 10 (phosphoenolpyruvate —> pyruvate).

17.6

$$^{2-}O_3POCH_2\overset{O}{\underset{\|}{C}}CH_2OH \longrightarrow \left[^{2-}O_3POCH_2\underset{\underset{H}{|}}{\overset{\overset{HO}{|}}{C}}=\underset{\underset{}{}}{\overset{\overset{OH}{|}}{CH}} \right] \longrightarrow ^{2-}O_3POCH_2\underset{\underset{H}{|}}{\overset{\overset{HO}{|}}{C}}\overset{O}{\underset{\|}{C}}H$$

Enol

Enzyme-catalyzed enolization is followed by formation of glyceraldehyde 3-phosphate.

17.7 Citrate and isocitrate are tricarboxylic acids.

17.8

$$\underset{\text{Citrate}}{\begin{array}{c} \text{COO}^- \\ | \\ \text{CH}_2 \\ | \\ \text{HO-C-COO}^- \\ | \\ \text{HC-H} \\ | \\ \text{COO}^- \end{array}} \longrightarrow \underset{\text{Aconitate}}{\begin{array}{c} \text{COO}^- \\ | \\ \text{CH}_2 \quad \text{O} \\ | \quad \quad \diagdown \\ \text{C=C} \\ || \quad \diagup \\ \text{HC} \quad \text{O}^- \\ | \\ \text{COO}^- \quad :\ddot{\text{O}}\text{H}_2 \end{array}} \longrightarrow \left[\begin{array}{c} \text{COO}^- \\ | \\ \text{CH}_2 \quad \text{O}^- \\ | \quad \quad \diagdown \\ \text{C=C} \\ | \quad \quad \diagup \\ \quad \quad \text{O} \\ \text{HC-}\overset{+}{\text{O}}\text{H}_2 \\ | \\ \text{COO}^- \end{array} \right] \longrightarrow \underset{\text{Isocitrate}}{\begin{array}{c} \text{COO}^- \\ | \\ \text{CH}_2 \\ | \\ \text{HC-COO}^- \\ | \\ \text{HC-OH} \\ | \\ \text{COO}^- \end{array}}$$

Enzyme-catalyzed E2 elimination of H₂O is followed by nucleophilic conjugate addition of water to produce isocitrate.

17.9

$$\underset{\text{Leucine}}{\begin{array}{c} \text{CH}_3 \\ | \\ \text{CHCH}_3 \\ | \\ \text{CH}_2 \\ | \\ \text{HC-}\overset{+}{\text{NH}}_3 \\ | \\ \text{COO}^- \end{array}} + \underset{\alpha\text{-Ketoglutarate}}{\begin{array}{c} \text{CH}_3 \\ | \\ \text{CH}_2 \\ | \\ \text{CH}_2 \\ | \\ \text{C=O} \\ | \\ \text{COO}^- \end{array}} \longrightarrow \begin{array}{c} \text{CH}_3 \\ | \\ \text{CHCH}_3 \\ | \\ \text{CH}_2 \\ | \\ \text{C=O} \\ | \\ \text{COO}^- \end{array} + \underset{\text{Glutamate}}{\begin{array}{c} \text{CH}_3 \\ | \\ \text{CH}_2 \\ | \\ \text{CH}_2 \\ | \\ \text{HC-}\overset{+}{\text{NH}}_3 \\ | \\ \text{COO}^- \end{array}}$$

17.10 Digestion is the breakdown of bulk food in the stomach and small intestine. Hydrolysis of amide, ester and acetal bonds yields amino acids, fatty acids, and simple sugars.

17.11 Metabolism refers to all reactions that take place inside cells. Digestion is a part of metabolism in which food is broken down into small organic molecules.

17.12 Metabolic processes that break down large food molecules are known as catabolism. Metabolic processes that assemble larger biomolecules from smaller ones are known as anabolism.

17.13

AMP

17.14 ATP transfers a phosphate group to another molecule in anabolic reactions.

286   Chapter 17

17.15 NAD⁺ is a biochemical oxidizing agent that converts alcohols to aldehydes or ketones, yielding NADH as a byproduct.

17.16 FAD is an oxidizing agent that introduces a conjugated double bond into a biomolecule, yielding FADH$_2$ as a byproduct.

17.17 Oxaloacetate is the starting point for the citric acid cycle.

17.18

$$\underset{\text{Lactate}}{CH_3\overset{OH}{\underset{|}{C}}HCOO^-} \xrightarrow[]{NAD^+ \quad NADH/H^+} \underset{\text{Pyruvate}}{CH_3\overset{O}{\overset{\|}{C}}COO^-}$$

NAD⁺ is needed to convert lactate to pyruvate.

17.19 a) One mole of glucose is catabolized to two moles of pyruvate, each of which yields one mole of acetyl CoA. Thus,

   1.0 mol glucose —> 2.0 mol acetyl CoA

b) Maltose is a disaccharide that yields two moles of glucose on hydrolysis. Since each mole of glucose yields two moles of acetyl CoA,

   1.0 mol maltose —> 2.0 mol glucose —> 4.0 mol acetyl CoA

c) A fatty acid with *n* carbons yields *n*/2 moles of acetyl CoA per mole of fatty acid. For palmitic acid (C$_{15}$H$_{31}$COOH),

   1.0 mol palmitic acid x $\dfrac{\text{8 mol acetyl CoA}}{\text{1 mol palmitic acid}}$ —> 8.0 mol acetyl CoA

17.20

	a) Glucose	b) Maltose	c) Palmitic acid
Molecular weight	180.2 amu	342.3 amu	256.4 amu
Moles in 100.0 g	0.5549 mol	0.2921 mol	0.3900 mol
Moles of acetyl CoA produced	2 x 0.5549 mol = 1.110 mol	4 x 0.2921 mol = 1.168 mol	8 x 0.3900 mol = 3.120 mol
Grams acetyl CoA produced	898.6 grams	945.6 grams	2526 grams

17.21 Palmitic acid is the most efficient precursor of acetyl CoA on a weight basis.

17.22

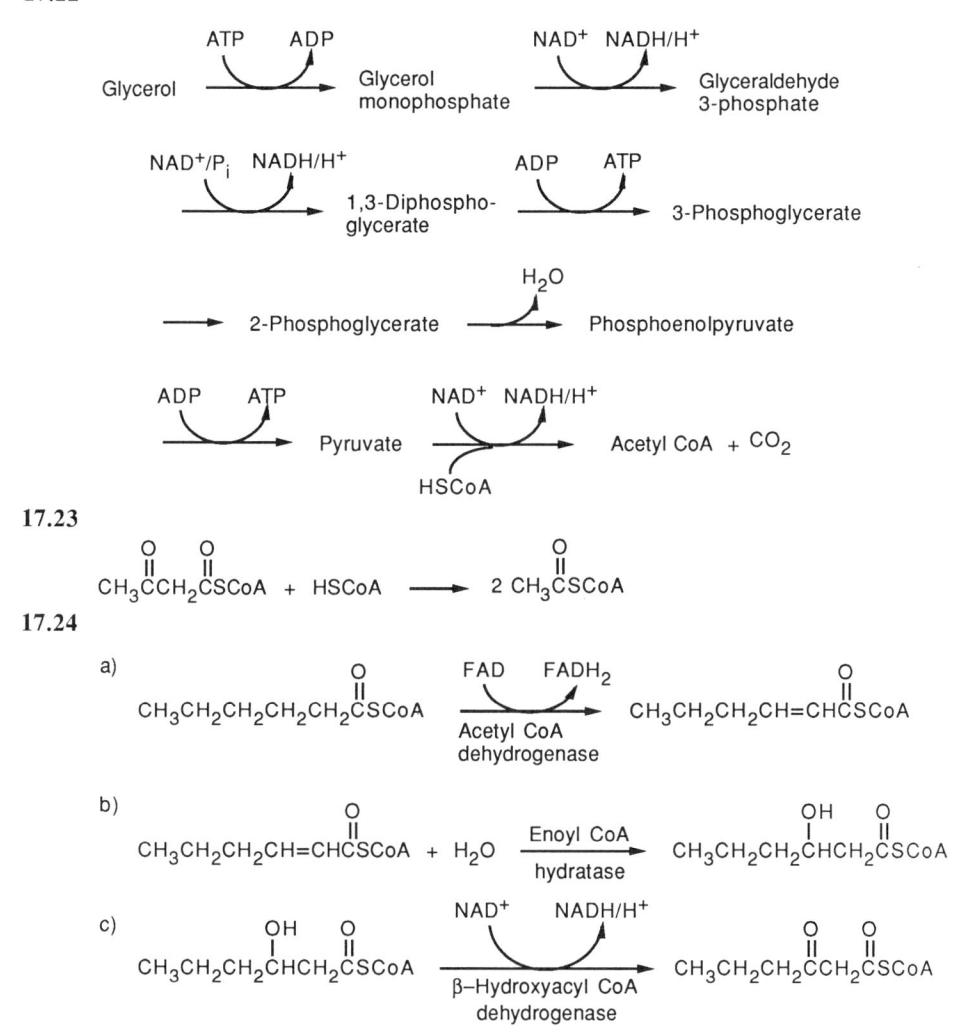

Glycerol $\xrightarrow{\text{ATP} \quad \text{ADP}}$ Glycerol monophosphate $\xrightarrow{\text{NAD}^+ \quad \text{NADH/H}^+}$ Glyceraldehyde 3-phosphate

$\xrightarrow{\text{NAD}^+/P_i \quad \text{NADH/H}^+}$ 1,3-Diphosphoglycerate $\xrightarrow{\text{ADP} \quad \text{ATP}}$ 3-Phosphoglycerate

$\longrightarrow$ 2-Phosphoglycerate $\xrightarrow{\text{H}_2\text{O}}$ Phosphoenolpyruvate

$\xrightarrow{\text{ADP} \quad \text{ATP}}$ Pyruvate $\xrightarrow[\text{HSCoA}]{\text{NAD}^+ \quad \text{NADH/H}^+}$ Acetyl CoA + CO_2

17.23

$$CH_3\overset{O}{\underset{\|}{C}}CH_2\overset{O}{\underset{\|}{C}}SCoA + HSCoA \longrightarrow 2\ CH_3\overset{O}{\underset{\|}{C}}SCoA$$

17.24

a) $CH_3CH_2CH_2CH_2CH_2\overset{O}{\underset{\|}{C}}SCoA \xrightarrow[\text{Acetyl CoA dehydrogenase}]{\text{FAD} \quad \text{FADH}_2} CH_3CH_2CH_2CH=CH\overset{O}{\underset{\|}{C}}SCoA$

b) $CH_3CH_2CH_2CH=CH\overset{O}{\underset{\|}{C}}SCoA + H_2O \xrightarrow{\text{Enoyl CoA hydratase}} CH_3CH_2CH_2\overset{OH}{\underset{|}{C}}HCH_2\overset{O}{\underset{\|}{C}}SCoA$

c) $CH_3CH_2CH_2\overset{OH}{\underset{|}{C}}HCH_2\overset{O}{\underset{\|}{C}}SCoA \xrightarrow[\beta\text{-Hydroxyacyl CoA dehydrogenase}]{\text{NAD}^+ \quad \text{NADH/H}^+} CH_3CH_2CH_2\overset{O}{\underset{\|}{C}}CH_2\overset{O}{\underset{\|}{C}}SCoA$

17.25

Amino acid	α–Keto acid
a) CH$_3$CH(CH$_3$)CH(+NH$_3$)COO$^-$	CH$_3$CH(CH$_3$)C(=O)COO$^-$
b) C$_6$H$_5$–CH$_2$CH(+NH$_3$)COO$^-$	C$_6$H$_5$–CH$_2$C(=O)COO$^-$
c) CH$_3$SCH$_2$CH$_2$CH(+NH$_3$)COO$^-$	CH$_3$SCH$_2$CH$_2$C(=O)COO$^-$

17.26 Pyridoxal phosphate is the cofactor involved in transamination.

17.27 The reactions of lipogenesis are the reverse of the reactions that occur in the fatty acid spiral.

$$CH_3\overset{O}{\underset{\|}{C}}CH_2\overset{O}{\underset{\|}{C}}SCoA \xrightarrow{\text{reduction of ketone}} CH_3\overset{OH}{\underset{|}{C}}HCH_2\overset{O}{\underset{\|}{C}}SCoA \xrightarrow{\text{dehydration}} CH_3CH=CH\overset{O}{\underset{\|}{C}}SCoA$$

$$\xrightarrow{\text{reduction of double bond}} CH_3CH_2CH_2\overset{O}{\underset{\|}{C}}SCoA$$

17.28

Ribulose 5-phosphate → [Enol] → Ribose 5-phosphate

The isomerization of ribulose 5-phosphate to ribose 5-phosphate occurs by way of an intermediate enol.

Organic Chemistry of Metabolic Pathways 289

17.29 This is a reverse aldol reaction, similar to step 4 of glycolysis.

$$\begin{array}{c} H-C=O \\ H-C-OH \\ H-C-OH \\ H-C-OH \\ CH_2OPO_3^{2-} \end{array} \longrightarrow \left[\begin{array}{c} H-C-O^- \\ H-C-OH \\ + \\ H-C=O \\ H-C-OH \\ CH_2OPO_3^{2-} \end{array} \right] \longrightarrow \begin{array}{c} CHO \\ CH_2OH \\ + \\ CHO \\ H-C-OH \\ CH_2OPO_3^{2-} \end{array}$$

17.30 The conversion of pyruvate to oxaloacetate is a mixed aldol condensation reaction between CO_2 and pyruvate.

$$O=C=O \ + \ H_2\ddot{C}-\overset{O}{\underset{\underset{O}{\|}}{C}}-C-O^- \longrightarrow {}^-OCH_2-\overset{O}{\underset{\|}{C}}-\overset{O}{\underset{\|}{C}}-CO^-$$

Pyruvate   Oxaloacetate

17.31 Formation of phosphoenolpyruvate occurs by way of a reverse aldol reaction, followed by esterification of the enolate anion.

$$\overset{O}{\underset{\|}{{}^-O-C}}-CH_2-\overset{O}{\underset{\|}{C}}-CO^- \longrightarrow \left[H_2\ddot{C}-\overset{O}{\underset{\|}{C}}-CO^- \longleftrightarrow H_2C=\overset{{}^-O}{\underset{\|}{C}}-\overset{O}{\underset{\|}{C}}O^- \right]$$

Oxaloacetate

ATP ↓ ADP

$$H_2C=\overset{{}^{2-}O_3PO}{\underset{\|}{C}}-\overset{O}{\underset{\|}{C}}O^-$$

Phosphoenol pyruvate

17.32

$$2\ CH_3\overset{O}{\underset{\|}{C}}SCoA \xrightarrow{HSCoA} CH_3\overset{O}{\underset{\|}{C}}CH_2\overset{O}{\underset{\|}{C}}SCoA \xrightarrow{H_2O\ \ HSCoA} CH_3\overset{O}{\underset{\|}{C}}CH_2\overset{O}{\underset{\|}{C}}O^-$$

Acetyl CoA   Acetoacetyl CoA   Acetoacetate

From acetoacetate:
- CO₂ → $CH_3\overset{O}{\underset{\|}{C}}CH_3$ (Acetone)
- NADH/H⁺ → NAD⁺ → $CH_3\overset{OH}{\underset{|}{C}}HCH_2\overset{O}{\underset{\|}{C}}O^-$ (3-Hydroxybutyrate)

17.33

$$H_3C\overset{O}{\underset{\|}{C}}SCoA + H_2\overset{\cdot\cdot}{\underset{\|}{C}}\overset{O}{\underset{\|}{C}}SCoA \longrightarrow \left[H_3C\overset{:\ddot{O}:^-}{\underset{CoAS}{C}}CH_2\overset{O}{\underset{\|}{C}}SCoA \right]$$

$$\downarrow$$

$$CH_3\overset{O}{\underset{\|}{C}}CH_2\overset{O}{\underset{\|}{C}}SCoA + {}^-SCoA$$

The reaction is a Claisen condensation.

Chapter Outline

I. An overview of metabolism and biochemical energy (Section 17.1)
   A. Digestion
   B. Citric acid cycle and respiratory chain
   C. Phosphorylation
II. Fat catabolism (Section 17.2)
   A. Hydrolysis of triglycerides to glycerol and fatty acids
   B. Glycerol enters the carbohydrate catabolic pathway
   C. Fatty acids enter the fatty acid spiral, where they are cleaved to acetyl CoA molecules
III. Carbohydrate Catabolism. Glycolysis (Section 17.3)
   Glucose is converted to two acetyl CoA molecules
IV. The citric acid cycle (Section 17.4)
   Acetyl CoA from fat and carbohydrate catabolism is converted to CO_2
V. Catabolism of proteins. Transamination (Section 17.5)
   The amino group of an amino acid is transferred to α-ketoglutarate

Study Skills for Chapter 17

After studying this chapter, you should be able to:
1. Predict the products of the following metabolic pathways:
    a. Fatty acid spiral (Problems 17.3, 17.4, 17.23, 17.24, 17.27, 17.32)
    b. Glycolysis (Problems 17.5, 17.18, 17.19, 17.22)
    c. Citric acid cycle (Problem 17.17)
    d. Transamination (Problems 17.19, 17.25)
2. Understand the terms and concepts of this chapter (Problems 17.7, 17.10, 17.11, 17.12, 17.13, 17.14, 17.15, 17.16, 17.26)
3. Formulate mechanisms of biochemical reactions (Problems 17.1, 17.2, 17.6, 17.8, 17.28, 17.29, 17.30, 17.31, 17.33)

Summary of Reaction Mechanisms

The following table summarizes the most important types of polar reactions in organic chemistry. Much of organic chemistry is accounted for by just these eleven mechanisms.

1. **Electrophilic Addition to Alkenes** (Sections 3.7 - 3.11)
   Alkenes reacts with electrophiles such as HBr to yield saturated addition products. The reaction occurs in two steps: The electrophile first reacts with the alkene double bond to yield a carbocation intermediate that reacts further to yield the addition product.

   Alkene → Carbocation → Addition product

2. **Electrophilic Aromatic Substitution** (Sections 5.5 - 5.10)
   Aromatic compounds react with electrophiles such as Br_2 to yield substitution products rather than alkene-addition-type products. The reaction occurs in two steps: The electrophile first reacts with the aromatic ring to yield a carbocation intermediate that then loses H^+ to form a substituted aromatic-ring product.

   Benzene → Carbocation → Bromobenzene + HBr

3. **Nucleophilic Substitution of Alkyl Halides**

   A. S_N2 Reaction (Section 7.7)
   Alkyl halides undergo substitution when treated with nucleophiles. Primary and secondary alkyl halides react by the S_N2 mechanism, which occurs in a single step involving attack of the incoming nucleophile from a direction 180° away from the leaving group. This results in an umbrella-like inversion of stereochemistry (Walden inversion).

   Alkyl halide → Substitution product

B. S_N1 Reaction (Section 7.8)

Tertiary alkyl halides undergo nucleophilic substitution by the S$_N$1 mechanism, which occurs in two step. Spontaneous dissociation of the alkyl halide into an anion and a carbocation intermediate takes place, followed by reaction of the carbocation with a nucleophile. The dissociation step is the slower of the two.

$$\underset{\text{Alkyl halide}}{H_3C-\underset{\underset{CH_3}{|}}{\overset{\overset{CH_3}{|}}{C}}-Br} \longrightarrow \left[\underset{\text{Carbocation}}{H_3C-\underset{CH_3}{\overset{CH_3}{C^+}} + Br^-}\right] \xrightarrow{H_2O} \underset{\text{Substitution product}}{H_3C-\underset{\underset{CH_3}{|}}{\overset{\overset{CH_3}{|}}{C}}-OH} + HBr$$

4. Elimination Reaction of Alkyl Halides

A. E2 Reaction (Section 7.9)

Alkyl halides undergo elimination of HX to yield alkenes on treatment with base, in addition to nucleophilic substitution (Reaction 3 above). When a strong base such as hydroxide ion (HO⁻) or alkoxide ion (RO⁻) is used, alkyl halides react by the E2 mechanism. The E2 reaction occurs in a single step in which base removes a neighboring hydrogen at the same time that the halide ion leaves.

$$\underset{\text{Alkyl halide}}{\overset{H}{\underset{Br}{C-C}}} \xrightarrow{Na^+ \; OH^-} \underset{\text{Alkene}}{C=C} + H_2O + NaBr$$

B. E1 Reaction (Section 7.10)

Tertiary alkyl halides undergo elimination by the E1 mechanism in competition with S$_N$1 substitution when a nonbasic nucleophile is used. The reaction takes place in two steps: Spontaneous dissociation of the alkyl halide leads to a carbocation intermediate that then loses H⁺.

$$\underset{\text{Alkyl halide}}{H_3C-\underset{\underset{CH_3}{|}}{\overset{\overset{CH_3}{|}}{C}}-Br} \longrightarrow \left[\underset{\text{Carbocation}}{H_3C-\underset{CH_3}{\overset{CH_3}{C^+}} + Br^-}\right] \xrightarrow{H_2O} \underset{\text{Alkene}}{\underset{H_3C}{\overset{H_3C}{>}}C=C\underset{CH_3}{\overset{CH_3}{<}}} + HBr$$

294 Reaction Mechanisms

5. **Nucleophilic Addition to Ketones and Aldehydes** (Sections 9.7 -9.11)

   Ketones and aldehydes react with nucleophiles to yield addition products. The reaction occurs by addition of the nucleophile to the carbonyl group, producing a tetrahedrally hybridized intermediate anion that can be protonated to yield an alcohol.

   Ketone → (CH$_3^-$ MgBr$^+$) → Tetrahedral intermediate → (H$_2$O) → Alcohol

6. **Nucleophilic Acyl Substitution** (Sections 10.5 - 10.10)

   Carboxylic acid derivatives (acid chlorides, acid anhydrides, esters, amides) undergo a substitution reaction on treatment with nucleophiles. A nucleophilic acyl substitution reaction takes place in two steps: Addition of the nucleophile to the carbonyl group produces a tetrahedrally hybridized intermediate that expels a leaving group to generate a new carbonyl compound.

   Acid chloride → (Na$^+$ CH$_3$O$^-$) → Tetrahedral intermediate → Ester + NaCl

7. **Carbonyl alpha-Substitution Reaction** (Sections 11.2 - 11.6)

   Carbonyl compounds having alpha hydrogens (hydrogens on carbon next to the carbonyl group) are in equilibrium with their enol tautomers. An α-substitution reaction takes place when an enol or enolate ion reacts with an electrophile to produce a substitution product.

   Ketone → (H$^+$) → Enol intermediate → (Br$_2$) → α–substitution product + HBr

8. Carbonyl condensation Reaction

A. Aldol condensation of ketones and aldehydes (Sections 11.7 - 11.9)

Ketones and aldehydes with alpha hydrogens undergo a base-catalyzed dimerization reaction leading to formation of β-hydroxy ketone/aldehyde products. The reaction occurs in two step when one molecule of ketone or aldehyde is converted into its enolate ion, which then does a nucleophilic addition reaction to the carbonyl group of a second molecule.

$$H_3C-\overset{O}{\underset{}{\overset{\|}{C}}}-H \quad \xrightarrow{:\bar{C}H_2-\overset{O}{\underset{}{\overset{\|}{C}}}-H} \quad \left[H_3C\cdots\underset{H}{\overset{O^-}{\overset{|}{C}}}\underset{H}{\overset{}{\underset{H}{C}}}\overset{O}{\underset{}{\overset{\|}{C}}}-H \right] \quad \xrightarrow{H_2O} \quad CH_3-\underset{H}{\overset{OH}{\overset{|}{C}}}-CH_2-\overset{O}{\underset{}{\overset{\|}{C}}}-H$$

Aldehyde   Tetrahedral intermediate   β-Hydroxy aldehyde

B. Claisen condensation of esters (Section 11.10)

Esters with alpha hydrogens undergo a base-catalyzed dimerization reaction leading to formation of a β-keto ester product. The reaction occurs in two steps when one molecule of ester is converted into its enolate ion that does a nucleophilic acyl substitution reaction on a second ester molecule.

$$H_3C-\overset{O}{\underset{}{\overset{\|}{C}}}-OCH_3 \quad \xrightarrow{:\bar{C}H_2-\overset{O}{\underset{}{\overset{\|}{C}}}-OCH_3} \quad \left[\underset{H_3CO}{\overset{H_3C}{}}\overset{O^-}{\underset{}{\overset{|}{C}}}\underset{H\ \ H}{\overset{}{C}}\overset{O}{\underset{}{\overset{\|}{C}}}-OCH_3 \right]$$

Ester   Tetrahedral intermediate

$$\xrightarrow{H_2O} \quad CH_3-\overset{O}{\underset{}{\overset{\|}{C}}}-CH_2-\overset{O}{\underset{}{\overset{\|}{C}}}-OCH_3$$

β-Keto ester

Summary of Functional Group Preparations

The following table summarizes the synthetic methods used to prepare important functional groups. The functional groups are listed alphabetically, followed by reference to the appropriate text section and a brief description of each synthetic method.

Acetals
(Sec. 9.9) from ketones and aldehydes by acid-catalyzed reaction with alcohols

Acid anhydrides
(Sec. 10.8) from acid chlorides by reaction with carboxylate salts

Acid chlorides
(Secs. 10.7) from carboxylic acids by reaction with either $SOCl_2$ or PCl_3

Alcohols
(Sec. 4.7) from alkenes by hydroxylation with $KMnO_4$
(Sec. 4.4) from alkenes by hydration with aqueous acid
(Sec. 7.7) from alkyl halides by S_N2 reaction with hydroxide ion
(Sec. 8.9) from ethers by acid-induced cleavage
(Sec. 8.11) from epoxides by acid-catalyzed ring opening with H_2O
(Sec. 8.5) from ketones and aldehydes by reduction with $NaBH_4$ or $LiAlH_4$
(Sec. 9.11) from ketones and aldehydes by addition of Grignard reagents
(Sec. 8.5) from carboxylic acids by reduction with $LiAlH_4$
(Sec. 10.7) from acid chlorides by reaction with Grignard reagents
(Secs. 8.5, 10.9) from esters by reduction with $LiAlH_4$
(Sec. 10.9) from esters by reaction with Grignard reagents

Aldehydes
(Sec. 4.7) from disubstituted alkenes by ozonolysis
(Secs. 8.7, 9.4) from primary alcohols by oxidation

Alkanes
(Sec. 4.6) from alkenes by catalytic hydrogenation
(Sec. 7.4) from alkyl halides protonolysis of Grignard reagents
(Sec. 9.10) from ketones and aldehydes by Wolff-Kishner reaction

Alkenes
   (Secs. 4.9, 7.9)   from alkyl halides by treatment with strong base (E2 reaction)
   (Secs. 4.9, 8.7)   from alcohols by dehydration
   (Sec. 4.14)   from alkynes by catalytic hydrogenation using the Lindlar catalyst
   (Sec. 11.3)   from α-bromo ketones by heating with pyridine

Amides
   (Sec. 10.6)   from carboxylic acids by heating with ammonia
   (Sec. 10.7)   from acid chlorides by treatment with an amine or ammonia
   (Sec. 10.8)   from acid anhydrides by treatment with an amine or ammonia
   (Sec. 10.9)   from esters by treatment with an amine or ammonia

Amines
   (Secs. 7.7, 12.5)   from primary alkyl halides by treatment with ammonia
   (Sec. 10.10, 12.5) from amides by reduction with LiAlH$_4$
   (Sec. 10.11, 12.5) from nitriles by reduction with LiAlH$_4$

Arenes
   (Sec. 5.8)   from arenes by Friedel-Crafts alkylation with a primary alkyl halide
   (Sec. 12.6)   from arenediazonium salts by treatment with hypophosphorous acid

Arylamines
   (Sec. 12.5)   from nitroarenes by reduction with either Fe, Sn, or H$_2$/Pd

Arenediazonium salts
   (Sec. 12.6)   from arylamines by reaction with nitrous acid

Arenesulfonic acids
   (Sec. 5.7)   from arenes by electrophilic aromatic substitution with SO$_3$/H$_2$SO$_4$

Carboxylic acids
   (Sec. 4.7)   from mono- and 1,2-disubstituted alkenes by ozonolysis
   (Sec. 5.11)   from arenes by side-chain oxidation with Na$_2$Cr$_2$O$_7$ or KMnO$_4$
   (Sec. 9.6)   from aldehydes by oxidation
   (Sec. 9.6)   from alcohols by oxidation
   (Sec. 10.4)   from alkyl halides by conversion into Grignard reagents followed by reaction with CO$_2$
   (Sec. 10.4, 10.11) from nitriles by vigorous acid or base hydrolysis
   (Sec. 10.7)   from acid chlorides by reaction with aqueous base

(Sec. 10.8)	from acid anhydrides by reaction with aqueous base
(Sec. 10.9)	from esters by hydrolysis with aqueous base
(Sec. 10.10)	from amides by hydrolysis with aqueous base

Cycloalkanes

(Sec. 5.11)	from arenes by hydrogenation over a PtO_2 catalyst

Epoxides

(Sec. 8.10)	from alkenes by treatment with a peroxyacid

Esters

(Secs. 7.7, 10.6)	from carboxylic acid salts by S_N2 reaction with primary alkyl halides
(Sec. 10.6)	from carboxylic acids by acid-catalyzed reaction with an alcohol (Fischer esterification)
(Sec. 10.7)	from acid chlorides by base-induced reaction with an alcohol
(Sec. 10.8)	from acid anhydrides by base-induced reaction with an alcohol
(Sec. 11.6)	from alkyl halides by alkylation with diethyl malonate

Ethers

(Sec. 8.6)	from primary alkyl halides by S_N2 reaction with alkoxide ions (Williamson ether synthesis)
(Sec. 8.10)	from alkenes by epoxidation with peroxyacids
(Sec. 8.8)	from phenols by reaction of phenoxide ions with primary alkyl halides

Halides, alkyl

(Secs. 4.1-4.2)	from alkenes by electrophilic addition of HX
(Sec. 4.5)	from alkenes by addition of halogen
(Sec. 4.14)	from alkynes by addition of halogen or HX
(Secs. 7.3, 8.7)	from alcohols by reaction with HX
(Secs. 7.3, 8.7)	from alcohols by reaction with $SOCl_2$
(Secs. 7.4, 8.7)	from alcohols by reaction with PBr_3
(Sec. 8.9)	from ethers by cleavage with HX
(Sec. 11.3)	from ketones by alpha-halogenation with bromine

Halides, aryl

(Secs. 5.6, 5.7)	from arenes by electrophilic aromatic substitution with halogen
(Sec. 12.6)	from arenediazonium salts by reaction with cuprous halides

Imines

(Sec. 9.10)	from ketones or aldehydes by reaction with primary amines

Ketones

(Sec. 4.7)	from alkenes by ozonolysis
(Sec. 4.15)	from alkynes by acid-catalyzed hydration
(Sec. 5.8)	from arenes by Lewis-acid-catalyzed reaction with an acid chloride (Friedel-Crafts acylation)
(Secs. 8.7, 9.5)	from secondary alcohols by oxidation
(Sec. 10.11)	from nitriles by reaction with Grignard reagents

Nitriles

(Secs. 7.7, 10.13)	from primary alkyl halides by S_N2 reaction with cyanide ion
(Sec. 12.6)	from arenediazonium ions by treatment with CuCN

Nitroarenes

(Sec. 5.7)	from arenes by electrophilic aromatic substitution with nitric/sulfuric acids

Organometallics

(Sec. 7.4)	formation of Grignard reagents from organohalides by treatment with magnesium

Phenols

(Secs. 5.7, 8.4)	from arenesulfonic acids by fusion with KOH
(Sec. 12.6)	from arenediazonium salts by reaction with aqueous acid

Quinones

(Sec. 8.8)	from phenols by oxidation with $Na_2Cr_2O_7$

Sulfides

(Sec. 8.12)	from thiols by S_N2 reaction of thiolate ions with primary alkyl halides

Thiols

(Sec. 8.12)	from primary alkyl halides by S_N2 reaction with hydrosulfide anion

Reagents Used in Organic Synthesis

The following table summarizes the uses of some important reagents in organic chemistry. The reagents are listed alphabetically, followed by a brief description of the uses of each and references to the appropriate text sections.

Acetic anhydride: Reacts with alcohols to yield acetate esters (Section 10.8).

Aluminum chloride: Acts as a Lewis acid catalyst in Friedel-Crafts alkylation and acylation reactions of aromatic compounds (Section 5.8).

Bromine: Adds to alkenes yielding 1,2-dibromides (Section 4.5).
- Adds to alkynes yielding either 1,2-dibromoalkenes or 1,1,2,2-tetrabromoalkanes (Section 4.14).
- Reacts with arenes in the presence of ferric bromide catalyst to yield bromoarenes (Section 5.6).
- Reacts with ketones in acetic acid solvent to yield α-bromo ketones (Section 11.3).

Di-*tert*-butoxy dicarbonate: Reacts with amino acids to give BOC protected amino acids suitable for use in peptide synthesis (Section 15.8).

Carbon dioxide: Reacts with Grignard reagents to yield carboxylic acids (Section 10.4).

Chlorine: Adds to alkenes to yield 1,2-dichlorides (Section 4.4).
- Reacts with alkanes in the presence of light to yield chloroalkanes by a radical chain reaction pathway (Section 7.2).
- Reacts with arenes in the presence of ferric chloride catalyst to yield chloroarenes (Section 5.7).

Chromium trioxide: Oxidizes alcohols in aqueous sulfuric acid to yield carbonyl-containing products. Primary alcohols yield carboxylic acids and secondary alcohols yield ketones (Sections 8.7, 9.4, 9.5).

Cuprous bromide: Reacts with arenediazonium salts to yield bromoarenes (Sandmeyer reaction; Section 12.6).

Cuprous chloride: Reacts with arenediazonium salts to yield chloroarenes (Sandmeyer reaction; Section 12.6).

Cuprous cyanide: Reacts with arenediazonium salts to yield substituted benzonitriles (Sandmeyer reaction; Section 12.6).

Dicyclohexylcarbodiimide (DCC): Couples an amine with a carboxylic acid to yield an amide. DCC is often used in peptide synthesis (Section 15.8).

2,4-Dinitrophenylhydrazine: Reacts with ketones and aldehydes to yield 2,4-DNPs that serve as useful crystalline derivatives (Section 9.10).

Ethylene glycol: Reacts with ketones or aldehydes in the presence of a acid catalyst to yield acetals that serve as useful carbonyl-protecting groups (Section 9.9).

Ferric bromide: Acts as a catalyst for the reaction of arenes with bromine to yield bromoarenes (Section 5.6).

Ferric chloride: Acts as a catalyst for the reaction of arenes with chlorine to yield chloroarenes (Section 5.7).

Grignard reagent: Adds to carbonyl-containing compounds (ketones, aldehydes, esters) to yield alcohols (Sections 9.11, 10.9).

Hydrazine: Reacts with ketones or aldehydes in the presence of potassium hydroxide to yield the corresponding alkanes (Wolff-Kishner reaction; Section 9.10).

Hydrogen bromide: Adds to alkenes to yield alkyl bromides. Markovnikov regiochemistry is observed (Sections 4.1, 4.2).
- Adds to alkynes to yield either bromoalkenes or 1,1-dibromoalkanes (Section 4.14).
- Reacts with alcohols to yield alkyl bromides (Section 7.3).

Hydrogen chloride: Adds to alkenes to yield alkyl chlorides. Markovnikov regiochemistry is observed (Section 4.2).
- Adds to alkynes to yield either chloroalkenes or 1,1-dichloroalkanes (Section 4.14).
- Reacts with alcohols to yield alkyl chlorides (Section 7.3).

Hydroxylamine: Reacts with ketones and aldehydes to yield oximes (Section 9.10).

Hypophosphorous acid: Reacts with arenediazonium salts to yield arenes (Section 12.6).

Iodomethane: Reacts with alkoxide anions to yield methyl ethers (Section 8.6).
- Reacts with carboxylate anions to yield methyl esters (Section 10.6).
- Reacts with amines to yield methylated amines (Section 12.5).

Iron: Reacts with nitroarenes in the presence of mineral acid to yield anilines (Section 12.5).

Lindlar catalyst: Acts as a catalyst for the hydrogenation of alkynes to yield cis alkenes (Section 4.14).

Lithium aluminum hydride: Reduces ketones, aldehydes, esters, and carboxylic acids to yield alcohols (Section 8.5, 10.9).
  - Reduces amides to yield amines (Sections 10.10, 12.5).
  - Reduces nitriles to yield amines (Sections 10.11, 12.5).

Magnesium: Reacts with organohalides to yield Grignard reagents (Section 7.4).

Mercuric sulfate: Acts as a catalyst for the addition of water to alkynes in the presence of aqueous sulfuric acid, yielding ketones (Section 4.15).

Methyl sulfate: A reagent used to methylate heterocyclic amine bases during Maxam-Gilbert DNA sequencing (Section 16.13).

Nitric acid: Reacts with arenes in the presence of sulfuric acid to yield nitroarenes (Section 5.7).
  - Oxidizes aldoses to yield aldaric acids (Section 14.8).

Nitrous acid: Reacts with amines to yield diazonium salts (Section 12.6).

Ozone: Adds to alkenes to cleave the carbon-carbon double bond and give ozonides. The ozonides can then be reduced with zinc in acetic acid to yield carbonyl compounds (Section 4.7).

Palladium on carbon: Acts as a hydrogenation catalyst in the reduction of carbon-carbon multiple bonds. Alkenes and alkynes are reduced to yield alkanes (Section 4.6).
  - Acts as a hydrogenation catalyst in the reduction of nitroarenes to yield anilines (Section 12.5).

Phenyl isothiocyanate: A reagent used in the Edman degradation of peptides to identify N-terminal amino acids (Section 15.7).

Phosphorus tribromide: Reacts with alcohols to yield alkyl bromides (Sections 7.3, 8.7).

Phosphorus trichloride: Reacts with carboxylic acids to yield acid chlorides (Sections 10.6).

Platinum oxide: Acts as a hydrogenation catalyst in the reduction of alkenes and alkynes to yield alkanes (Section 4.6).

Potassium hydroxide: Reacts with alkyl halides to yield alkenes by an elimination reaction (Sections 4.9, 7.9).

Potassium permanganate: oxidizes alkenes under alkaline conditions to yield 1,2-diols (Section 4.7).
- Oxidizes alkenes under neutral or acidic conditions to give carboxylic acid double-bond cleavage products (Sections 4.7).
- Oxidizes arenes to yield benzoic acids (Section 5.11).

Pyridine: Reacts with α-bromo ketones to yield α,β-unsaturated ketones (Section 11.3).
- Acts as a catalyst for the reaction of alcohols with acid chlorides to yield esters (Section 10.7).
- Acts as a catalyst for the reaction of alcohols with acetic anhydride to yield acetate esters (Section 10.8).

Pyridinium chlorochromate: Oxidizes primary alcohols to yield aldehydes and secondary alcohols to yield ketones (Sections 9.4, 9.5).

Silver oxide: Oxidizes primary alcohols in aqueous ammonia solution to yield aldehydes (Tollens oxidation; Sections 14.8).

Sodium borohydride: Reduces ketones and aldehydes to yield alcohols (Section 8.5).
- Reduces quinones to yield hydroquinones (Section 8.8).

Sodium cyanide: Reacts with alkyl halides to yield alkanenitriles (Section 10.11).

Sodium dichromate: Oxidizes primary alcohols to yield carboxylic acids and secondary alcohols to yield ketones (Sections 8.7, 9.5).
- Oxidizes alkylbenzenes to yield benzoic acids (Section 5.11).

Sodium hydroxide: Reacts with arenesulfonic acids at high temperature to yield phenols (Sections 5.7, 8.4).

Sodium iodide: Reacts with arenediazonium salts to yield aryl iodides (Section 12.6).

Stannous chloride: Reduces nitroarenes to yield arylamines (Section 12.5).
- Reduces quinones to yield hydroquinones (Section 8.8).

Sulfur trioxide: Reacts with arenes in sulfuric acid solution to yield arenesulfonic acids (Section 5.7).

Sulfuric acid: Reacts with alcohols to yield alkenes (Sections 4.9, 8.7).
- Reacts with alkynes in the presence of water and mercuric acetate to yield ketones (Section 4.15).

Thionyl chloride: Reacts with primary and secondary alcohols to yield alkyl chlorides (Sections 7.3, 8.7).
- Reacts with carboxylic acids to yield acid chlorides (Section 10.6).

Trifluoroacetic acid: Acts as a catalyst for cleaving the BOC protecting group from amino acids in peptide synthesis (Section 15.8).

Zinc: Reduces ozonides, produced by addition of ozone to alkenes, to yield ketones and aldehydes (Section 4.7).

Organic Name Reactions

Aldol condensation reaction (Section 11.8): the nucleophilic addition of an enol or enolate ion to a ketone or aldehyde, yielding a β-hydroxy ketone.

$$2 \ R\text{-}\underset{\underset{\,}{|}}{C}\text{-}\overset{\overset{O}{\|}}{C}\text{-}H \quad \xrightarrow{\text{NaOH}} \quad R\text{-}\overset{\overset{O}{\|}}{C}\text{-}\underset{\underset{R}{|}}{C}\text{-}\overset{\overset{OH}{|}}{C}\text{-}\underset{\underset{\,}{|}}{C}\text{-}H$$

Claisen condensation reaction (Section 11.10): a nucleophilic acyl substitution reaction that occurs when an ester enolate ion attacks the carbonyl group of a second ester molecule. The product is a β-keto ester.

$$2 \ R\text{-}CH_2\text{-}\overset{\overset{O}{\|}}{C}\text{-}OCH_3 \quad \xrightarrow[\text{2. } H_3O^+]{\text{1. Na}^+ \ ^-OCH_3} \quad R\text{-}CH_2\text{-}\overset{\overset{O}{\|}}{C}\text{-}\underset{\underset{R}{|}}{CH}\text{-}\overset{\overset{O}{\|}}{C}\text{-}OCH_3 \ + \ CH_3OH$$

Edman degradation (Section 15.7): a method for cleaving the N-terminal amino acid from a peptide by treatment of the peptide with N-phenylisothiocyanate.

$$Ph\text{-}N{=}C{=}S \ + \ H_2N\text{-}\underset{\underset{R}{|}}{CH}\text{-}\overset{\overset{O}{\|}}{C}\text{-}NH\text{-}\xi \quad \longrightarrow \quad \text{(hydantoin)} \ + \ H_2N\text{-}\xi$$

Fehling's test (Section 14.8): a chemical test for aldehydes, involving treatment with cupric ion.

Fischer esterification reaction (Section 10.6): the acid-catalyzed reaction between a carboxylic acid and an alcohol to yield an ester.

$$R\text{-}\overset{\overset{O}{\|}}{C}\text{-}OH \ + \ R'\text{-}OH \quad \xrightarrow{H^+, \text{ heat}} \quad R\text{-}\overset{\overset{O}{\|}}{C}\text{-}OR' \ + \ H_2O$$

Friedel-Crafts reaction (Section 5.8): the alkylation or acylation of an aromatic ring by treatment with an alkyl- or acyl chloride in the presence of a Lewis-acid catalyst.

$$\text{benzene} \xrightarrow{\text{R-Cl, AlCl}_3} \text{Ar-R} \quad \text{Alkylation}$$

$$\text{benzene} \xrightarrow{\text{R-C(=O)-Cl, AlCl}_3} \text{Ar-C(=O)-R} \quad \text{Acylation}$$

Grignard reaction (Section 9.11): the nucleophilic addition reaction of an alkylmagnesium halide to a ketone, aldehyde, or ester carbonyl group.

$$\text{R-Mg-X} + \text{R'}_2\text{C=O} \xrightarrow[\text{2. H}_3\text{O}^+]{\text{1. mix}} \text{R'}_2\text{C(OH)-R}$$

Grignard reagent (Section 7.4): an organomagnesium halide, RMgX, prepared by reaction between an organohalide and magnesium metal.

Malonic ester synthesis (Section 11.6): a multi-step sequence for converting an alkyl halide into a carboxylic acid with the addition of two carbon atoms to the chain.

$$\text{R-CH}_2\text{-X} + \text{Na}^+ \text{ }^-\text{:CH(COOCH}_3)\text{-C(=O)-OCH}_3 \xrightarrow[\text{2. H}_3\text{O}^+, \text{ heat}]{\text{1. heat}} \text{RCH}_2\text{-CH}_2\text{COCH}_3 + \text{CO}_2 + \text{CH}_3\text{OH}$$

Maxam-Gilbert DNA sequencing (Section 16.13): a rapid and efficient method for sequencing long chains of DNA by employing selective cleavage reactions.

Sandmeyer reaction (Section 12.6): a method for converting aryldiazonium salts into aryl halides by treatment with cuprous halide.

$$\text{Ar-N}\equiv\text{N}^+ + \text{CuBr} \longrightarrow \text{Ar-Br} \quad (\text{or Cl, I})$$

Tollen's test (Section 14.8): a chemical test for detecting aldehydes by treatment with ammoniacal silver nitrate. A positive test is signaled by formation of a silver mirror on the walls of the reaction vessel.

Walden inversion (Sections 7.5, 7.7): the inversion of stereochemistry at a chiral center that occurs during S_N2 reactions.

$$Nu:^- + \;\;C-X \longrightarrow Nu-C \;\; + :X^-$$

Williamson ether synthesis (Section 8.6): a method for preparing ethers by treatment of a primary alkyl halide with an alkoxide ion.

$$R-O^- \; Na^+ + R'CH_2Br \longrightarrow R-O-CH_2R' + NaBr$$

Wolff-Kishner reaction (Section 9.10): a method for converting a ketone or aldehyde into the corresponding hydrocarbon by treatment with hydrazine and strong base.

$$R-\underset{\underset{R'}{}}{\overset{\overset{O}{\|}}{C}}-R' \xrightarrow{N_2H_4, \; KOH} R-CH_2-R'$$

Glossary

Absorption spectrum (Section 13.1): a plot of wavelength of incident light versus amount of light absorbed. Organic molecules show absorption spectra in both the infrared and ultraviolet regions of the electromagnetic spectrum. By interpreting these spectra, useful structural information about the sample can be obtained. (See: infrared spectrum, ultraviolet spectrum)

Acetal (Section 9.9): a functional group consisting of two ether-type oxygen atoms bound to the same carbon, $R_2C(OR')_2$. Acetals are often used as protecting groups for ketones and aldehydes since they are stable to basic and nucleophilic reagents but can be easily removed by acidic hydrolysis.

Achiral (Section 6.2): lacking handedness. A molecule is achiral if it has a plane of symmetry and is thus superimposable on its mirror image. (See chiral)

Acidity constant, K_a (Section 1.13): a value that expresses the strength of an acid in water solution. The larger the K_a, the stronger the acid.

Activating group (Section 5.9): an electron-donating group such as hydroxyl (–OH) or amino (–NH$_2$) that increases the reactivity of an aromatic ring toward electrophilic aromatic substitution. All activating groups are ortho/para directing.

Activation energy (Section 3.10): the difference in energy levels between ground state and transition state. The amount of activation energy required by a reaction determines the rate at which the reaction proceeds. The majority of organic reactions have activation energies of 10 - 25 kcal/mol.

Acyl group (Section 5.8): A name for the $-\overset{\overset{\displaystyle O}{\|}}{C}-R$ group.

Acylation (Section 5.8): the introduction of an acyl group, –COR, onto a molecule. For example, acylation of an alcohol yields an ester (R'OH → R'OCOR), acylation of an amine yields an amide (R'NH$_2$ → R'NHCOR), and acylation of an aromatic ring yields an alkyl aryl ketone (ArH → ArCOR).

Aldaric acid (Section 14.8): the dicarboxylic acid resulting from oxidation of an aldose.

Alditol (Section 14.8): the polyalcohol resulting from reduction of the carbonyl group of a sugar.

Aldol reaction (Section 11.8): A carbonyl-condensation reaction between two ketones or aldehydes leading to a β-hydroxy ketone or aldehyde product.

Aldonic acid (Section 14.8): the monocarboxylic acid resulting from mild oxidation of an aldose.

Aldose (Section 14.1): a simple sugar with an aldehyde carbonyl group.

Alicyclic (Section 2.7): referring to an aliphatic cyclic hydrocarbon such as a cycloalkane or cycloalkene.

Aliphatic (Section 2.2): referring to a nonaromatic hydrocarbon such as a simple alkane, alkene, or alkyne.

Alkaloid (Section 12.8): a naturally occurring compound that contains a basic amine functional group. Morphine is an example of an alkaloid.

Alkane (Section 2.2): a compound that contains only carbon and hydrogen and has only single bonds.

Alkyl group (Section 2.2): a part structure, formed by removing a hydrogen from an alkane.

Alkylation (Sections 5.8, 8.6, and 11.6): introduction of an alkyl group onto a molecule. For example, aromatic rings can be alkylated to yield arenes (ArH → ArR), alkoxide anions can be alkylated to yield ethers (R'O$^-$ → R'OR), and enolate anions can be alkylated to yield α-substituted carbonyl compounds.

Allylic (Section 4.11): referring to the position next to a double bond. For example, CH_2=CHCH$_2$Br is an allylic bromide, and an allylic carbocation is a conjugated, resonance-stabilized species containing a vacant p orbital next to a double bond – (C=C-C$^+$ ↔ $^+$C-C=C).

Amine (Section 12.1): an organic derivative of ammonia, RNH$_2$, R$_2$NH, or R$_3$N.

α-Amino acid (Section 15.1): a compound with an amino group attached to the carbon atom next to the carboxyl group, RCH(NH$_2$)COOH.

Amplitude (Section 13.1): The height of a wave from midpoint to peak.

Anabolism (Section 17.1): Metabolic reactions that synthesize larger molecules from smaller precursors.

Androgen (Section 16.5): a steroidal male sex hormone such as testosterone.

Angle strain (Section 2.9): the strain introduced into a molecule when a bond angle is deformed from its ideal value. Angle strain is particularly important in small-ring cycloalkanes where it results from compression of bond angles to less than their ideal tetrahedral values. For example, cyclopropane has approximately 22 kcal/mol of angle strain because of bond deformations from the 109° tetrahedral angle to 60°.

Anomers (Section 14.6): cyclic stereoisomers of sugars that differ only in their configurations at the hemiacetal (anomeric) carbon.

Anti periplanar geometry (Section 7.9): reaction geometry in which all reacting atoms line in a plane, with one group on top and another on the bottom of the molecule. The E2 reaction of alkyl halides, for example, has anti periplanar geometry.

Anti stereochemistry (Section 4.5): referring to opposite sides of a double bond or molecule. An anti addition reaction is one in which the two ends of the double bond are attacked from different sides. For example, addition of Br$_2$ to cyclohexene yields *trans*-1,2-dibromocyclohexane, the product of anti addition. An anti elimination reaction is one in which the two groups leave from opposite sides of the molecule. (See syn stereochemistry)

Anticodon (Section 16.11): a sequence of three bases on tRNA that read the codons on mRNA and bring the correct amino acids into position for protein synthesis.

Aromaticity (Chapter 5): the special characteristics of cyclic conjugated pi electron systems that result from their electronic structures. These characteristics include unusual stability and a tendency to undergo substitution reactions rather than addition reactions on treatment with electrophiles.

Axial bond (Section 2.10): a bond to chair cyclohexane that lies along the ring axis perpendicular to the rough plane of the ring. (See equatorial bond)

Axial bonds

Azo compound (Section 12.6): a compound containing the –N=N– functional group.

Basicity constant, K_b (Section 12.3): a value that expresses the strength of a base in water solution. The larger the K_b, the stronger the base.

Benzylic (Section 5.11): referring to the position next to an aromatic ring. For example, a benzylic cation is a resonance-stabilized, conjugated carbocation having its positive charge located on a carbon atom next to the benzene ring in a pi orbital that overlaps the aromatic pi system.

β-oxidation pathway (Section 17.2): A series of four enzyme-catalyzed reactions that cleave two carbon atoms at a time from the end of a fatty acid chain.

Bimolecular reaction (Section 7.7): a reaction that occurs between two reagents.

Bond angle (Section 1.8): the angle formed between two adjacent bonds.

Bond-dissociation energy (Section 1.6): see Bond strength

Bond length (Section 1.6): the equilibrium distance between the nuclei of two atoms that are bonded to each other.

Bond strength (Section 1.6): the amount of energy needed to break a bond to produce two radical fragments.

Brønsted-Lowry acid (Section 1.13): a substance that donates a hydrogen ion (proton, H^+) to a base.

Brønsted-Lowry base (Section 1.13): a substance that accepts a hydrogen ion, H^+ from an acid.

Bromonium ion (Section 4.5): a species with a positively charged, divalent bromine atom, R_2Br^+. Bromonium ions are intermediates in the addition reaction of bromine with alkenes.

Carbanion (Section 7.4): a carbon-anion, or substance that contains a trivalent, negatively charged carbon atom ($R_3C:^-$). Carbanions are sp^3 hybridized and have eight electrons in the outer shell of the negatively charged carbon.

Carbocation (Section 3.8): a carbon-cation, or substance that contains a trivalent, positively charged carbon atom having six electrons in its outer shell (R_3C^+). Carbocations are planar and sp^2 hybridized.

Carbocycle (Section 12.7): a cyclic molecule that has only carbon atoms in the ring. (See heterocycle)

Carbohydrate (Chapter 14): a polyhydroxy aldehyde or polyhydroxy ketone. The name derives from the fact that glucose, the most abundant carbohydrate, has the formula $C_6H_{12}O_6$ and was originally thought to be a "hydrate of carbon". Carbohydrates can be either simple sugars such as glucose or complex sugars such as cellulose.

Carbonyl condensation reaction (Section 11.7): a reaction between two carbonyl compounds in which the α carbon of one partner bonds to the carbonyl carbon of the other.

Carbonyl group (Section 2.1): the carbon-oxygen double bond functional group, $-\overset{\overset{\displaystyle O}{\|}}{C}-$

Catabolism (Section 17.1): Metabolic reactions that break down large molecules.

Chain reaction (Section 7.2): a reaction that, once initiated, sustains itself in an endlessly repeating cycle of propagation steps. The radical chlorination of alkanes in an example of a chain reaction that is initiated by irradiation with light and that then continues in a series of propagation steps.

Step 1:	Initiation:	$Cl_2 \rightarrow 2\ Cl\cdot$
Steps 2 and 3:	Propagation:	$Cl\cdot + CH_4 \rightarrow HCl + \cdot CH_3$
		$\cdot CH_3 + Cl_2 \rightarrow CH_3Cl + Cl\cdot$
Step 4:	Termination:	$R\cdot + R\cdot \rightarrow R\text{-}R$

Chain-growth polymer (Sections 4.8 and 10.12): a polymer produced by a chain reaction procedure in which an initiator adds to a carbon-carbon double bond to yield a reactive intermediate. The chain is then built as more monomers add successively to the reactive end of the growing chain.

Chair cyclohexane (Section 2.9): a three-dimensional conformation of cyclohexane that resembles the rough shape of a chair. The chair form of cyclohexane, which has neither angle strain nor eclipsing strain, represents the lowest energy conformation of the molecule.

Chair cyclohexane

Chemical shift (Section 13.7): the position on the NMR chart where a nucleus absorbs. By convention, the chemical shift of tetramethylsilane is set at zero and all other absorptions usually occur downfield (to the left on the chart). Chemical shifts are expressed in delta units, δ, where one delta equals one part per million of the spectrometer operating frequency. For example, one delta on a 60 megahertz instrument equals 60 Hertz. The chemical shift of a given nucleus is related to the chemical environment of that nucleus in the molecule, thus allowing one to obtain structural information by interpreting the NMR spectrum.

Chiral (Section 6.2): having handedness. Chiral molecules do not have a plane of symmetry and are therefore not superimposable on their mirror image. A chiral molecule thus exists in two forms, one right handed and one left handed. The most common (though not the only) cause of chirality in a molecule is the presence of a carbon atom that is bonded to four different substituents. (See achiral)

Cis-trans isomers (Section 2.8 and 3.3): stereoisomers that differ in their stereochemistry about a double bond or a ring. Cis-trans isomers are also called geometric isomers.

Citric acid cycle (Section 17.4): The third stage of catabolism, in which acetyl groups are degraded to CO_2.

Claisen condensation reaction (Section 11.10): A carbonyl-condensation reaction between two esters leading to formation of a β-keto ester product.

Codon (Section 16.11): a three-base sequence on the messenger RNA chain that encodes the genetic information necessary to cause specific amino acids to be incorporated into proteins. Codons on mRNA are read by complementary anticodons on tRNA.

Coenzyme (Section 15.12): a small organic molecule that acts as an enzyme cofactor.

Complex carbohydrate (Section 14.1): a carbohydrate composed of two or more simple sugars linked together by acetal bonds.

Condensed structure (Section 2.2): a shorthand way of drawing structures in which bonds are understood rather than shown.

Configuration (Section 6.6): the three-dimensional arrangement of atoms bonded to a chiral center relative to the stereochemistry of other chiral centers in the same molecule.

Conformation (Section 2.5): the exact three-dimensional shape of a molecule at any given instant, assuming that rotation around single bonds is frozen.

Conjugate acid (Section 1.13): The product that results when a bases accepts H^+.

Conjugate base (Section 1.13): the anion that results from dissociation of a Brønsted acid.

Conjugation (Section 4.10): a series of alternating single and multiple bonds with overlapping *p* orbitals. For example, 1,3-butadiene is a conjugated diene, benzene is a cyclic conjugated triene, and 3-buten-2-one is a conjugated enone.

Constitutional isomers (Section 2.2): isomers that have their atoms connected in a different order. For example, butane and 2-methylpropane are constitutional isomers.

Coupling constant (Section 13.10): the magnitude (expressed in Hertz) of the spin-spin splitting interaction between nuclei whose spins are coupled. Coupling constants are denoted *J*.

Covalent bond (Section 1.5): a bond formed by sharing electrons between two nuclei. (See ionic bond)

D-sugar (Section 14.3): A sugar whose hydroxyl group at the chiral carbon farthest from the carbonyl group points to the right when the molecule is drawn in Fischer projection.

Deactivating group (Section 5.9): an electron-withdrawing substituent that decreases the reactivity of an aromatic ring towards electrophilic aromatic substitution. Most deactivating groups, such as nitro, cyano, and carbonyl are meta-directors, but halogen substituents are ortho/para directors.

Decarboxylation (Section 11.6): a reaction that involves loss of carbon dioxide from the starting material. β-Keto acids decarboxylate particularly readily on heating.

Dehydration (Section 4.9): the loss of water. Most alcohols can be dehydrated to yield alkenes, but aldol condensation products (β-hydroxy ketones) dehydrate particularly readily.

Dehydrohalogenation (Sections 4.9 and 7.9): a reaction that involves loss of HX from the starting material. Alkyl halides undergo dehydrohalogenation to yield alkenes on treatment with strong base.

Delocalization (Section 4.11): a spreading out of electron density over a conjugated pi electron system. For example allylic cations and allylic anions are delocalized because their charges are spread out by resonance stabilization over the entire pi-electron system.

Delta (δ) scale (Section 13.7): The arbitrary scale used for defining the position of NMR absorptions. One delta unit is equal to one part-per-million of spectrometer frequency.

Deshielding (Section 13.7): an effect observed in NMR that causes a nucleus to absorb downfield (to the left) of tetramethylsilane standard. Deshielding is caused by a withdrawal of electron density from the nucleus and is responsible for the observed chemical shifts of vinylic and aromatic protons.

Dextrorotatory (Section 6.3): a word used to describe an optically active substance that rotates the plane of polarization of plane-polarized light in a right-handed (clockwise) direction. The direction of rotation is not related to the absolute configuration of the molecule. (See levorotatory)

Diastereomer (Section 6.7): a term that indicates the relationship between non-mirror-image stereoisomers. Diastereomers are stereoisomers that have the same configuration at one or more chiral centers, but differ at other chiral centers.

Diazotization (Section 12.6): the conversion of a primary amine, RNH_2 into a diazonium salt, RN_2^+ by treatment with nitrous acid. Aryl diazonium salts are stable, but alkyl diazonium salts are extremely reactive and are rarely isolable.

Digestion (Section 17.1): The first stage of catabolism, in which food molecules are hydrolyzed to yield fatty acids, amino acids, and monosaccharides.

Disulfide link (Section 15.5): A sulfur-sulfur link between two cysteine residues in a peptide.

DNA (Section 16.7): deoxyribonucleic acid, the biopolymer consisting of deoxyribonucleotide units linked together through phosphate-sugar bonds. DNA, which is found in the nucleus of cells, contains an organism's genetic information.

Doublet (Section 13.10): a two-line NMR absorption caused by spin-spin splitting when the spin of the nucleus under observation couples with the spin of a neighboring magnetic nucleus.

Downfield (Section 13.7): used to refer to the left hand portion of the NMR chart. (See deshielding)

Eclipsed conformation (Section 2.5): the geometric arrangement around a carbon-carbon single bond in which the bonds to substituents on one carbon are parallel to the bonds to substituents on the neighboring carbon as viewed in a Newman projection. For example, the eclipsed conformation of ethane has the C-H bonds on one carbon lined up with the C-H bonds on the neighboring carbon.

Eclipsed conformation

E2 reaction (Section 7.9): An elimination reaction that takes place in a single step through a bimolecular mechanism.

Edman degradation (Section 15.7): a method for selectively cleaving the *N*-terminal amino acid from a peptide.

Electromagnetic spectrum (Section 13.1): the range of electromagnetic energy, including infrared, ultraviolet and visible radiation.

Electronegativity (Section 1.12): the ability of an atom to attract electrons and thereby polarize a bond. As a general rule, electronegativity increases in going across the periodic table from right to left and in going from bottom to top.

Electrophile (Section 3.6): an "electron-lover", or substance that accepts an electron pair from a nucleophile in a polar bond-forming reaction.

Electrophilic aromatic substitution reaction (Section 5.5): the substitution of an electrophile for a hydrogen atom on an aromatic ring.

Electrophoresis (Section 15.3): a technique used for separating charged organic molecules, particularly proteins and amino acids. The mixture to be separated is placed on a buffered gel or paper and an electric potential is applied across the ends of the apparatus. Negatively charged molecules migrate towards the positive electrode and positively charged molecules migrate towards the negative electrode.

Glossary 317

Enantiomers (Section 6.5): stereoisomers of a chiral substance that have a mirror-image relationship. Enantiomers must have opposite configurations at all chiral centers.

Endothermic (Section 3.9): a term used to describe reactions that absorb energy and that therefore have positive ΔH changes. In reaction energy diagrams, the products of endothermic reactions have higher energy levels than the starting materials.

Enol (Sections 4.15 and 11.1): a vinylic alcohol, C=C–OH

Enolate ion (Section 11.4): the anion of an enol; a resonance-stabilized α-keto carbanion.

Entgegen (*E*) (Section 3.4): a term used to describe the stereochemistry of a carbon-carbon double bond. The two groups on each carbon are assigned priorities according to sequence rules, and the two carbons are then compared. If the high priority groups on each carbon are on opposite sides of the double bond, the bond has *E* geometry.

Enzyme (Section 15.11): a biological catalyst. Enzymes are large proteins that catalyze specific biochemical reactions.

Epoxide (Section 8.10): a three-membered ring ether functional group.

Equatorial bond (Section 2.10): a bond to cyclohexane that lies along the rough equator of the ring. (See axial bond)

Equatorial bonds

Essential amino acid (Section 15.1): an amino acid that must be obtained in the diet.

Ether (Section 8.1): a compound with two organic groups bonded to the same oxygen atom, R-O-R'.

Exothermic (Section 3.9): a term used to describe reactions that release energy and that therefore have negative enthalpy changes. On reaction energy diagrams, the products of exothermic reactions have energy levels lower than those of starting materials.

Fat (Section 16.2): a solid triacylglycerol derived from animal sources.

Fatty acid spiral (Section 17.2): A series of four enzyme-catalyzed reactions that cleave two carbon atoms at a time from the end of a fatty acid chain.

Fibrous protein (Section 15.9): proteins that consist of polypeptide chains arranged side by side in long threads. These proteins are tough, insoluble in water, and are used in nature for structural materials such as hair, hooves, and fingernails.

Fingerprint region (Section 13.2): the complex region of the infrared spectrum from 1500 cm^{-1} to 400 cm^{-1}. If two substances have identical absorption patterns in the fingerprint region of the IR, they are almost certainly identical.

Fischer esterification reaction (Section 10.6): the conversion of a carboxylic acid into an ester by acid-catalyzed reaction with an alcohol.

Fischer projection (Section 14.2): a means of depicting the configuration of chiral molecules on a flat page. A Fischer projection employs a cross to represent the chiral center; the horizontal arms of the cross represent bonds coming out of the plane of the page, and the vertical arms of the cross represent bonds going back into the plane of the page.

Frequency (Section 13.1): the number of electromagnetic wave cycles that travel past a fixed point in a given unit of time. Frequencies are usually expressed in units of reciprocal seconds, s^{-1}, or Hertz.

Friedel-Crafts reaction (Section 5.8): the introduction of an alkyl or acyl group onto an aromatic ring by an electrophilic substitution reaction.

Functional group (Section 2.1): an atom or group of atoms that is part of a larger molecule and that has a characteristic chemical reactivity. Functional groups display the same chemistry in all molecules where they occur.

Geometric isomers: an old term for cis-trans isomers.

Globular protein (Section 15.9): proteins that are coiled into compact, nearly spherical shapes. These proteins, which are generally water soluble and mobile within the cell, are the structural class to which enzymes belong.

Glycol (Section 8.11): a 1,2-diol such as ethylene glycol, HOCH₂CH₂OH.

Glycolysis (Section 17.3): A series of ten enzyme-catalyzed reactions that break down a glucose molecule into two pyruvates.

Glycoside (Section 14.8): a cyclic acetal formed by reaction of a sugar with another alcohol.

Grignard reagent (Section 7.4): An organomagnesium halide, RMgX.

Haworth projection (Section 14.5): a means of viewing stereochemistry in cyclic hemiacetal forms of sugars. Haworth projections are drawn so that the ring is flat and is viewed from an oblique angle with the hemiacetal oxygen at the upper right.

Haworth projection of glucose

Heat of reaction, ΔH (Section 3.9): the amount of heat released or absorbed in a reaction.

Hemiacetal (Section 9.9): a compound that has one –OR group and one –OH group bonded to the same carbon atom.

Heterocycle (Section 12.7): a cyclic molecule whose ring contains more than one kind of atom. For example, pyridine is a heterocycle that contains five carbon atoms and one nitrogen atom in its ring.

Heterogenic bond formation (Section 3.6): what occurs when one partner donates both electrons in forming a new bond. Polar reactions always involve heterogenic bond formation:

$$A^+ + B:^- \rightarrow A:B$$

Heterolytic bond breakage (Section 3.6): the kind of bond breaking that occurs in polar reactions when one fragment leaves with both of the bonding electrons:

$$A:B \rightarrow A^+ + B:^-$$

Holoenzyme (Section 15.12): the combination of enzyme and cofactor.

Homogenic bond formation (Section 3.6): what occurs in radical reactions when each partner donates one electron to the new bond:

A· + B· → A:B

Homolytic bond breakage (Section 3.6): the kind of bond breaking that occurs in radical reactions when each fragment leaves with one bonding electron:

A:B → A· + B·

Hybrid orbital (Section 1.7): an orbital that is derived from a combination of ground-state (s, p, d) atomic orbitals. Hybrid orbitals, such as the sp^3, sp^2, and sp hybrids of carbon, are strongly directed and form stronger bonds than ground-state atomic orbitals.

Hydration (Section 4.4): addition of water to a molecule, such as occurs when alkenes are treated with strong sulfuric acid.

Hydrogenation (Section 4.6): addition of hydrogen to a double or triple bond to yield a saturated product.

Hydrogen bond (Section 8.2): a weak attraction between a hydrogen atom bonded to an electronegative element and an electron lone pair on another atom. Hydrogen bonding plays an important role in determining the secondary structure of proteins and in stabilizing the DNA double helix.

Hydrophobic (Section 16.3): repelled by water (and attracted to hydrocarbons).

Inductive effect (Section 1.12): the electron-attracting or electron-withdrawing effect that is transmitted through sigma bonds as the result of a nearby dipole. Electronegative elements have an electron-withdrawing inductive effect, and electropositive elements have an electron-donating inductive effect.

Infrared spectroscopy (Section 13.1): a kind of optical spectroscopy that uses infrared energy. IR spectroscopy is particularly useful in organic chemistry for determining the kinds of functional groups present in molecules.

Initiator (Section 7.2): a substance with an easily broken bond that is used to initiate radical chain reactions. For example, radical chlorination of alkanes is initiated when light energy breaks the weak chlorine-chlorine bond to form chlorine radicals.

Intermediate (Section 3.11): a species that is formed during the course of a multistep reaction but is not the final product. Intermediates are more stable than transition states, but may or may not be stable enough to isolate.

Intramolecular, intermolecular (Section 8.11): reactions that occur within the same molecule are intramolecular, whereas reactions that occur between two molecules are intermolecular.

Ionic bond (Section 1.4): a bond between two ions due to the electrical attraction of unlike charges. Ionic bonds are formed between strongly electronegative elements (such as the halogens) and strongly electropositive elements (such as the alkali metals).

Isoelectric point (Section 15.3): the pH at which the number of positive charges and the number of negative charges on a protein or amino acid are exactly balanced.

Isomers (Section 2.2): compounds with the same molecular formula but different structures.

Kekulé structure (Section 1.5): a representation of a molecule in which a line between atoms represents a covalent bond.

L-sugar (Section 14.3): a sugar whose hydroxyl group at the chiral carbon farthest from the carbonyl group points to the left when the molecule is drawn in Fischer projection.

Leaving group (Section 7.6): the group that is replaced in a substitution reaction. The best leaving groups in nucleophilic substitution reactions are those that form the most stable, least basic, anions.

Levorotatory (Section 6.3): used to describe an optically active substance that rotates the plane of polarization of plane-polarized light in a left-handed (counterclockwise) direction. (See dextrorotatory)

Lewis acid (Section 1.13): a substance with a vacant low-energy orbital that can accept an electron pair from a base. All electrophiles are Lewis acids, but transition metal salts such as $AlCl_3$ and $ZnCl_2$ are particularly good ones. (See Lewis base)

Lewis base (Section 1.13): a substance that donates an electron lone pair to an acid. All nucleophiles are Lewis bases. (See Lewis acid)

Lewis structure (Section 1.5): a representation of a molecule showing covalent bonds as a pair of electron dots between atoms.

Line-bond structure (Section 1.5): a representation of a molecule showing covalent bonds as lines between atoms. (See Kekulé structure)

Lipid (Chapter 16.1): a naturally occurring substance isolated from cells and tissues by extraction with nonpolar solvents. Lipids belong to many different structural classes, including fats, prostaglandins, and steroids.

Lipophilic (Section 16.3): fat-loving. Long non-polar hydrocarbon chains tend to cluster together in polar solvents because of their lipophilic properties.

Lone-pair electrons (Section 1.5): nonbonding electron pairs that occupy valence orbitals. It is the lone-pair electrons that are used by nucleophiles in their reactions with electrophiles.

Major groove (Section 16.8): The large groove in double helical DNA.

Markovnikov's rule (Section 4.2): a guide for determining the regiochemistry (orientation) of electrophilic addition reactions. In the addition of HX to an alkene, the hydrogen atom becomes bonded to the alkene carbon that has fewer alkyl substituents. A modern statement of this same rule is that electrophilic addition reactions proceed via the most stable carbocation intermediate.

Mechanism (Section 3.6): a complete description of how a reaction occurs. A mechanism must account for all starting materials and all products, and must describe the details of each individual step in the overall reaction process.

Meso (Section 6.8): A meso compound contains chiral centers but is nevertheless achiral by virtue of a symmetry plane. For example, (2R,3S)-butanediol has two chiral carbon atoms, but is achiral because of a symmetry plane between carbons 2 and 3.

Metabolism (Section 17.1): The total of all reactions in living organisms.

Micelle (Section 16.3): a spherical cluster of soap-like molecules that aggregate in aqueous solution. The ionic heads of the molecules lie on the outside where they are solvated by water, and the organic tails bunch together on the inside of the micelle.

Minor groove (Section 16.8): The small groove in double helical DNA.

Monomer (Sections 4.8 and 10.12): the simple starting units from which polymers are made.

Multiplet (Section 13.10): a symmetrical pattern of peaks in an nmr spectrum that arises by spin-spin splitting of a single absorption because of coupling between neighboring magnetic nuclei.

Mutarotation (Section 14.6): the spontaneous change in optical rotation observed when a pure anomer of a sugar is dissolved in water. Mutarotation is caused by the reversible opening and closing of the acetal linkage, which yields an equilibrium mixture of anomers.

Newman projection (Section 2.5): a way indicating stereochemical relationships between substituent groups on neighboring carbons. The carbon-carbon bond is viewed end-on, and the carbons are indicated by a circle. Bonds radiating from the center of the circle are attached to the front carbon, and bonds radiating from the edge of the circle are attached to the rear carbon.

Newman projection

Nitrile (Section 10.11): a compound that contains the -C≡N functional group.

Nonbonding electron (Section 1.5): a valence electron not used for bonding.

Normal alkane (Section 2.2): a straight-chain alkane, as opposed to a branched alkane. Normal alkanes are denoted by the suffix *n*, as in *n*-C_4H_{10} (*n*-butane)..

Nuclear magnetic resonance, NMR (Section 13.5): a spectroscopic technique that provides information about the carbon-hydrogen framework of a molecule. NMR works by detecting the energy absorption accompanying the transition between nuclear spin states that occurs when a molecule is placed in a strong magnetic field and irradiated with radio-frequency waves. Different nuclei within a molecule are in slightly different magnetic environments and therefore show absorptions at slightly different frequencies.

Nucleophile (Section 3.5): a "nucleus-lover", or species that donates an electron pair to an electrophile in a polar bond-forming reaction. Nucleophiles are also Lewis bases. (See electrophile)

Nucleoside (Section 16.6): a nucleic acid constituent, consisting of a sugar residue bonded to a heterocyclic purine or pyrimidine base.

Nucleotide (Section 16.6): a nucleic acid constituent, consisting of a sugar residue bonded both to a heterocyclic purine or pyrimidine base and to a phosphoric acid. Nucleotides are the monomer units from which DNA and RNA are constructed.

Nylons (Section 10.12): polyamides prepared by reaction between a diacid and a diamine.

Olefin: an alternative name for an alkene.

Optical isomers (Section 6.5): enantiomers. Optical isomers are isomers that have a mirror-image relationship.

Optically active (Section 6.3): a substance that rotates the plane of polarization of plane-polarized light. Note that an optically active sample must contain chiral molecules, but that not all samples with chiral molecules are optically active. For instance, a racemic sample is optically inactive even though the individual molecules are chiral. (See chiral)

Orbital (Section 1.1) the volume of space in which an electron is most likely to be found. Orbitals are described mathematically by wavefunctions, which describe the energies of electrons around nuclei.

Oxidation (Section 4.7): The addition of oxygen to a molecule or removal of hydrogen from it.

Oxirane (Section 8.10): an alternative name for an epoxide.

Ozonide (Section 4.7): the product formed by addition of ozone to a carbon-carbon double bond. Ozonides are usually treated with a reducing agent such as zinc in acetic acid to produce carbonyl compounds.

Paraffins (Section 2.4): a common name for alkanes.

Peptides (Section 15.4): amino-acid polymers in which the individual amino acid residues are linked by amide bonds. (See proteins)

Periplanar (Section 7.9): a conformation in which bonds to neighboring atoms have a parallel arrangement. In an eclipsed conformation, the neighboring bonds are syn-periplanar; in a staggered conformation, the bonds are anti-periplanar.

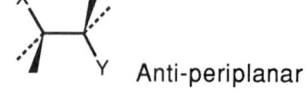

Phenyl (Section 5.4): the -C₆H₅ group.

Phospholipid (Section 16.4): lipids that contain a phosphate residue. For example, phosphoglycerides contain a glycerol backbone linked to two fatty acids and a phosphoric acid.

Phosphoric acid anhydride (Section 17.1): A functional group containing the P–O–P linkage.

Phosphorylation (Section 17.1): a reaction that transfers a phosphate group from a phosphoric anhydride to an alcohol.

Pi bond (Section 1.10): the covalent bond formed by sideways overlap of atomic orbitals. For example, carbon-carbon double bonds contain a pi bond formed by sideways overlap of two p orbitals.

Plane of symmetry (Section 6.2): an imaginary plane that bisects a molecule such that one half of the molecule is the mirror image of the other half. Molecules containing a plane of symmetry are achiral.

Plane-polarized light (Section 6.3): ordinary light that has its electric vectors in a single plane rather than in random planes. The plane of polarization is rotated when the light is passed through a solution of a chiral substance.

Polar reaction (Section 3.6): a reaction in which bonds are made when a nucleophile donates two electrons to an electrophile, and in which bonds are broken when one fragment leaves with both electrons from the bond. Polar reactions are the most common class of reactions. (See heterogenic and heterolytic reactions)

Polarity (Sections 1.12): the unsymmetrical distribution of electrons in molecules that results when one atom attracts electrons more strongly than another.

Polycyclic aromatic hydrocarbon (Section 5.12): a molecule that has two or more benzene rings fused together.

Polymer (Sections 4.8 and 10.12): a large molecule made up of repeating smaller units. For example, polyethylene is a synthetic polymer made from repeating ethylene units, and DNA is a biopolymer made of repeating deoxyribonucleotide units.

Polysaccharide (Section 14.10): a complex carbohydrate having many simple sugars bonded together by acetal links.

Primary, secondary, tertiary, quaternary (Section 2.2): terms used to describe the substitution pattern at a specific site. A primary site has one organic substituent attached to it, a secondary site has two organic substituents, a tertiary site has three, and a quaternary site has four.

	Carbon	Hydrogen	Alcohol	Amine
primary	RCH_3	RCH_3	RCH_2OH	RNH_2
secondary	R_2CH_2	R_2CH_2	R_2CHOH	R_2NH
tertiary	R_3CH	R_3CH	R_3COH	R_3N
quaternary	R_4C			

Primary structure (Section 15.10): the amino acid sequence in a protein. (See secondary structure, tertiary structure)

Propagation step (Section 7.2): the step or series of steps in a radical chain reaction that carry on the chain. The propagation steps must yield both product and a reactive intermediate to carry on the chain.

Protein (Section 15.9): a large biological polymer containing fifty or more amino acid residues. Proteins serve both as structural materials (hair, horns, fingernails) and as enzymes that control an organism's chemistry. (See peptide)

Protic solvent: a solvent such as water or alcohol that can serve as a proton donor. Protic solvents are particularly good at stabilizing anions by hydrogen bonding, thereby lowering their reactivity.

Quartet (Section 13.10): a set of four peaks in the NMR, caused by spin-spin splitting of a signal by three adjacent nuclear spins.

Quaternary (See primary)

Quaternary ammonium salt (Section 12.1): a compound with four organic substituents bonded to a positively charged nitrogen, $R_4N^+\ X^-$.

Quaternary structure (Section 15.10): the highest level of protein structure, involving a specific aggregation of individual proteins into a larger cluster.

Quinone (Section 8.8): a compound that contains the cyclohexadienone functional group.

***R,S* convention** (Section 6.6): a method for defining the absolute configuration around chiral centers. Sequence rules are used to assign relative priorities to the four substituents on the chiral center and the center is oriented such that the group of lowest (fourth) priority faces directly away from the viewer. If the three remaining substituents have a right-handed or clockwise relationship in going from first to second to third priority, then the chiral center is denoted R (rectus, right). If the three remaining substituents have a left-handed or counterclockwise relationship, the chiral center is denoted S (sinister, left). (see sequence rules)

R configuration   *S* configuration

Racemic mixture (Section 6.10): a mixture consisting of equal parts (+) and (-) enantiomers of a chiral substance. Even though the individual molecules are chiral, racemic mixtures are optically inactive.

Racemization (Section 6.10): the process whereby one enantiomer of a chiral molecule becomes converted into a 50:50 mixture of enantiomers, thus losing its optical activity. For example, this might happen during an S_N1 reaction of a chiral alkyl halide.

Radical (Section 3.6): When used in organic nomenclature, the word radical refers to a part of a molecule that appears in its name – for example the "phenyl" in phenyl acetate. When used chemically, however, a radical is a species that has an odd number of electrons, such as the chlorine radical, Cl·.

Radical reaction (Section 3.6): a reaction in which bonds are made by donation of one electron from each of two reagents, and in which bonds are broken when each fragments leaves with one electron. (See homogenic, homolytic)

Reaction energy diagram (Section 3.10): a pictorial representation of the course of a reaction, in which potential energy is graphed as a function of reaction progress. Starting materials, transition states, intermediates, and final products are all represented, and their appropriate energy levels are indicated.

Reducing sugar (Section 14.8): any sugar that reduces silver ion in the Tollens test or cupric ion in the Fehling's or Benedict's tests. All sugars that are aldehydes or that can be readily converted into aldehydes are reducing. Glycosides, however, are not reducing sugars.

Regiochemistry (Section 4.9): a term describing the orientation of a reaction that occurs on an unsymmetrical substrate. Markovnikov's rule, for example, predicts the regiochemistry of electrophilic addition reactions.

Regioselective (Section 4.2): a term describing the orientation of a reaction that occurs with a specific regiochemistry to give primarily a single product, rather than a mixture of products.

Replication (Section 16.10): the process by which double-stranded DNA uncoils and is replicated to produce two new copies.

Resolution (Sections 6.10 and 12.4): the process by which a racemic mixture is separated into its two pure enantiomers. For example, a racemic carboxylic acid might be converted by reaction with a chiral amine base into a diastereomeric mixture of salts, which could be separated by fractional crystallization. Regeneration of the free acids would then yield the two pure enantiomeric acids.

Resonance hybrid (Sections 4.11 and 4.12): a molecule, such as benzene, that cannot be represented adequately by a single Kekulé structure but must instead be considered as an average of two or more resonance structures. The resonance structures themselves differ only in the positions of their electrons, not their nuclei.

Respiratory chain (Section 17.1): The fourth stage of catabolism, in which ATP is synthesized.

Restriction endonuclease (Section 16.13): an enzyme that is able to cut a DNA strand at a specific base sequence in the chain.

Ring-flip (Section 2.11): the molecular motion that converts one chair conformation of cyclohexane into another chair conformation. The effect of a ring-flip is to convert an axial substituent into an equatorial substituent.

RNA (Section 16.6): ribonucleic acid, the biopolymer found in cells that serves to transcribe the genetic information found in DNA and uses that information to direct the synthesis of proteins.

Saccharide (Section 14.1): a sugar.

Salt bridge (Section 15.10): the ionic attraction between charged amino acid side chains that helps stabilize a protein's tertiary structure.

Sandmeyer reaction (Section 12.6): the conversion of an arenediazonium salt into an aryl halide by reaction with a cuprous halide.

Saponification (Section 10.9): an old term for the base-induced hydrolysis of an ester to yield a carboxylic acid salt.

Saturated (Section 2.2): having only single bonds and thus not being able to undergo addition reactions. Alkanes, for example, are saturated, but alkenes are unsaturated.

Sawhorse structure (Section 2.5): a stereochemical manner of representation that portrays a molecule using a stick drawing and gives a perspective view of the conformation around single bonds.

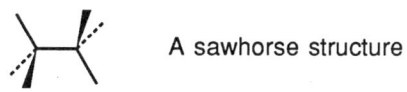

A sawhorse structure

Secondary (See primary)

Secondary structure (Section 15.10): the level of protein substructure that involves organization of chain sections into ordered arrangements such as β-pleated sheets or α-helices.

Sequence rules (Sections 3.4 and 6.6): a series or rules for assigning relative priorities to substituent groups on a double-bond carbon atom or on a chiral center. Once priorities have been established, E,Z-double bond geometry and R,S-configurational assignments can be made. (See entgegen, R,S convention, zusammen)

Shielding (Section 13.8): an effect observed in NMR that causes a nucleus to absorb toward the right (upfield) side of the chart. Shielding is caused by donation of electron density to the nucleus. (See deshielding)

Sigma bond (Section 1.6): a covalent bond formed by head-on overlap of atomic orbitals.

Soap (Section 16.3): the mixture of long-chain fatty acid salts obtained by base hydrolysis of animal fat.

***sp* Orbital** (Section 1.11): a hybrid orbital derived from the combination of an *s* and a *p* atomic orbital. The two *sp* orbitals that result from hybridization are oriented at an angle of 180° to each other.

***sp²* Orbital** (Section 1.10): a hybrid orbital derived by combination of an *s* atomic orbital with two *p* atomic orbitals. The three *sp²* hybrid orbitals that result lie in a plane at angles of 120° to each other.

***sp³* Orbital** (Section 1.7): a hybrid orbital derived by combination of an *s* atomic orbital with three *p* atomic orbitals. The four *sp³* hybrid orbitals that result are directed towards the corners of a tetrahedron at angles of 109° to each other.

Specific rotation, $[\alpha]_D$ (Section 6.4): The specific rotation of a chiral compound is a physical constant defined by the equation:

$$[\alpha]_D = \frac{\text{observed rotation}}{\text{path length} \times \text{concentration}} = \frac{\alpha}{l \times c}$$

where *l* is the path length of the sample solution expressed in decimeters and *c* is the concentration of the sample solution expressed in g/mL.

Sphingolipid (Section 16.4): a phospholipid based on the sphingosine backbone rather than on glycerol.

Spin-spin splitting (Section 13.10): the splitting of an NMR signal into a multiplet caused by an interaction between nearby magnetic nuclei whose spins are coupled. The magnitude of spin-spin splitting is given by the coupling constant, *J*.

Staggered conformation (Section 2.5): the three-dimensional arrangement of atoms around a carbon-carbon single bond in which the bonds on one carbon exactly bisect the bond angles on the second carbon as viewed end-on. (See eclipsed conformation)

Staggered conformation

Step-growth polymer (Section 10.12): a polymer produced by a series of polar reactions between two difunctional monomers. The polymer normally has the two monomer units in alternating order and usually has other atoms in addition to carbon in the polymer backbone. Nylon, a polyamide produced by reaction between a diacid and a diamine, is an example.

Stereochemistry (Chapter 6): the branch of chemistry concerned with the three-dimensional arrangement of atoms in molecules.

Stereogenic center (Section 6.2): an atom (usually carbon) that is bonded to four different groups and is therefore chiral. (See chiral)

Stereoisomers (Section 2.8): isomers that have their atoms connected in the same order but that have different three-dimensional arrangements. The term stereoisomer includes both enantiomers and diastereomers but does not include constitutional isomers.

Stereoselective: a term indicating that only a single stereoisomer is produced in a given reaction, rather than a mixture.

Steric strain (Section 2.10): the strain imposed on a molecule when two groups are too close together and try to occupy the same space. Steric strain is responsible both for the greater stability of trans versus cis alkenes, and for the greater stability of equatorially substituted versus axially substituted cyclohexanes.

Steroid (Section 16.5): a lipid whose structure is based on the tetracyclic carbon skeleton:

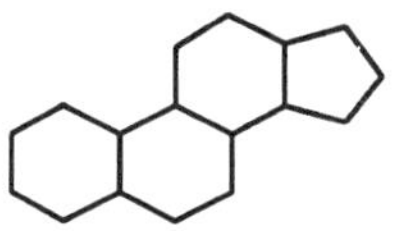

steroid skeleton

Steroids occur in both plants and animals and have many important hormonal functions.

Sulfide (Section 8.12): a compound that has two organic groups bonded to a sulfur atom, R–S–R'.

Syn stereochemistry (Section 4.6): A syn addition reaction is one in which the two ends of the double bond are attacked from the same side. For example, hydrogenation of alkenes has syn stereochemistry. A syn elimination is one in which the two groups leave from the same side of the molecule.

Tautomers (Sections 4.15 and 11.1): isomers that are rapidly interconverted. For example, enols and ketones are tautomers since they are rapidly interconverted on treatment with either acid or base catalysts.

Tertiary (see primary)

Tertiary structure (Section 15.10): the level of protein structure that involves the manner in which the entire protein chain is folded into a specific three-dimensional arrangement.

Thiol (Section 8.12): A compound with the –SH functional group.

Transamination (Section 17.5): A reaction in which the –NH$_2$ group of an amine changes place with the keto group of an α-keto acid.

Transcription (Section 16.11): the process by which the genetic information encoded in DNA is read and used to synthesize RNA in the nucleus of the cell. A small portion of double-stranded DNA uncoils, and complementary ribonucleotides line up in the correct sequence for RNA synthesis.

Transition state (Section 3.10): an imaginary activated complex between reagents, representing the highest energy point on a reaction curve. Transition states are unstable complexes that cannot be isolated.

Translation (Section 16.12): the process by which the genetic information transcribed from DNA onto mRNA is read by tRNA and used to direct protein synthesis.

Triacylglycerol (Section 16.2): lipids such as animal fat and vegetable oil consisting chemically of triesters of glycerol with long-chain fatty acids.

Triplet (Section 13.10): a symmetrical three-line splitting pattern observed in the ^{1}H NMR when a proton has two equivalent neighbor protons.

Ultraviolet (UV) spectroscopy (Section 13.3): an optical spectroscopy employing ultraviolet irradiation. UV spectroscopy provides structural information about the extent of pi-electron conjugation in organic molecules.

Unsaturated (Section 3.1): An unsaturated molecule is one that has multiple bonds and can undergo addition reactions. Alkenes and alkynes, for example, are unsaturated. (See saturated)

Upfield (Section 13.7): used to refer to the right-hand portion of the NMR chart. (See shielding)

Vitamin (Section 15.11): a small organic molecule that must be obtained in the diet and that is required for proper growth.

Vinylic (Section 7.7): a term that refers to a substituent at a double-bond carbon atom. For example, chloroethylene is a vinylic chloride, and enols are vinylic alcohols.

Wavelength (Section 13.1): the length of a wave from peak to peak. The wavelength of electromagnetic radiation is inversely proportional to frequency and inversely proportional to energy. (See frequency)

Wavenumber (Section 13.1): The wavenumber is the reciprocal of the wavelength in centimeters. Thus, wavenumbers are expressed in cm^{-1}.

Williamson ether synthesis (Section 8.6): the reaction of an alkoxide ion with an alkyl halide to yield an ether.

Wolff-Kishner reaction (Section 9.10): a reaction for reducing a ketone or aldehyde to an alkane by reaction with hydrazine and KOH.

Zaitsev's rule (Section 4.9): a rule stating that E2 elimination reactions normally yield the more highly substituted alkene as major product.

Zusammen (Z) (Section 3.4): a term used to describe the stereochemistry of a carbon-carbon double bond. The two groups on each carbon are assigned priorities according to a series of sequence rules, and the two carbons are compared. If the high priority groups on each carbon are on the same side of the double bond, the bond has Z geometry. (See Entgegen, sequence rules)

Zwitterion (Section 15.2): a neutral dipolar molecule in which the positive and negative charges are not adjacent. For example, amino acids exist as zwitterions.

$$H_3\overset{+}{N}-\underset{R}{CH}-\overset{O}{\overset{\|}{C}}O^-$$ An amino acid zwitterion

Abbreviations

Å	symbol for Angstrom unit (10^{-8} cm)
Ac-	acetyl group, $CH_3\overset{\overset{O}{\|}}{C}-$
Ar-	aryl group
at. no.	atomic number
at. wt.	atomic weight
$[\alpha]_D$	specific rotation
BOC	tert-butoxycarbonyl group, $(CH_3)_3CO\overset{\overset{O}{\|}}{C}-$
bp	boiling point
n-Bu	n-butyl group, $CH_3CH_2CH_2CH_2$-
sec-Bu	sec-butyl group, $CH_3CH_2CH(CH_3)$-
t-Bu	tert-butyl group, $(CH_3)_3C$-
cm	centimeter
cm^{-1}	wavenumber or reciprocal centimeter
D	stereochemical designation of carbohydrates and amino acids
DCC	dicyclohexylcarbodiimide, C_6H_{11}-N=C=N-C_6H_{11}
δ	chemical shift in ppm downfield from TMS
Δ	symbol for heat; also symbol for change
ΔH	heat of reaction
dm	decimeter (0.1 m)
DMF	dimethylformamide, $(CH_3)_2NCHO$
DMSO	dimethyl sulfoxide, $(CH_3)_2SO$
DNA	deoxyribonucleic acid
DNP	dinitrophenyl group, as in 2,4-DNP (2,4-dinitrophenylhydrazone)
(E)	entgegen, stereochemical designation of double bond geometry
E_{act}	activation energy
E1	unimolecular elimination reaction
E2	bimolecular elimination reaction
Et	ethyl group, CH_3CH_2-
g	gram

hv	symbol for light
Hz	Hertz, or cycles per second
i-	iso
IR	infrared
J	Joule
J	symbol for coupling constant
K	Kelvin temperature
K_a	symbol for acid dissociation constant
kcal	kilocalories
L	stereochemical designation of carbohydrates and amino acids
LAH	lithium aluminum hydride, $LiAlH_4$
Me	methyl group, CH_3-
mg	milligram (0.001 g)
MHz	megahertz (10^6 cycles per second)
mL	milliliter (0.001 L)
mm	millimeter (0.001 m)
mol. wt.	molecular weight
mp	melting point
μγ	microgram (10^{-6} gram)
mμ	millimicron (nanometer, 10^{-9} meter)
n-	normal, straight chain alkane or alkyl group
ng	nanogram (10^{-9} gram)
nm	nanometer (10^{-9} meter)
NMR	nuclear magnetic resonance
-OAc	acetate group, $-O\overset{\overset{O}{\|\|}}{C}CH_3$
PCC	pyridinium chlorochromate
Ph	phenyl group, $-C_6H_5$
pH	measure of acidity of aqueous solution
pK_a	measure of acid strength (= -log K_a)
pm	picometer (10^{-12}) meter)
ppm	parts per million
n-Pr	*n*-propyl group, $CH_3CH_2CH_2$-
i-Pr	isopropyl group, $(CH_3)_2CH$-

prim	primary
R-	symbol for a generalized alkyl group
(R)	rectus, stereochemical designation of chiral centers
RNA	ribonucleic acid
(S)	sinister, stereochemical designation of chiral centers
sec-	secondary
S_N1	unimolecular substitution reaction
S_N2	bimolecular substitution reaction
tert-	tertiary
THF	tetrahydrofuran
TMS	tetramethylsilane nmr standard, $(CH_3)_4Si$
UV	ultraviolet
yr	year
X-	halogen group (–F, –Cl, –Br, –I)
(Z)	zusammen, stereochemical designation of double bond geometry
⟶	chemical reaction in direction indicated
⇌	reversible chemical reaction
↔	resonance symbol
⤴	curved arrow indicating direction of electron flow
≡	is equivalent to
>	greater than
<	less than
≈	approximately equal to
R⌇	indicates that the organic fragment shown is a part of a larger molecule
▬	single bond coming out of the plane of the paper
-----	single bond receding into the plane of the paper
......	partial bond
δ^+, δ^-	partial charge
*	isotopically labeled atom
‡	denoting the transition state

Proton NMR Chemical Shifts

Type of Proton		Chemical Shift (δ)
Alkyl, primary	R-CH$_3$	0.7 – 1.3
Alkyl, secondary	R-CH$_2$-R	1.2 – 1.4
Alkyl, tertiary	R$_3$C-H	1.4 – 1.7
Allylic	-C=C-C-H	1.6 – 1.9
Alpha to carbonyl	-C(=O)-C-H	2.0 – 2.3
Benzylic	Ar-C-H	2.3 – 3.0
Acetylenic	R-C≡C-H	2.5 – 2.7
Alkyl chloride	Cl-C-H	3.0 – 4.0
Alkyl bromide	Br-C-H	2.5 – 4.0
Alkyl iodide	I-C-H	2.0 – 4.0
Amine	N-C-H	2.2 – 2.6
Epoxide	(epoxide C-H)	2.5 – 3.5
Alcohol	HO-C-H	3.5 – 4.5
Ether	RO-C-H	3.5 – 4.5
Vinylic	-C=C-H	5.0 – 6.5
Aromatic	Ar-H	6.5 – 8.0
Aldehyde	R-C(=O)-H	9.7 – 10.0
Carboxylic acid	R-C(=O)-O-H	11.0 – 12.0
Alcohol	R-O-H	3.5 – 4.5
Phenol	Ar-O-H	2.5 – 6.0

Infrared Absorption Frequencies

Functional Group Class		Frequency (cm^{-1})
Alcohol	—O—H	3300 - 3600 (s)
Aldehyde	—C—O—	1050 (s)
	—CO—H	2720, 2820 (m)
aliphatic	C=O	1725 (s)
aromatic	C=O	1705 (s)
Alkane	—C—H	2850 - 2960 (s)
	—C—C—	800 - 1300 (m)
Alkene	=C(H)	3020 - 3100 (m)
	C=C	1650 - 1670 (m)
	RCH=CH$_2$	910, 990 (m)
	R$_2$C=CH$_2$	890 (m)
Alkyne	≡C—H	3300 (s)
	—C≡C—	2100 - 2260 (m)
Alkyl bromide	—C—Br	500 - 600 (s)
Alkyl chloride	—C—Cl	600 - 800 (s)
Amine, *primary*	—N(H)(H)	3400, 3500 (s)

secondary	⟩N–H	3350 (s)
Ammonium salt	–⁺N–H	2200 - 3000 (broad)
Aromatic ring	Ar–H	3030 (m)
monosubstituted	Ar–R	690 - 710 (s)
		730 - 770 (s)
o-disubstituted		735 - 770 (s)
m-disubstituted		690 - 710 (s)
		810 - 850 (s)
p-disubstituted		810 - 840 (s)
Carboxylic acid	–O–H	2500 - 3300 (broad)
associated	⟩C=O	1710 (s)
free	⟩C=O	1760 (s)
Acid anhydride	⟩C=O	1820, 1760 (s)
Acid chloride		
aliphatic	⟩C=O	1810 (s)
aromatic	⟩C=O	1770 (s)
Amide, *aliphatic*	⟩C=O	1690 (s)
aromatic	⟩C=O	1675 (s)
N-substituted	⟩C=O	1680 (s)

N,N-disubstituted	\>C=O	1650 (s)
Ester, *aliphatic*	\>C=O	1735 (s)
aromatic	\>C=O	1720 (s)
Ether	−O−C\<	1050 - 1150 (s)
Ketone, *aliphatic*	\>C=O	1715 (s)
aromatic	\>C=O	1690 (s)
6-memb. ring	\>C=O	1715 (s)
5-memb. ring	\>C=O	1750 (s)
Nitrile, *aliphatic*	−C≡N	2250 (m)
aromatic	−C≡N	2230 (m)
Phenol	−O−H	3500 (s)

(s) = strong; (m) = medium intensity

Nobel Prizes in Chemistry

1901 **Jacobus H. van't Hoff** (Dutch):
"for the discovery of laws of chemical dynamics and of osmotic pressure"

1902 **Emil Fischer** (German):
"for syntheses in the groups of sugars and purines"

1903 **Svante A. Arrhenius** (Swedish):
"for his theory of electrolytic dissociation"

1904 **Sir William Ramsey** (British):
"for the discovery of gases in different elements in the air and for the determination of their place in the periodic system"

1905 **Adolf von Baeyer** (German):
"for his researches on organic dyestuffs and hydroaromatic compounds"

1906 **Henri Moissan** (French):
"for his research on the isolation of the element fluorine and for placing at the service of science the electric furnace that bears his name"

1907 **Eduard Buchner** (German):
"for his biochemical researches and his discovery of cell-less formation"

1908 **Ernest Rutherford** (British):
"for his investigation into the disintegration of the elements and the chemistry of radioactive substances"

1909 **Wilhelm Ostwald** (German):
"for his work on catalysis and on the conditions of chemical equilibrium and velocities of chemical reactions"

1910 **Otto Wallach** (German):
"for his services to organic chemistry and the chemical industry by his pioneer work in the field of alicyclic substances"

1911 **Marie Curie** (French):
"for her services to the advancement of chemistry by the discovery of the elements radium and polonium"

1912 **Victor Grignard** (French):
"for the discovery of the so-called Grignard reagent, which has greatly helped in the development of organic chemistry"

Paul Sabatier (French):
"for his method of hydrogenating organic compounds in the presence of finely divided metals"

1913 **Alfred Werner** (Swiss):
"for his work on the linkage of atoms in molecules by which he has thrown new light on earlier investigations and opened up new fields of research especially in inorganic chemistry"

1914 **Theodore W. Richards** (U.S.):
"for his accurate determinations of the atomic weights of a great number of chemical elements"

1915 **Richard M. Willstätter** (German):
"for his research on plant pigments, principally on chlorophyll"

1916 No award

1917 No award

1918 **Fritz Haber** (German):
"for the synthesis of ammonia from its elements, nitrogen and hydrogen"

1919 No award

1920 **Walther H. Nernst** (German):
"for his thermochemical work"

1921 **Frederick Soddy** (British):
"for his contributions to the chemistry of radioactive substances and his investigations into the origin and nature of isotopes"

1922 **Francis W. Aston** (British):
"for his discovery, by means of his mass spectrograph, of the isotopes of a large number of nonradioactive elements, as well as for his discovery of the whole-number rule"

1923 **Fritz Pregl** (Austrian):
"for his invention of the method of microanalysis of organic substances"

1924 No award

1925 **Richard A. Zsigmondy** (German):
for his demonstration of the heterogeneous nature of colloid solutions, and for the methods he used, which have since become fundamental in modern colloid chemistry"

1926 **Theodor Svedberg** (Swedish):
"for his work on disperse systems"

1927 **Heinrich O. Wieland** (German):
"for his research on bile acids and related substances"

1928 **Adolf O. R. Windaus** (German):
"for his studies on the constitution of the sterols and their connection with the vitamins"

1929 **Arthur Harden** (British):
Hans von Euler-Chelpin (Swedish):
"for their investigation on the fermentation of sugar and of fermentative enzymes"

1930 **Hans Fischer** (German):
"for his researches into the constitution of hemin and chlorophyll, and especially for his synthesis of hemin"

1931 **Frederich Bergius** (German):
Carl Bosch (German):
"for their contributions to the invention and development of chemical high-pressure methods"

1932 **Irving Langmuir** (U.S.):
"for his discoveries and investigations in surface chemistry"

1933 No award

1934 **Harold C. Urey** (U.S.):
"for his discovery of heavy hydrogen"

1935 **Frederic Joliot** (French):
Irene Joliot-Curie (French):
"for their synthesis of new radioactive elements"

1936 **Peter J. W. Debye** (Dutch/U.S.):
"for his contributions our knowledge of molecular structure through his investigations on dipole moments and on the diffraction of X-rays and electrons in gases"

1937 **Walter N. Haworth** (British):
"for his researches into the constitution of carbohydrates and vitamin C"

Paul Karrer (Swiss):
"for his researches into the constitution of carotenoids, flavins, and vitamins A and B"

1938 **Richard Kuhn** (German):
"for his work on carotenoids and vitamins"

1939 **Adolf F. J. Butenandt** (German):
"for his work on sex hormones"

Leopold Ruzicka (Swiss):
"for his work on polymethylenes and higher terpenes"

1940 No award

1941 No award

1942 No award

1943 **Georg de Hevesy** (Hungarian):
"for his work on the use of isotopes as tracer elements in researches on chemical processes"

1944 **Otto Hahn** (German):
"for his discovery of the fission of heavy nuclei"

1945 **Artturi I. Virtanen** (Finnish):
"for his researches and inventions in agricultural and nutritive chemistry, expecially for his fodder preservation method"

1946 **James B. Sumner** (U.S.):
"for his discovery that enzymes can be crystallized"

John H. Northrop (U.S.):
Wendell M. Stanley (U.S.):
for their preparation of enzymes and virus proteins in a pure form"

1947 **Sir Robert Robinson** (British):
"for his investigations on plant products of biological importance, particularly the alkaloids"

1948 **Arne W. K. Tiselius** (Swedish):
"for his researches on electrophoresis and adsorption analysis, especially for his discoveries concerning the complex nature of the serum proteins"

1949 **William F. Giauque** (U.S.):
"for his contributions in the field of chemical thermodynamics, particularly concerning the behavior of substances at extremely low temperatures"

1950 **Kurt Alder** (German):
Otto P. H. Diels (German):
"for their discovery and development of the diene synthesis"

1951 **Edwin M. McMillan** (U.S.):
Glenn T. Seaborg (U.S.):
"for their discoveries in the chemistry of the transuranium elements"

1952 **Archer J. P. Martin** (British):
Richard L. M. Synge (British):
"for their development of partition chromatography"

1953 **Hermann Staudinger** (German):
"for his discoveries in the field of macromolecular chemistry"

1954 **Linus C. Pauling** (U.S.):
"for his research into the nature of the chemical bond and its application to the elucidation of the structure of complex substances"

1955 **Vincent du Vigneaud** (U.S.):
"for his work on biochemically important sulfur compounds, especially for the first synthesis of a polypeptide hormone"

1956 **Sir Cyril N. Hinshelwood** (British):
Nikolai N. Semenov (U.S.S.R.):
"for their research in clarifying the mechanisms of chemical reactions in gases"

1957 **Sir Alexander R. Todd** (British):
"for his work on nucleotides and nucleotide coenzymes"

1958 **Frederick Sanger** (British):
"for his work on the structure of proteins, particularly insulin"

1959 **Jaroslav Heyrovsky** (Czechoslovakian):
"for his discovery and development of the polarographic method of analysis"

1960 **Willard F. Libby** (U.S.):
"for his method to use carbon-14 for age determination in archaeology, geology, geophysics, and other branches of science"

1961 **Melvin Calvin** (U.S.):
"for his research on the carbon dioxide assimilation in plants"

1962 **John C. Kendrew** (British):
Max F. Perutz (British):
"for their studies of the structures of globular proteins"

1963 **Giulio Natta** (Italian):
Karl Ziegler (German):
"for their work in the controlled polymerization of hydrocarbons through the use of organometallic catalysts"

1964 **Dorothy C. Hodgkin** (British):
"for her determinations by X-ray techniques of the structures of important biochemical substances, particularly vitamin B-12 and penicillin"

1965 **Robert B. Woodward** (U.S.):
"for his outstanding achievements in the 'art' of organic synthesis"

1966 **Robert S. Mulliken** (U.S.):
"for his fundamental work concerning chemical bonds and the electronic structure of molecules by the molecular orbital method"

1967 **Manfred Eigen** (German):
Ronald G. W. Norrish (British):
George Porter (British):
"for their studies of extremely fast chemical reactions, effected by disturbing the equilibrium with very short pulses of energy"

1968 **Lars Onsager** (U.S.):
"for his discovery of the reciprocal relations bearing his name, which are fundamental for the thermodynamics of irreversible processes"

1969 **Sir Derek H. R. Barton** (British):
Odd Hassel (Norwegian):
"for their contributions to the development of the concept of conformation and its application in chemistry"

1970 **Luis F. Leloir** (Argentinian):
"for his discovery of sugar nucleotides and their role in the biosynthesis of carbohydrates"

1971 **Gerhard Herzberg** (Canadian):
"for his contributions to the knowledge of electronic structure and geometry of molecules, particularly free radicals"

1972 **Christian B. Anfinsen** (U.S.):
"for his work on ribonuclease, especially concerning the connection between the amino acid sequence and the biologically active conformation"

Stanford Moore (U.S.):
William H. Stein (U.S.):
"for their contribution to the understanding of the connection between chemical structure and catalytic activity of the active center of the ribonuclease molecule"

1973 **Ernst Otto Fischer** (German):
Geoffrey Wilkinson (British):
"for their pioneering work, performed independently, on the chemistry of the organo-metallic sandwich compounds"

1974 **Paul J. Flory** (U.S.):
"for his fundamental achievements, both theoretical and experimental, in the physical chemistry of macromolecules"

1975 **John Cornforth** (Australian/British):
"for his work on the stereochemistry of enzyme-catalyzed reactions"

Vladimir Prelog (Yugoslavian/Swiss):
"for his work on the stereochemistry of organic molecules and reactions"

1976 **William N. Lipscomb** (U.S.):
"for his studies on the structures of boranes illuminating problems of chemical bonding"

1977 **Ilya Pregogine** (Belgian):
"for his contributions to nonequilibrium thermodynamics, particularly the theory of dissipative structures"

1978 **Peter Mitchell** (British):
"for his contribution to the understanding of biological energy transfer through the formulation of the chemiosmotic theory"

1979 **Herbert C. Brown** (U.S.):
"for his application of boron compounds to synthetic organic chemistry"

Georg Wittig (German):
"for developing phosphorus reagents, presently bearing his name"

1980 **Paul Berg** (U.S.):
"for his fundamental studies of the biochemistry of nucleic acids, with particular regard to recombinant DNA"

Walter Gilbert (U.S.)
Frederick Sanger (British):
"for their contributions concerning the determination of base sequences in nucleic acids"

1981 **Kenichi Fukui** (Japanese)
Roald Hoffmann (U.S.):
for their theories, developed independently, concerning the course of chemical reactions"

1982 **Aaron Klug** (British):
"for his development of crystallographic electron microscopy and his structural elucidation of biologically important nucleic acid - protein complexes"

1983 **Henry Taube** (U.S.):
"for his work on the mechanisms of electron transfer reactions, especially in metal complexes"

1984 **R. Bruce Merrifield** (U.S.):
"for his development of methodology for chemical synthesis on a solid matrix"

1985 **Herbert A. Hauptman** (U.S.):
Jerome Karle (U.S.):
"for their outstanding achievements in the development of direct methods for the determination of crystal structures"

1986 **John C. Polanyi** (Canadian):
"for his pioneering work in the use of infrared chemiluminescence in studying the dynamics of chemical reactions"

Dudley R. Herschbach (U.S.):
Yuan T. Lee (U.S.):
"for their contributions concerning the dynamics of chemical elementary processes"

1987 **Donald J. Cram** (U.S.):
Jean-Marie Lehn (French):
Charles J. Pedersen (U.S.):
"for their development and use of molecules with structure-specific interactions of high selectivity"

1988 **Johann Deisenhofer** (German):
Robert Huber (German):
Hartmut Michel (German):
"for their determination of the structure of the photosynthetic reaction center of bacteria"

1989 **Sidney Altman** (U.S.):
Thomas R. Cech (U.S.):
"for their discovery of catalytic properties of RNA"

1990 **Elias J. Corey** (U.S.):
"for his development of the theory and methodology of organic synthesis"

1991 **Richard R. Ernst** (Swiss):
"for his contributions to the development of the methodology of high resolution NMR spectroscopy"

1992 **Rudolph A. Marcus** (U.S.):
"for his contributions to the theory of electron-transfer reactions in chemical systems"

Top 40 Organic Chemicals
U.S. Chemical Industry – 1992

Ethylene (20,203,000 tons/yr):
prepared by thermal cracking of ethane and propane during petroleum refining; used as starting material for manufacture of polyethylene, ethylene oxide, ethylene glycol, ethylbenzene, 1,2-dichloroethane, and other bulk chemicals.

Propylene (11,297,000 tons/yr):
prepared by steam cracking of light hydrocarbon fractions during petroleum refining; used as starting material for the manufacture of polypropylene, acrylonitrile, propylene oxide, and isopropyl alcohol.

1,2-Dichloroethane (Ethylene dichloride; 7,968,000 tons/yr):
prepared by addition of chlorine to ethylene in the presence of $FeCl_3$ catalyst at 50°C; used as a chlorinated solvent and as starting material for the manufacture of vinyl chloride.

Vinyl chloride (6,613,000 tons/yr):
prepared by addition of chlorine to ethylene followed by elimination of HCl; used as starting material for preparation of poly(vinyl chloride) polymers (hoses, pipes, molded objects).

Benzene (5,984,000 tons/yr):
obtained from petroleum by catalytic reforming of hexane and cyclohexane over a platinum catalyst; used as starting material for the synthesis of ethylbenzene, cumene, cyclohexane, and aniline.

Methyl *tert*-butyl ether (MTBE; 5,430,000 tons/yr):
prepared by acid-catalyzed addition of methanol to isobutylene; used as an octane enhancer in gasoline.

Ethylbenzene (4,495,000 tons/yr):
prepared during catalytic reforming in petroleum refining and by an acid-catalyzed Friedel-Crafts alkylation of benzene with ethylene; used almost exclusively for production of styrene.

Styrene (4,471,000 tons/yr):
prepared by high-temperature catalytic dehydrogenation of ethylbenzene; used in the manufacture of polystyrene polymers (thermoplastics, packaging materials).

Methanol (4,364,000 tons/yr):
prepared by high temperature reaction of a mixture of H_2, CO, and CO_2 ("synthesis gas") over a catalyst at 100 atmospheres pressure; used as a solvent and as starting material for the manufacture of formaldehyde, acetic acid, and methyl *tert*-butyl ether.

Toluene (3,013,000 tons/yr):
prepared during catalytic reforming of petroleum; used as a gasoline additive and as a degreasing solvent.

***p*-Xylene** (2,828,000 tons/yr):
prepared by separation from the mixed xylenes that result during catalytic reforming in gasoline refining; used as starting material for manufacture of the dimethyl terephthalate needed for polyester synthesis.

Dimethyl terephthalate (2,819,000 tons/yr):
   prepared from *p*-xylene by oxidation and esterification; used in the manufacture of polyester polymers (textiles, upholstery, recording tape, and film).

Ethylene oxide (2,780,000 tons/yr):
   prepared by high-temperature air oxidation of ethylene over a silver catalyst; used as starting material for the preparation of ethylene glycol and poly(ethylene glycol).

Ethylene glycol (2,562,000 tons/yr):
   prepared by high-temperature reaction between water and ethylene oxide at neutral pH; used as antifreeze and as a starting material for polymers and latex paints.

Cumene (2,283,000 tons/yr):
   prepared by a phosphoric-acid-catalyzed Friedel-Crafts reaction between benzene and propylene; used primarily for conversion into phenol and acetone.

Phenol (1,854,000 tons/yr):
   prepared from cumene by air oxidation to cumene hydroperoxide, followed by acid-catalyzed decomposition; used as starting material for preparing phenolic resins, epoxy resins, and caprolactam.

Acetic acid (1,800,000 tons/yr):
   prepared by metal-catalyzed air oxidation of acetaldehyde under pressure at 80°C and by reaction of methanol with carbon monoxide; used to make vinyl acetate polymers, ethyl acetate solvent, and cellulose acetate polymers.

1,3-Butadiene (1,589,000 tons/yr):
   prepared by steam cracking of gas oil during petroleum refining and by dehydrogenation of butane and butene; used primarily as a monomer component in the manufacture of styrene-butadiene rubber (SBR), polybutadiene rubber, and acrylonitrile-butadiene-styrene (ABS) copolymers.

Acrylonitrile (1,415,000 tons,yr):
   prepared by the Sohio ammoxidation process in which propylene, ammonia, and air are passed over a catalyst at 500°C; used in the preparation of acrylic fibers, nitrile rubber, and acrylonitrile-butadiene-styrene (ABS) copolymer.

Vinyl acetate (1,329,000 tons/yr):
   prepared from reaction of acetic acid, ethylene, and oxygen; used for manufacture of poly(vinyl acetate) (paint emulsions, plywood adhesives, textiles).

Formaldehyde (1,291,000 tons/yr):
   prepared by air oxidation of methanol over a silver or metal oxide catalyst; used in the manufacture of phenolic resins, melamine resins, and plywood adhesives.

Acetone (1,197,000 tons/yr):
   prepared by acid-catalyzed decomposition of cumene hydroperoxide and by air oxidation of isopropyl alcohol at 300°C over a metal oxide catalyst; used as a solvent and as starting material for synthesizing bisphenol A and methyl methacrylate.

Cyclohexane (1,103,000 tons/yr):
prepared by catalytic hydrogenation of benzene; used as starting material for synthesis of the caprolactam and adipic acid needed for nylon.

Caprolactam (689,000 tons/yr):
prepared from phenol by conversion into cyclohexanone, followed by formation and acid-catalyzed rearrangement of cyclohexanone oxime; used as starting material for the manufacture of nylon-6.

Isobutylene (643,000 tons/yr):
prepared from catalytic cracking of petroleum; used in the manufacture of methyl *tert*-butyl ether, isoprene, and butylated phenols.

1-Butanol (632,000 tons/yr):
prepared in the oxo process by reaction of propylene with carbon monoxide; used as solvent and as a starting material for synthesis of butyl acetate and dibutyl phthalate.

Isopropyl alcohol (621,000 tons/yr):
prepared by direct high-temperature addition of water to propylene; used in cosmetics formulations, as a solvent and deicer, and as starting material for manufacture of acetone.

Bisphenol A (608,000 tons/yr):
prepared by reaction of phenol with acetone; used in the manufacture of epoxy resins and adhesives, polycarbonates, and polysulfones.

Aniline (504,000 tons/yr):
prepared by catalytic reduction of nitrobenzene with hydrogen at 350°C; used as starting material for preparing toluene diisocyanate and for the synthesis of dyes and pharmaceuticals.

o-Xylene (459,000 tons/yr):
obtained by separation from the mixed xylenes that result during catalytic reforming in petroleum refining; used as starting material for preparation of phthalic acid and phthalic anhydride.

Phthalic anhydride (449,000 tons/yr):
prepared by oxidation of *o*-xylene at 400°C and by oxidation of naphthalene obtained from coal tar; used for the synthesis of polyesters and plasticizers.

Methyl methacrylate (419,000 tons/yr):
prepared by acetone cyanohydrin by treatment with sulfuric acid to effect dehydration, followed by esterification with methanol; used for the synthesis of methacrylate polymers such as Lucite.

Chloromethane (Methyl chloride; 360,000 tons/yr):
prepared by reaction of methanol with HCl at 0°C and by radical chlorination of methane; used in the manufacture of silicones, synthetic rubber, and methyl cellulose.

1,1,1-Trichloroethane (Methylchloroform; 360,000 tons/yr):
prepared by addition of HCl to vinyl chloride to give 1,1-dichloroethane, followed by radical chlorination with Cl_2; used as a solvent, industrial cleaner, and metal degreaser.

Ethanol (351,000 tons/yr):
  prepared by direct vapor phase hydration of ethylene at 300°C over an acidic catalyst; used as a solvent, as a constituent of cleaning preparations, and as starting material for ester synthesis.

Ethanolamines (347,000 tons/yr):
  prepared by reaction of ammonia with ethylene oxide at 100°C; used in soaps, detergents, cosmetics, and corrosion inhibitors.

2-Ethylhexanol (342,000 tons/yr):
  prepared from butanal by aldol condensation and catalytic hydrogenation (the Oxo Process); used in the manufacture of plasticizers, lubricating-oil additives, and detergents.

Propylene glycol (333,000 tons/yr):
  prepared by high temperature reaction of propylene oxide with water; used in the preparation of polyesters and as an additive in the food industry.

2-Butanone (Methyl ethyl ketone; 247,000 tons/yr):
  prepared by oxidation of 2-butanol over a ZnO catalyst at 400°C; used as a solvent for vinyl coatings, lacquers, rubbers, and paint removers.

Maleic anhydride (218,000 tons/yr):
  prepared by the vapor-phase air oxidation of hydrocarbons such as butane and butene over a solid catalyst; used in the manufacture of polyesters, lubricants, and plasticizers.